VHF/UHF FIRE, POLICE, HAM SCANNERS

Schematic/Servicing Manual

Dedication

This book is dedicated to my mother, Dorothy Dansby Norman, and my memory of her inspiration and imagination.

No. 835
$9.95

VHF/UHF FIRE, POLICE, HAM SCANNERS

Schematic/Servicing Manual

By David F. Norman

TAB BOOKS
Blue Ridge Summit, Pa. 17214

FIRST EDITION

FIRST PRINTING—JUNE 1976

Printed in the United States
of America

Hardbound Edition: International Standard Book No. 0-8306-6835-7

Paperbound Edition: International Standard Book No. 0-8306-5835-1

Library of Congress Card Number: 76-8625

Introduction

Ever since the first transmission of wireless messages, hobbyists and electronic enthusiasts have been "reading the mail," or listening in. The earliest monitoring receivers were crystal sets—crude RF detectors in series with an antenna and earth ground, with a pair of high-impedance headphones across the detector whereby the listener could barely hear the transmitted signals.

If the receivers were crude the transmitters were even more so; they were simply generators of electrical interference. For several years most of the progress made in radio was by improvement of the transmitting apparatus, and receivers remained basically the same. Even as transmitted signals began to occupy a smaller bandwidth, receiver sensitivity was still controlled largely by the length of the receiving antenna, and selectivity was mostly wishful thinking. Then advances in receiver design began to outstrip transmitter improvement as radio began to be taken seriously by government and business as a viable alternative to visual or landline communications.

Within a relatively short time, transmitters—at least those which used the Morse code—reached their apex. In fact a CW transmitter today is very much like that of 50 years ago, the difference being mainly in stability. The big difference in radio today as opposed to 50 years ago is in receiver design. Now receivers are readily available that can snatch a signal of only a fraction of a microvolt from between two much stronger signals and amplify it enough to fill large room.

Radio services placed into operation as short-range systems have sometimes exceeded all expectations. In some cases—marine VHF for example—systems designed as noninterfering have had such range as to necessitate clear channel regulations. In other words a marine telephone coastal station operating on, say, channel 26 has a buffer zone of nearly 100 miles between it and the nearest station operating on the same channel. The area in between is covered by a station operating on perhaps channel 28, which has its own buffer zone. Only a few years ago 10 miles was considered good VHF range. Coverage of a 60-mile radius required thousands of feet of antenna height. The unexpected range is almost wholly due to improvements in receiver design.

Not all the improvements have been in the areas of sensitivity and selectivity. Many changes in operational simplicity have occurred and more are being introduced almost daily. Development of fixed detectors eliminated the rather ticklish procedure of finding a "hot" spot on a galena crystal, and that was only the beginning. Even fairly late-model monitoring equipment required manual tuning, or "dial twirling," or switching of various tuned circuits or crystals to receive different frequencies. Now as many as 20 different channels in 3 or 4 different bands can be sampled by a single unit at a rate of 8 to 20 channels per second. Channels not in use are automatically bypassed, and only channels in use are locked in. As traffic is completed, the unit resumes its search for an active channel—and all of this takes place without the operator doing anything more than merely turning the unit on and setting the VOLUME control. These modern scanning monitors are a blend of state-of-the-art RF and digital techniques.

Only a few shops were really ready for their advent, and many TV and stereo technicians found that they had to hurriedly acquire a working knowledge of narrowband FM and digital logic.

Without going into much theory—each of the above subjects has warranted several books by itself—this book provides specific service information for technicians in this new field. Even if the unit being serviced is not covered in this manual, there are enough similarities between all scanners that some information contained herein is bound to be helpful.

Credit is due the following: Heath Co., Regency Electronics, E. F. Johnson Co., Unimetrics, Midland Electronics, and Lampkin Corp. Thanks also to Susan Broughton, my typist, and the others whose cooperation made this book possible.

David F. Norman

Contents

Chapter 1

Principles of Scanning Monitors

I have no intention of delving deeply into theory. However I have found that when I am stumped on a service problem, it often helps to go back and study a circuit similar to the one I am attempting to repair. Sometimes I gain a new perspective on the problem and suddenly what to do about it pops into mind. When I have a problem with audio outputs, for example, I look around in my technical library and find one or more discussions of audio outputs. Often merely reviewing what makes what work, how and when, is time well spent. It is with that in mind that this chapter is included.

While the following information is specifically pertinent to the Heath GR-110, it is generally applicable to most scanners. Study of the circuit description and reference to the schematic diagram should give you a good basic understanding of scanning monitors. Even if you feel that your knowledge of FM and digital circuits is adequate, you will find this brief review helpful.

HEATH GR-110 CIRCUITRY[1]

Refer to the block diagram (Fig. 1-1) and the schematic diagram (Fig. 1-2) while reading this circuit description. The part numbers on the schematic are arranged in the following groups to help you locate specific parts on the schematic, chassis, and circuit boards:

[1] The circuit descriptions and diagrams that follow are from the Heath GR-110 instruction manual and are reprinted by permission.

- 0–99 Parts mounted on the chassis
- 100–199 Parts mounted on the scan circuit board
- 200–299 Parts mounted on the IF circuit board
- 300–399 Parts mounted on the audio circuit board

RF CIRCUIT

Field-effect transistors Q101 and Q102 amplify the incoming signal at the received frequency. Coils L101 through L105 are tank circuits tuned to the frequency of incoming signals.

The output from transistor Q102 is coupled to gate G1 of mixer transistor Q103. Gate G2 receives a signal from the crystal oscillator circuit that is 10.7 MHz lower in frequency than the incoming signal. The 10.7 MHz IF output of transistor Q103 is capacitively coupled through capacitor C3 to the input of the IF circuit.

IF CIRCUIT

The signal from the RF amplifier is coupled through monolithic crystal filters Y201 and Y202. These filters and their matching coils, L201 and T201, help shape the filter bandpass, eliminating frequencies other than the 10.7 MHz IF.

After amplification by IC201 the signal is again filtered by ceramic filter Y203. Here the

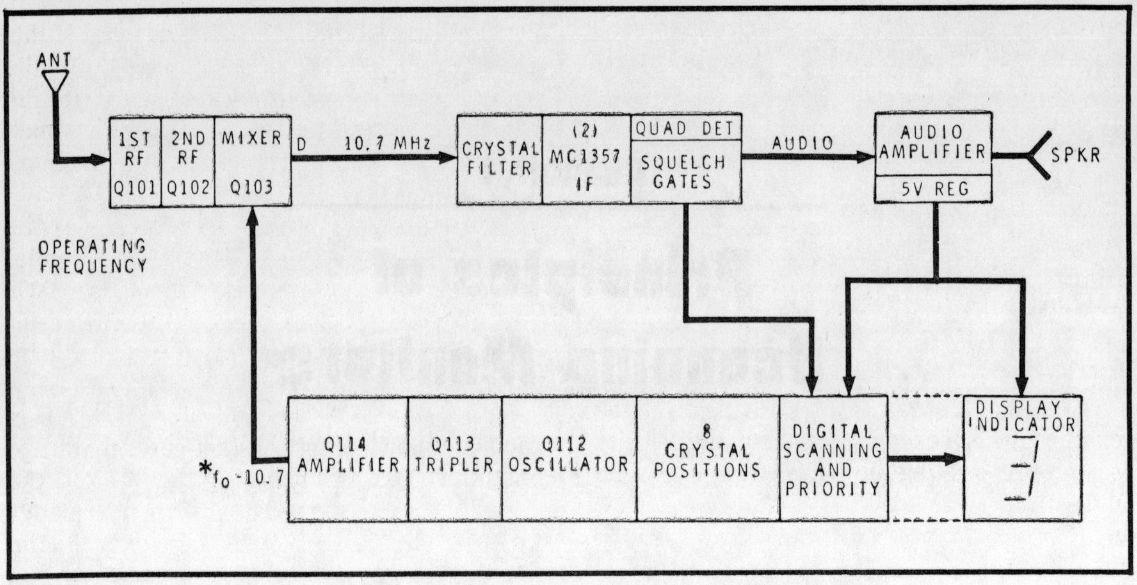

Fig. 1-1. Block diagram of Heath GR-110 VHF scanning monitor.

IF signal bandpass is shaped to give more attenuation to out-of-band signals.

From Y203 the signal is coupled through limiting amplifier IC202. In addition to amplifying the IF signal, IC202 is also the demodulator, or quadrature detector. Coil T202 is part of the associated circuit of IC202 and forms a phase shift network for the quadrature detector.

SQUELCH

The output at pin *1* of IC202 has two functions. One part of this output is capacitively coupled to the base of squelch gate transistor Q201.

Another part of the amplified IF signal is routed through the squelch circuit. The purpose of this circuit is to cut off the receiver output when no signal is coming through the unit's circuits. After passing through squelch control R2, the noise is amplified by transistors Q202 and Q203, rectified in a voltage doubler circuit, and then filtered. The resultant DC voltage is then amplified by transistor Q204, filtered, and coupled to transistor Q205.

At Q205 the signal is once again amplified and inverted. The output of transistor Q205 operates transistor gate Q206, which in turn operates audio squelch gate transistor Q201. The output from transistor Q205 also acts as a switch for the scanning circuits.

One major difference between AM receivers and modern FM receivers is in the area of squelch circuitry. While AM squelch circuits use AGC voltages (either as-is or inverted) to change bias on a stage and thus turn it on or off, FM units depend on separate circuitry which amplifies noise received through the demodulator.

The more noise that passes through the demodulator, the harder the squelch circuitry blocks the audio from being heard at the speaker; the less noise that is present, the less the squelch gate is blocked. The SQUELCH control determines how much of the noise present at the demodulator output is amplified and thus how hard the squelch gate is turned off. But regardless of the SQUELCH control setting, a received signal that is strong enough to develop much more than 20 dB of receiver quieting will open the squelch and be amplified by the audio stages. This means that, except for setting the basic quieting of the receiver when no signal is being received, the SQUELCH control has no effect on the receiver other than deciding how noisy a signal shall be permitted through the audio circuitry to the speaker.

SCAN CIRCUITS

The squelch gate signal from transistor Q205 is coupled to the base of transistor Q111. This transistor acts as a control switch, starting and stopping the action of scan oscillator transistors Q108 and Q109. A delay is generated in the circuitry of transistor Q110. When a received signal stops, a delay of approximately 4 sec begins. After this delay time, scan oscillator transistors Q108 and Q109 resume operation.

When a signal has the monitor locked on any channel other than priority channel 0, the priority oscillator will cause a pulsing noise in the audio signal. This occurs each time the priority channel is being scanned for the presence of a signal.

Priority Oscillator Operation

When integrated circuit IC101 has stopped counting and is locked on a channel, the priority oscillator consisting of transistors Q106 and Q107 will pulse an input of each gate in IC102 and IC103 to ground. This causes the outputs of three input AND gates to go low. Integrated circuit IC104 then inverts these signals to a high (logic 1) level and couples these signals to the inputs of IC105 and IC106. When all their inputs are high (logic 1), an indication of 0 will be presented by decoder drivers IC105 and IC106.

This sequence occurs approximately every 4 sec and has a duration of 20 msec. During this time the channel 0 crystal oscillates. If a signal is detected the scan circuit will lock onto the transmitted signal by using the squelch voltage to trigger pin 4 of IC103, the input of a 3-input NAND gate. A voltage is coupled from the 0 output through an inverter to produce a high on pin 4 of IC103. That insures that the monitor will stay locked onto the priority signal detected.

Crystal Oscillator Operation

The crystal oscillator is basically a Colpitts oscillator with a quarter-wave impedance-transforming network in series with the crystal and the base of transistor Q112. The associated diodes form an RF path between the crystal and transistor and are normally biased off. When a ground is placed at the series resistor by IC105, the associated diode will conduct and place the crystal in the circuit. Transistor Q113 is a frequency tripler. The output at coil L108 is rich in third harmonics, which are amplified by transistor Q114. This signal is then coupled to gate G2 of mixer transistor Q103.

AUDIO AMPLIFIER

The audio signal from the IF circuit board is routed through a deemphasis network consisting of C302, C303, and R301. The purpose of this circuit is to filter high frequencies from the audio signal before they are heard in the speaker. The signal is fed to control R1, which is adjusted to provide a comfortable listening level at the speaker.

From control R1 the signal is coupled to the base of transistor Q301, part of a differential amplifier, which also consists of transistor Q302 and common-emitter resistor R306. The result of this common connection is to combine the input and feedback signals.

The signal at the collector of Q301 is coupled to the base of constant-current voltage amplifier Q303. From the emitter of Q303 the signal is coupled to the base of transistor Q304 and through diode D301 to the base of Q305. The constant voltage drop across D301 provides a bias voltage for the driver and output stages. This voltage is applied across the two base—emitter junctions that are connected in parallel with diode D302 and transistors Q304 and Q305.

Resistor R308 drops the voltage enough to keep these three junction transistors biased on at all times. Diode D302 limits the amount of voltage that will develop across R308, thereby limiting the amount of current developed in the output stages.

Driver transistors Q304 and Q305 and output transistors Q306 and Q307 form a quasi-complementary output circuit. In this circuit a positive-going signal applied to the base of Q304 causes the transistor to conduct. This in turn causes transistor Q306 to conduct, increasing the current through the speaker voice coil. Negative-going signals cause Q305 and Q307 to conduct, decreasing the current through the speaker voice coil. This alternate increase and decrease of current through the speaker coil converts the electrical energy to audible signals.

Since the input and output signals are in phase, they tend to be coupled back to the input of the circuit. Resistor R316 and capacitor C308 at the output are for parasitic suppression and eliminate this feedback tendency.

POWER SUPPLY

The power supply consists of a dual-input power source. It provides for a 13.8V DC or 120V AC power source.

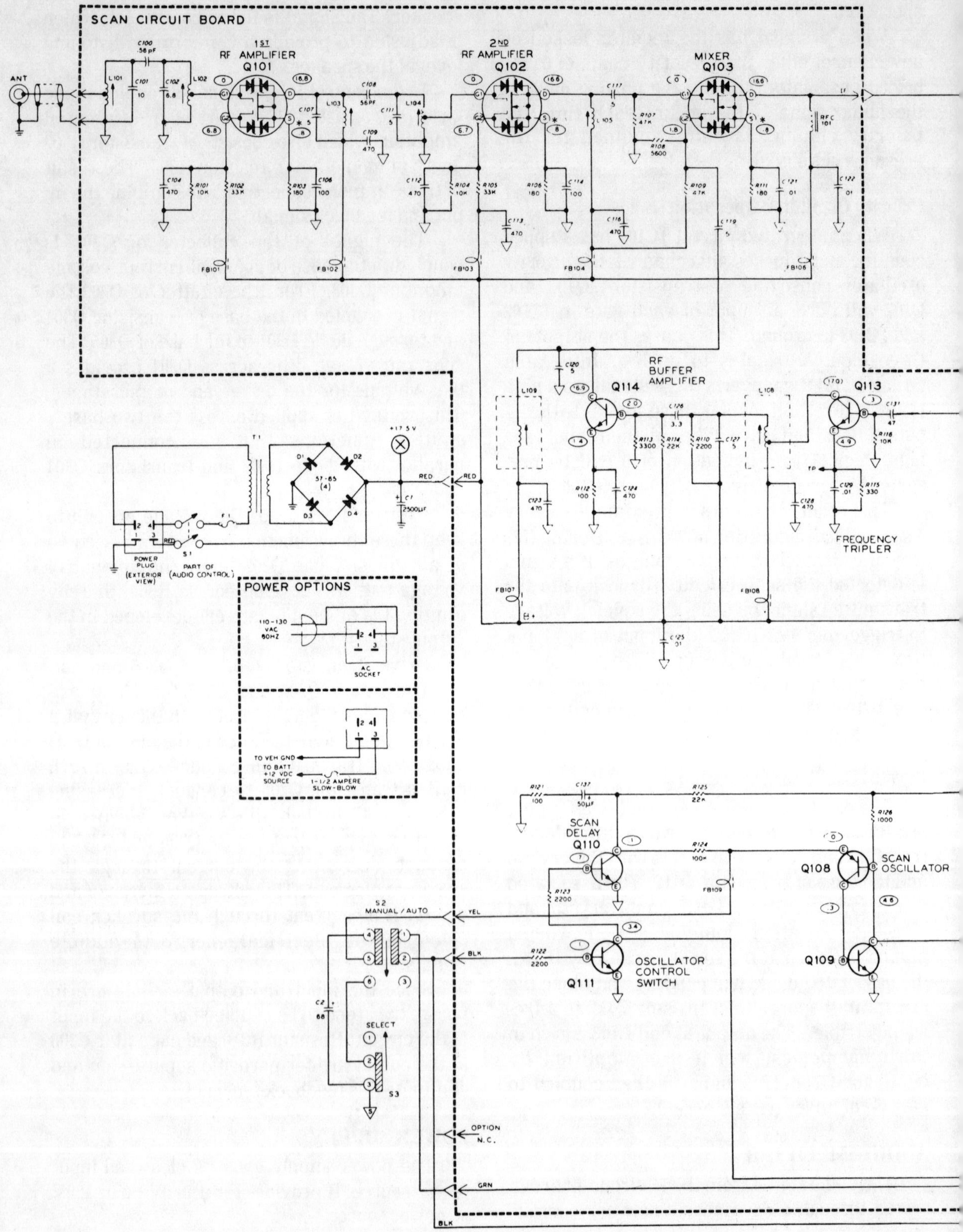
SCAN CIRCUIT BOARD
ANT
1ST RF AMPLIFIER Q101
2ND RF AMPLIFIER Q102
MIXER Q103
RF BUFFER AMPLIFIER
Q114
Q113
FREQUENCY TRIPLER
+17.5 V
RED
POWER PLUG (EXTERIOR VIEW)
PART OF (AUDIO CONTROL)
POWER OPTIONS
110-130 VAC 60HZ
AC SOCKET
TO VEH GND
TO BATT +12 VDC SOURCE
1-1/2 AMPERE SLOW-BLOW
SCAN DELAY Q110
Q108
SCAN OSCILLATOR
OSCILLATOR CONTROL SWITCH
Q111
Q109
S2 MANUAL / AUTO
SELECT
S3
YEL
BLK
OPTION N.C.
GRN

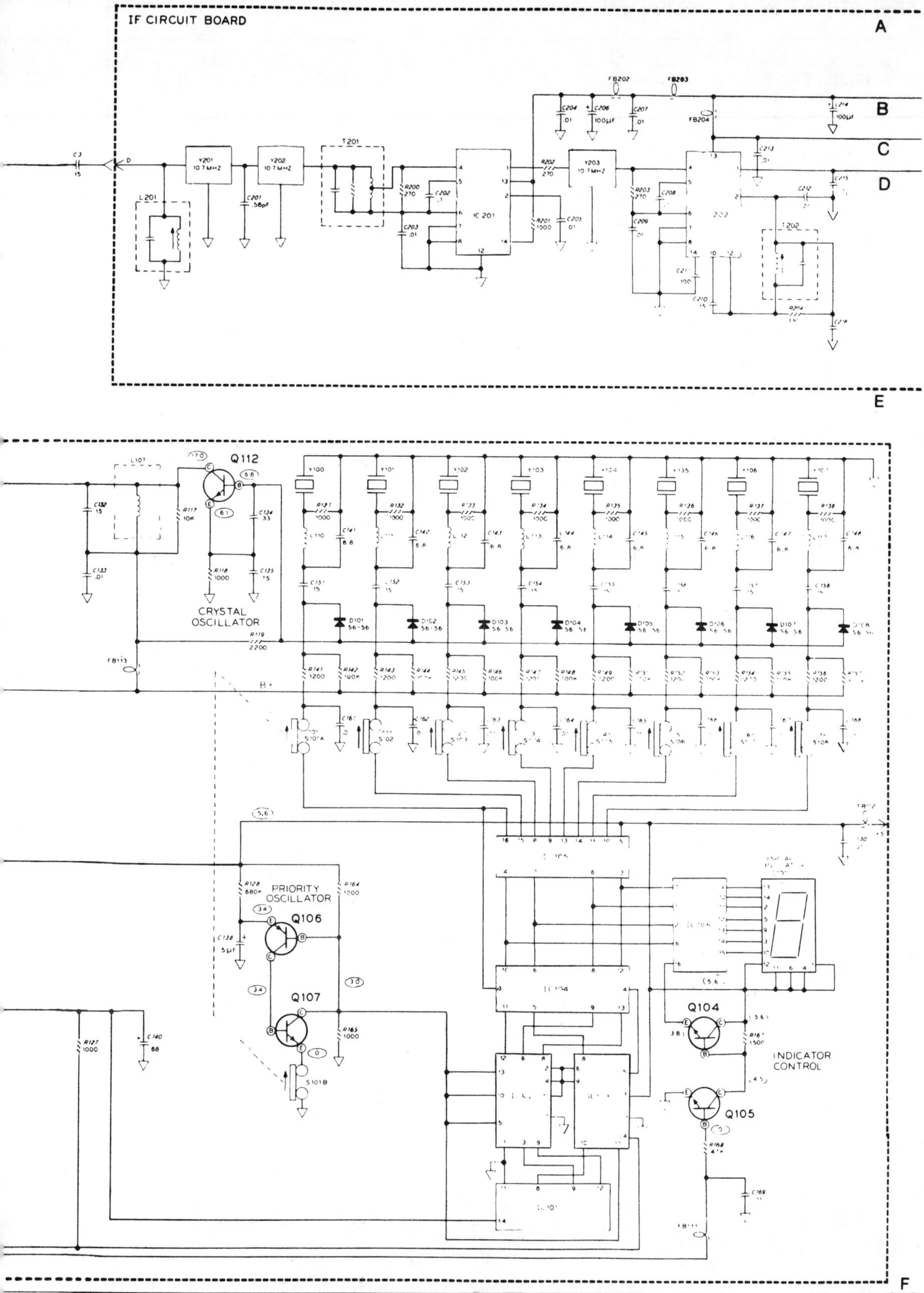

Fig. 1-2. Schematic of the Heath Model GR-110 VHF scanning monitor.

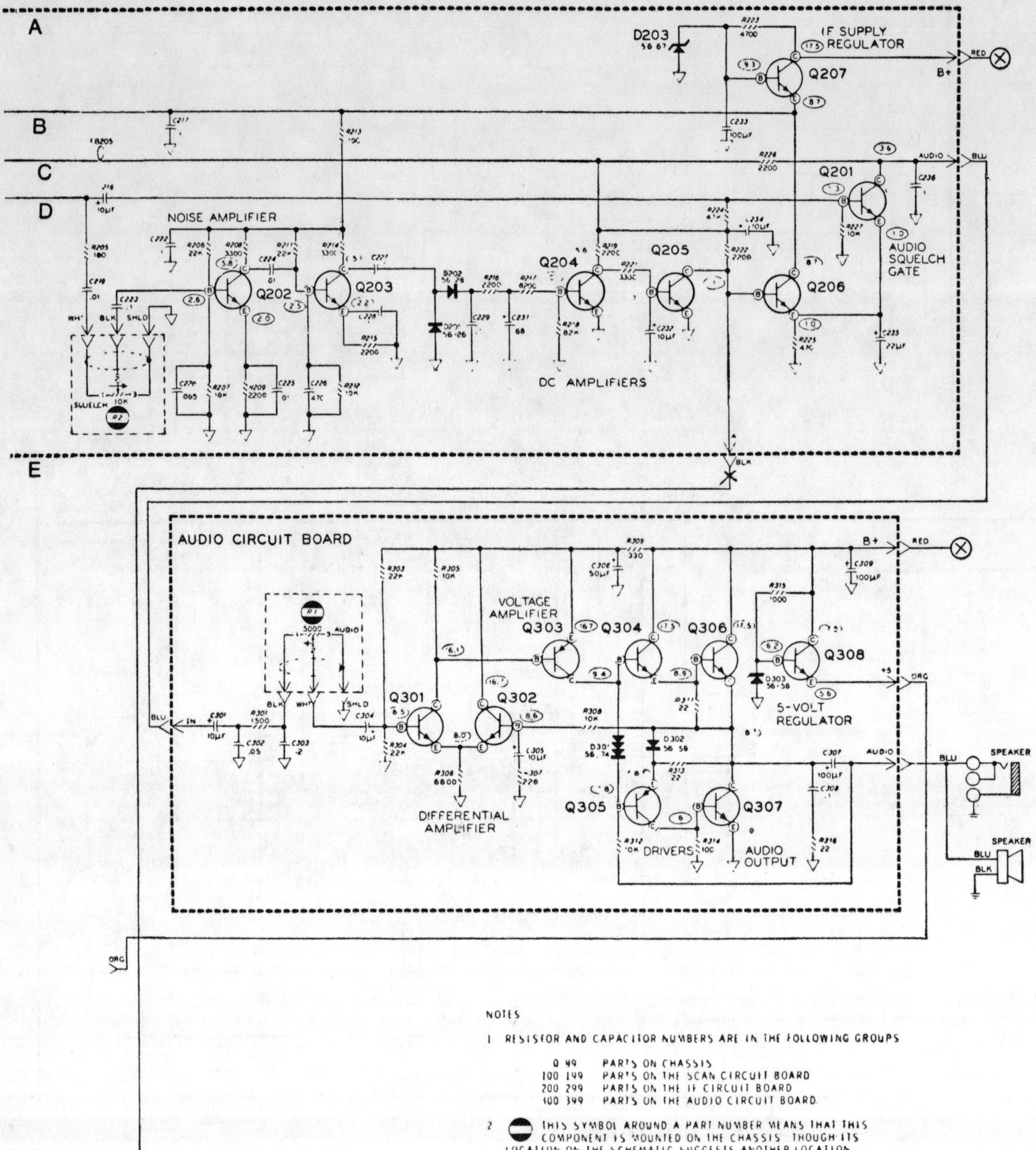

NOTES

1 RESISTOR AND CAPACITOR NUMBERS ARE IN THE FOLLOWING GROUPS

0 49	PARTS ON CHASSIS
100 199	PARTS ON THE SCAN CIRCUIT BOARD
200 299	PARTS ON THE IF CIRCUIT BOARD
300 399	PARTS ON THE AUDIO CIRCUIT BOARD

2 ⊖ THIS SYMBOL AROUND A PART NUMBER MEANS THAT THIS COMPONENT IS MOUNTED ON THE CHASSIS THOUGH ITS LOCATION ON THE SCHEMATIC SUGGESTS ANOTHER LOCATION

3 ALL RESISTORS ARE 1/2 WATT RESISTOR VALUES ARE IN OHMS K 1000

4 CAPACITORS LESS THAN 1 ARE IN μF (MICROFARADS) ALL OTHER CAPACITORS ARE IN pF (PICOFARADS) UNLESS OTHERWISE MARKED

5 ⬭ THIS SYMBOL INDICATES A POSITIVE DC VOLTAGE MEASUREMENT TAKEN WITH AN 11 MEGOHM VTVM FROM THE POINT INDICATED TO CHASSIS GROUND SQUELCH CONTROL FULL CLOCKWISE REFER TO THE VOLTAGE CHARTS ON PAGES 79 AND 80 FOR MEASUREMENTS UNDER OTHER CONDITIONS VOLTAGES MAY VARY ±20%

6 VOLTAGES MARKED WITH AN ASTERISK (*) MAY VARY BETWEEN 8 AND 1.8 VOLTS

7 ⊗ THIS SYMBOL INDICATES A COMMON B+ CONNECTION

8 THIS SYMBOL INDICATES A CIRCUIT BOARD GROUND

9 ⏚ THIS SYMBOL INDICATES A CHASSIS GROUND

10 THIS SYMBOL INDICATES A CIRCUIT BOARD GROUND THROUGH A SHIELD LEAD

11 ARROWS AT CONTROLS INDICATE CLOCKWISE ROTATION, VIEWED FROM THE SHAFT END OF THE CONTROL

12. REFER TO THE CHASSIS PHOTOGRAPHS AND CIRCUIT BOARD X-RAY VIEWS FOR THE PHYSICAL LOCATION OF PARTS

When a 13.8V DC source is used, the power is routed through ON-OFF switch S1, through fuse F1 to the primary winding of power transformer T1. Transformer T1 steps the voltage down to 17.5V AC in the secondary winding. Bridge rectifier diodes D1 through D4 convert this AC voltage to DC. It is then filtered by capacitor C1 and routed to the various B+ points on the circuit boards.

The 5V power for the scan circuits is regulated by transistor Q308 on the audio circuit board and its associated circuitry.

The 13.8V power supply is regulated by the action of a regulator and capacitance multiplier on the IF circuit board, transistor Q207, and its associated components. The primary purpose of this regulator is to keep the output of the IF circuits at a constant level.

Chapter 2

Antennas and Accessories

Antennas of all types are one of my favorite subjects, and it would be easy to get into an involved discussion of VHF antennas. However I will try to cover only a few basic points. If you are interested in pursuing the subject further—and all technicians should be—I suggest that you purchase or borrow one or more of the books devoted to antennas alone. I often refer to the "ARRL Antenna Handbook." Even if you don't have a lot of math (I don't), you can still find much food for thought. New developments are made public almost daily; if we don't watch it, our customers will hear about some new gimmick before we do.

When you hear about a new type of antenna that might improve the performance of some of your customers' systems, either find out more about it or buy one to try. Even if you operate on a budget that will not permit the purchase of many items on speculation, you probably have at least one customer who will try anything if you sell him the antenna at close to cost. Let me make one point clear: I am definitely opposed to discounting everything in the store; you don't have to. But it never hurts to let an affluent customer pay for part of the experimentation that is so necessary for you to keep up to date. If you think about it you can probably remember that you already have at least one customer who has asked you to let him in on anything new that hits the market. Just be honest with him and tell him that the only guarantee is whatever the manufacturer specifies—no more, no less. I would suggest that you make certain that he is not going to be piqued if it doesn't work out. This is an area where you will have to use your own judgment.

BASE STATION ANTENNAS

Antennas for monitoring police, taxi, government, and other FM transmissions take many forms. They can be as simple as a piece of wire or as elaborate as a multielement yagi or parabolic reflector. Usually, however, commonly used monitor antennas take one of three forms.

Dipoles

For single-band (VHF, low band, or UHF alone) local reception, a centerfed half-wave dipole (Fig. 2-1) cut to the center frequency will work quite well. This type of antenna is usually fed with 72Ω coaxial cable, but occasionally 50Ω coax is employed. The length of the coax is unimportant except for a linear loss factor (Fig. 2-2). If commercially built antennas are not available, a homebrew dipole may be easily constructed.

Dipole antennas may be either vertically polarized (Fig. 2-3A) or horizontally polarized (Fig. 2-3B). Since most stations in the VHF bands transmit vertically polarized signals, the receiving antenna must be similarly oriented for best reception.

Horizontally polarized antennas are usually used only for fixed point-to-point communica-

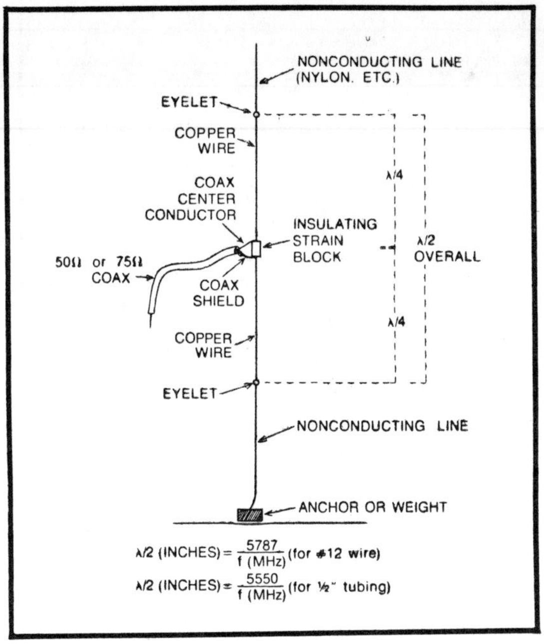

Fig. 2-1. Homemade doublet (half-wave dipole).

tions in the bands above 30 MHz. If an undesired signal is interfering with a desired signal, a little experimentation with antenna orientation may give sufficient isolation without resorting to a much more expensive directional antenna. All horizontally oriented dipoles are bidirectional, with the major lobes broadside to the antenna (Fig. 2-4) at low angles (non-skip) of radiation or reception.

Ground Planes

For improved omnidirectional reception at greater distances, a *ground plane* antenna is recommended. These antennas normally take two forms, conventional (Fig. 2-5A) and gain type (Fig. 2-5B).

The conventional ground plane is actually very similar to a vertical dipole, but it provides some small gain over the dipole. This gain—or increased signal delivered to the receiver for a given field strength at the antenna—is due to

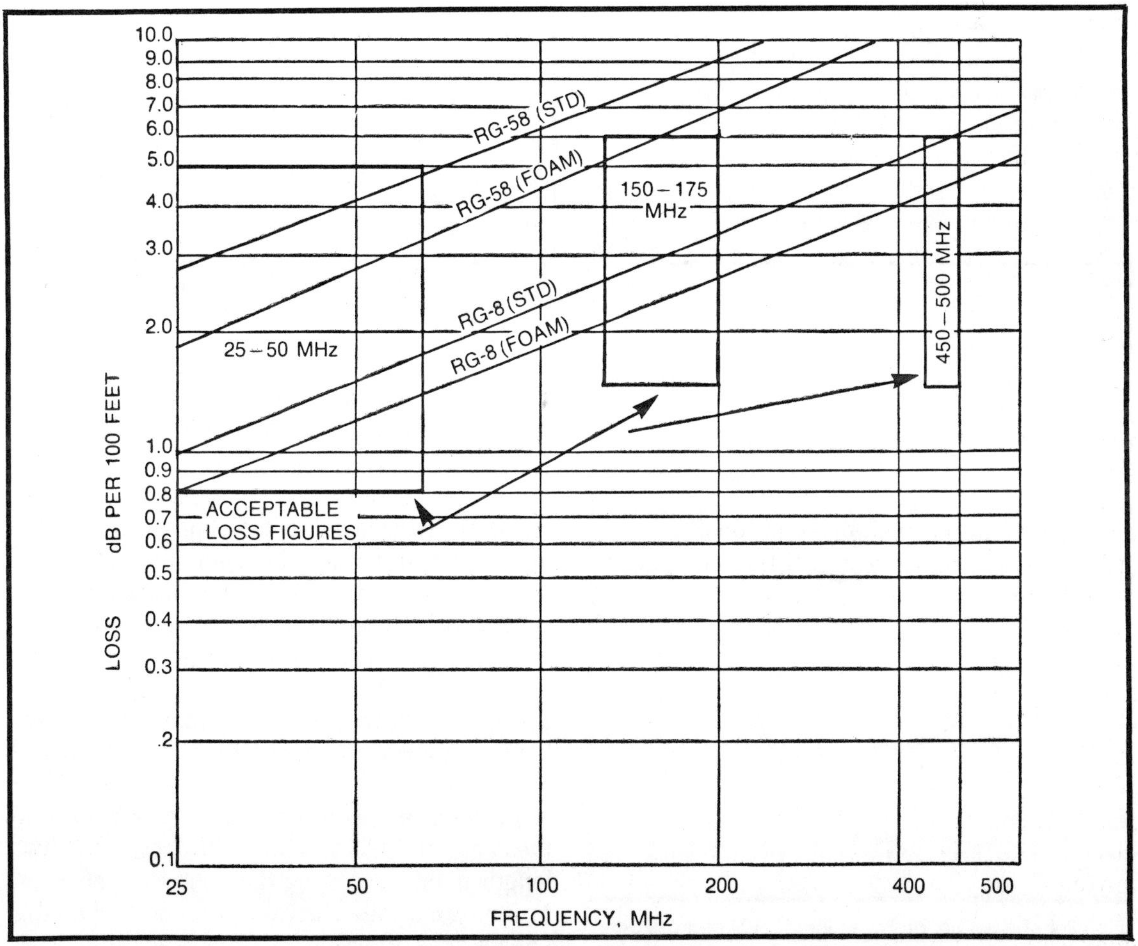

Fig. 2-2. Loss per 100 feet for selected coaxial cables. Blocks are acceptable loss figures for each band.

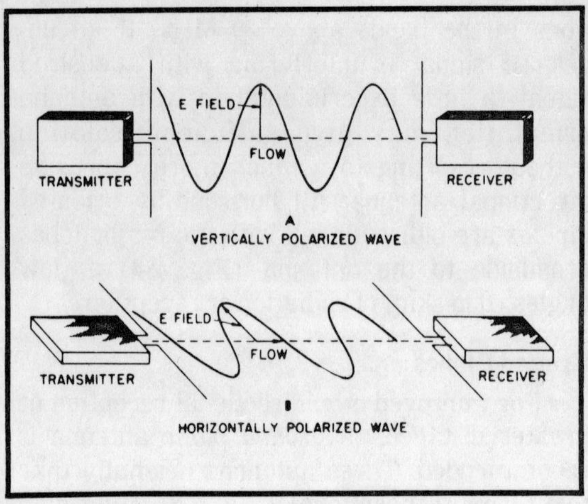

Fig. 2-3. Vertical and horizontal polarization.

the lower radiation (or reception) angle of the ground plane.

Additional gain may be provided by increasing the *captive area* of antenna. Any change in length must be compensated for by some matching device to maintain antenna impedance at the desired 50Ω to 75Ω (Fig. 2-6).

An antenna of the correct length acts as a resonant circuit and presents pure resistance to the excitation circuit (Fig. 2-6A). An antenna having other than the correct length displays both resistance and reactance to the excitation circuit. An antenna slightly longer than a half-

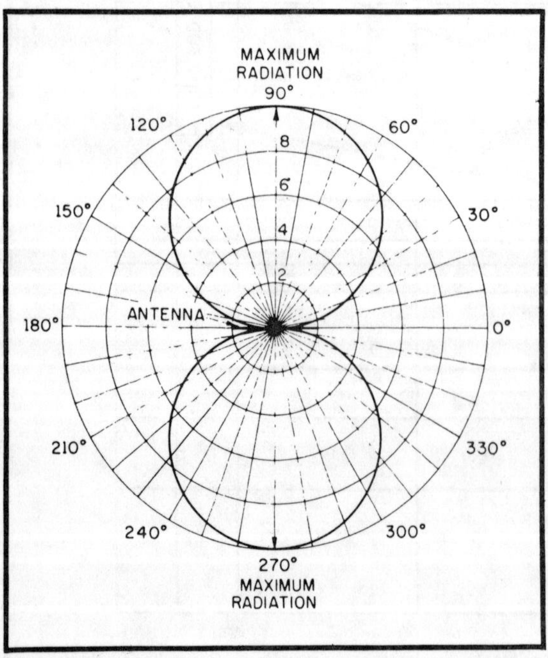

Fig. 2-4. Radiation or reception lobes of horizontal dipole.

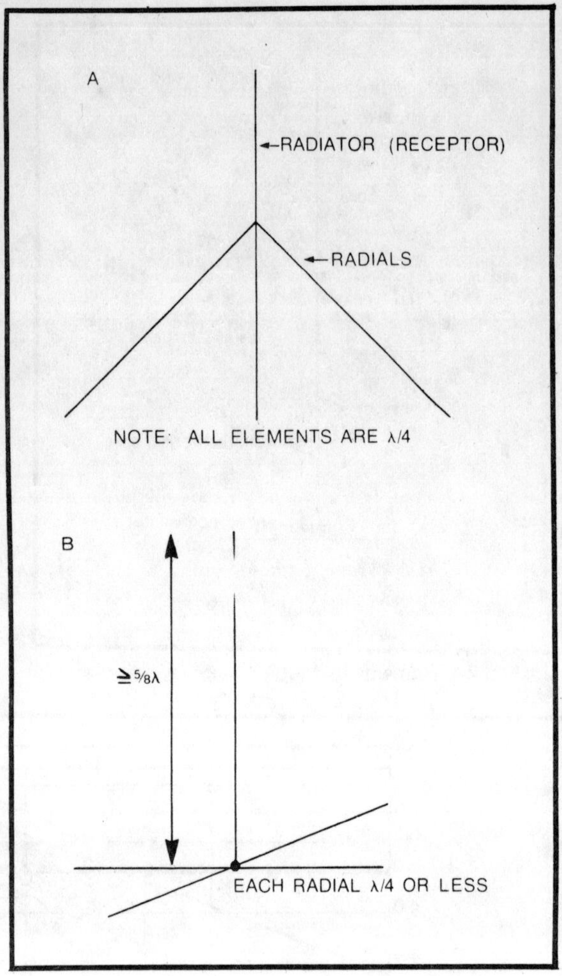

Fig. 2-5. Conventional λ/4 ground plane, in A; gain-type ground plane, in B.

wave, for example, acts as an inductive circuit. This is understandable if you think of the antenna in terms of a quarter-wave RF line. When the antenna is excited at the center, it is equivalent to a quarter-wave RF line, looking at it from the generator end. Any two-wire line longer than a quarter-wave (but less than a half-wave) is like a quarter-wave section with an open-circuited section (less than a quarter-wave long) added to it. The open section, which is capacitive, is inverted to inductance at the input terminals by the quarter-wave. In the same manner, a slightly long antenna looks inductive.

According to the equivalent lumped circuit at B, the inductance is not entirely balanced by the capacitance. The remedy for correcting the length is either cutting the conducter shorter, or tuning out the inductive reactance by adding capacitance in series.

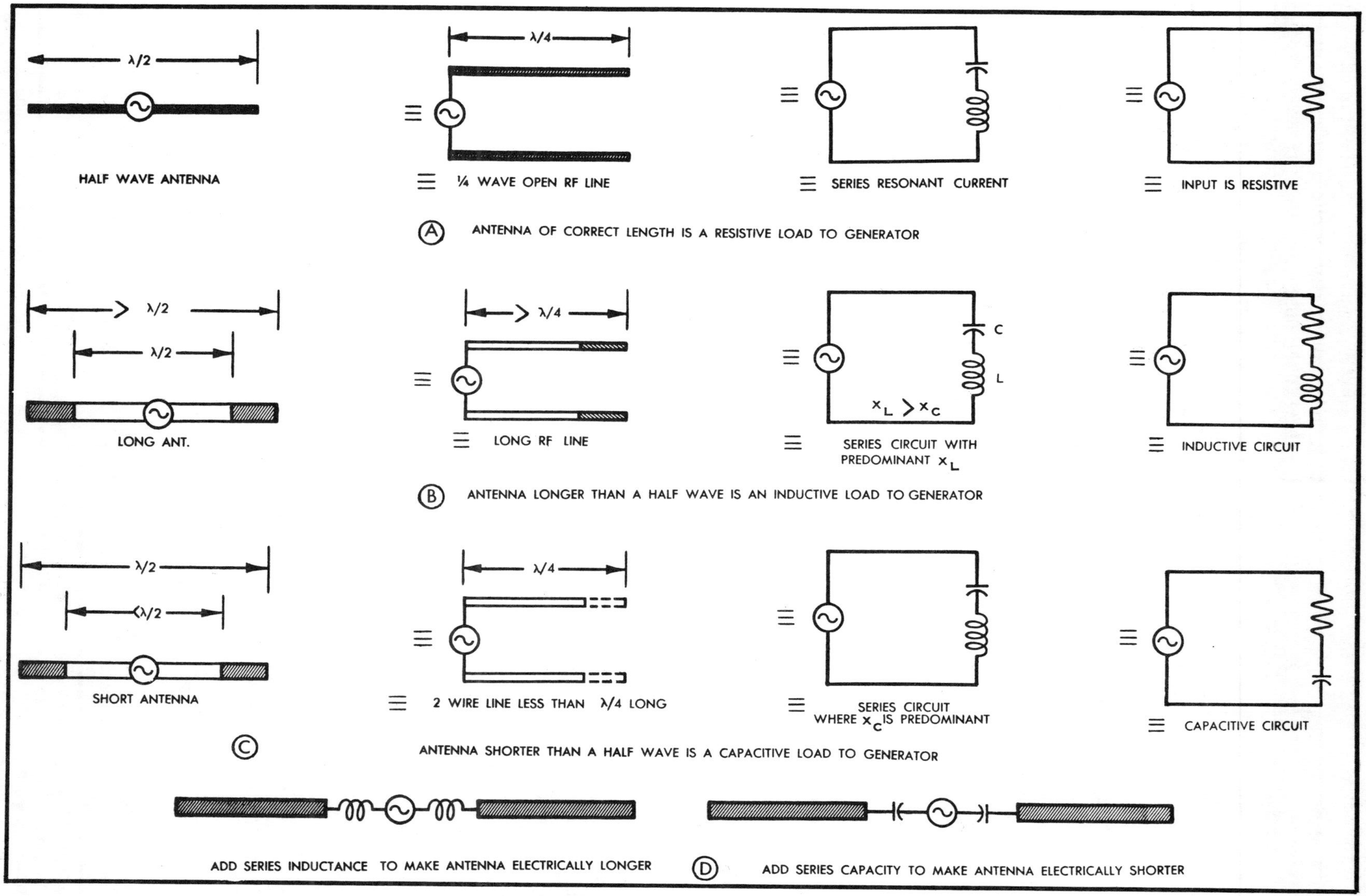

Fig. 2-6. Effect of length on antenna impedance.

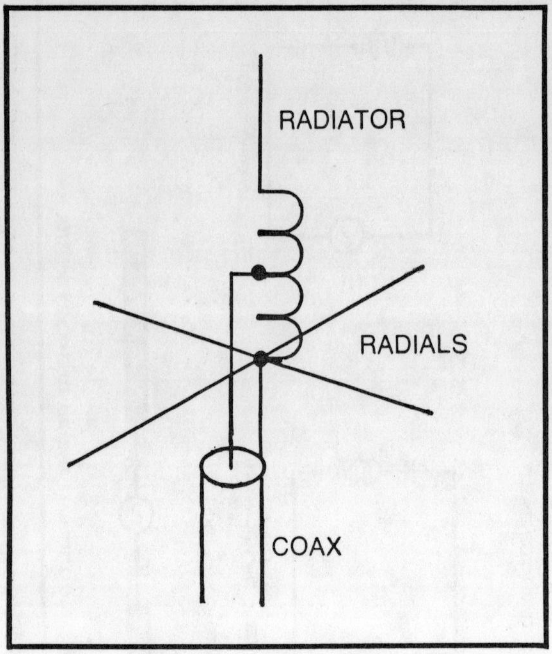

Fig. 2-7. Impedance-matching device in base of gain-type ground plane.

If the antenna is physically shorter than its resonant length, the input impedance becomes capacitively reactive. The two-wire line shorter than a quarter-wavelength displays capacitance at its input terminals. (See Fig. 2-6C.) The correction in this case is either to add inductance in series with the antenna (see Fig. 2-7) to bring it back to resonance or to add physical length to the antenna.

Figure 2-8 is an example of a gain-type ground plane antenna. Remember that the gain figure of any antenna is referenced to either a dipole cut to the same frequency or a theoretical antenna known as an *isotropic source*. While an isotropic source does not have any counterpart in fact, it could be defined as an antenna which radiates a signal of equal amplitude in all directions—a perfect sphere of radiation.

For the purposes of this explanation, you may assume that radiation angle and reception angle are identical, and for all practical purposes gain figures may be applied equally to antennas used for reception or transmission.

The physical construction of ground plane antennas is normally of rigid aluminum, steel, or fiberglass. This strong, light construction facilitates multibanding.

It is quite possible to design an antenna that will provide good performance and even gain on two or more bands without adjustment. This configuration (Fig. 2-9) uses traps to make the antenna appear one length at the center of the VHF band and another length at the center of the UHF band.

Another advantage of the multiband ground plane is that a single downlead will handle all signals. Splitting of signals may be accomplished either externally (as with MATV signals in a hotel) or internally, depending on the design of the monitors.

A disadvantage of multiband VHF antennas is that most are suitable only for reception. The relatively high reflected power of these receive-

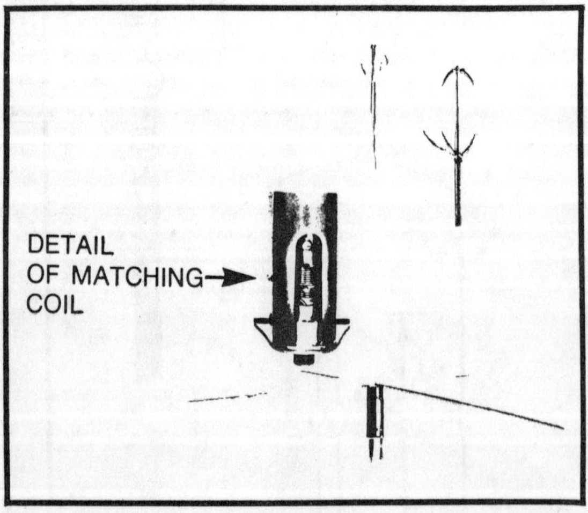

Fig. 2-8. This gain-type ground plane antenna is designed for CB but may be modified for low band by shortening elements.

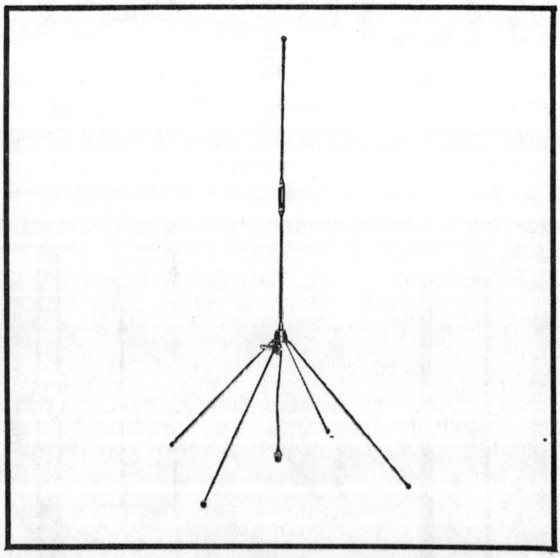

Fig. 2-9. Multiband ground plane.

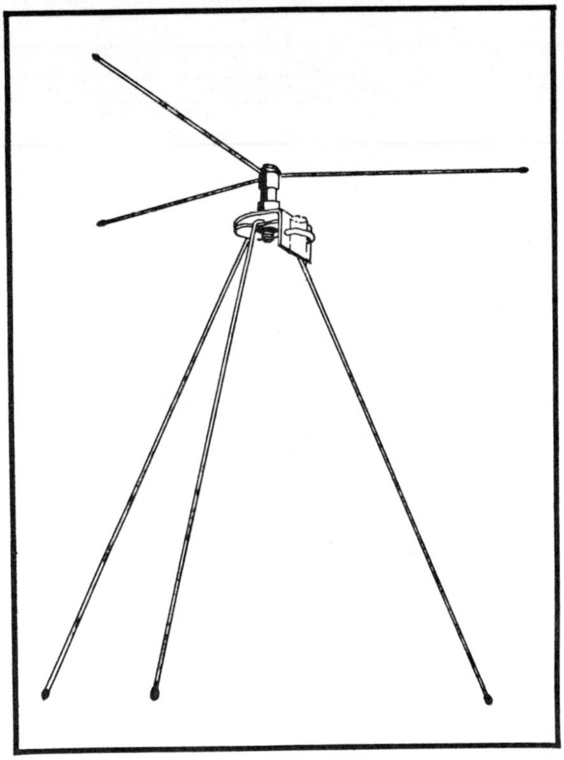

Fig. 2-10. Hustler discone antenna, for reception of 40–700 MHz.

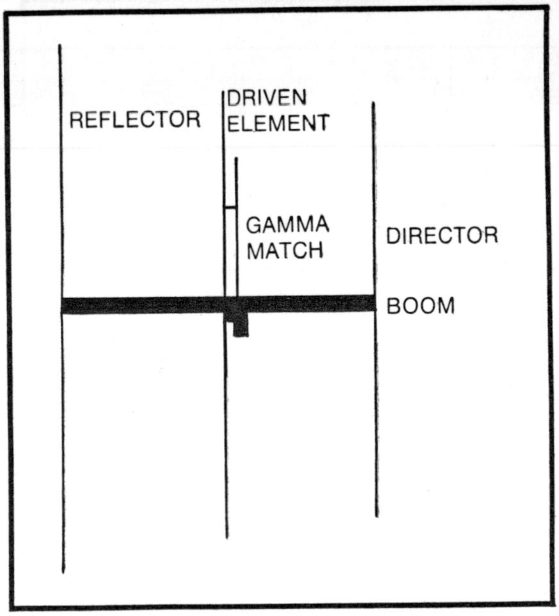

Fig. 2-11. Three-element yagi antenna.

only antennas could damage some transmitters. This is mentioned here because customers frequently ask if the antennas can be used for transmitting in the same bands or other bands such as CB. It seems that everyone wants to save a buck.

Money saved on an antenna system is usually no bargain. However, if reception of a wide range of frequencies with a single antenna is essential, a *discone* antenna (Fig. 2-10) may be the answer.

Beam Antennas

The third type of antenna commonly used for base station monitoring is the 2- or 3-element directional antenna, or *beam* (Fig. 2-11). Directivity and gain in a particular direction increase with the number of elements; but unless DX is the important factor, two or three elements are adequate.

An example of where a beam might be indicated would be an installation 20 or 30 mi from a city where the desired signals originated. A 3-element beam will provide between 6 and 8 dB forward gain and good front-to-back ratio and side rejection.

The 2-element beam gives a cardioid pattern (Fig. 2-12) sufficient for broadly directional suburban reception. It does not provide the forward gain of beams with more elements, but extreme directivity is not always desirable. This is especially true with respect to reception of mobiles working over several square miles.

MOBILE ANTENNAS FOR MONITORING

Since most scanning monitors are installed in vehicles, it is important that the installing technician be knowledgeable about the various mobile antennas.

There is almost an infinite variety of mounts available. This chapter will cover the antenna itself.

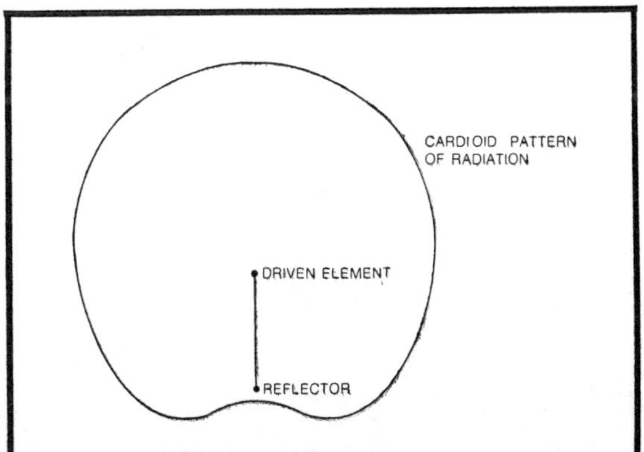

Fig. 2-12. Two-element beam and its associated cardioid pattern.

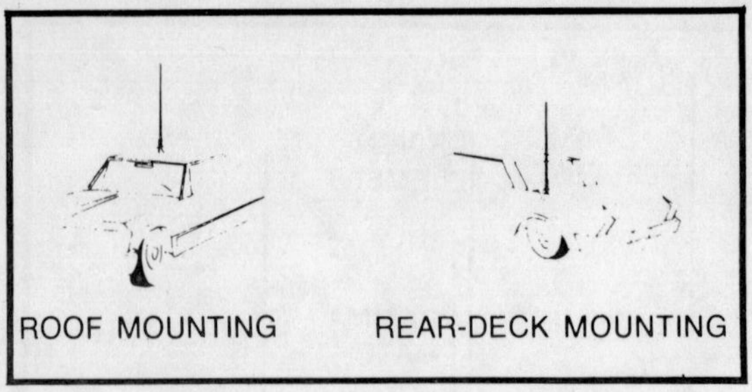

Fig. 2-13. Quarter-wave whips in typical mounting positions.

Almost without exception, mobile monitoring antennas are vertically polarized and omnidirectional, at least as far as the mounting position permits.

Quarter-Wave Whip

The basic antenna is the coaxially fed quarter-wave whip (Fig. 2-13). This antenna has *unity* gain (compared to a basic dipole) and has a feed point impedance of approximately 50Ω.

The basic dipole is a centerfed half-wave antenna, and the basic quarter-wave whip is electrically identical. The "other half" of the dipole is the mirror image presented by the ground plane surface of the vehicle (Fig. 2-14.)

The formula in Fig. 2-1 can be used for computing antenna length by using one-half of the total length for a half-wave dipole. For example, if the overall length of a half-wave antenna is calculated at 216 in., a quarter-wave whip at the same frequency would be equal to 108 in. Always allow an extra few inches before final trimming: It's much easier to cut off than to add on.

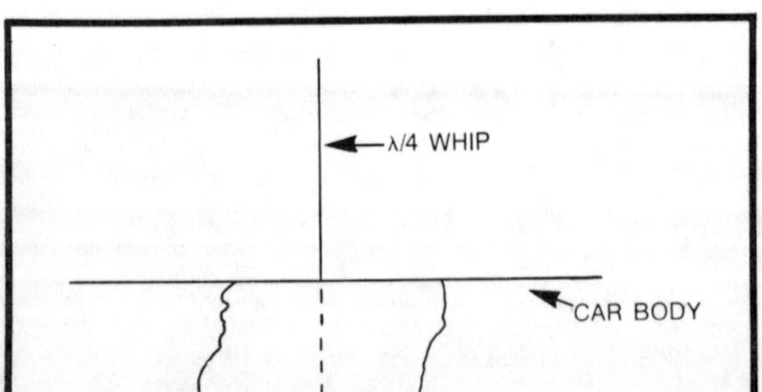

Fig. 2-14. With mirror image, quarter-wave whip is electrically identical to half-wave dipole.

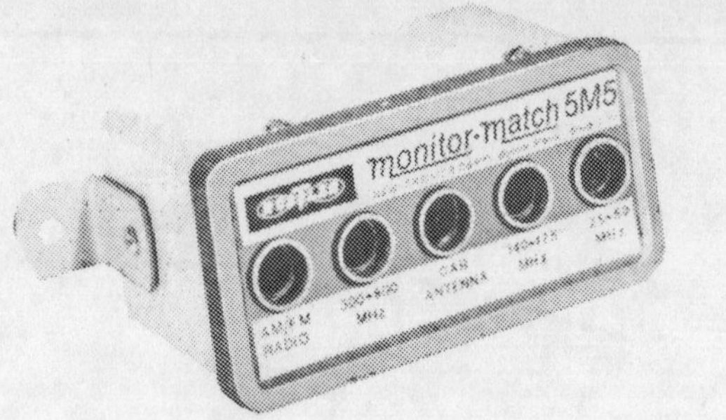

Fig. 2-15. Hustler's Monitor Match 5M5 is an inexpensive signal splitter for feeding several monitors from one antenna and downlead.

Multiband Monitoring Antennas

Several telescoping antennas that will permit optimizing of any one band of frequencies are currently manufactured. Their biggest drawback is that using them is the same as using a single whip. Performance on other bands suffers drastically without retuning. For maximum reception each monitor or band should have its own antenna. However not everyone wants that kind of performance badly enough to make a vehicle look like a porcupine.

If some compromise of performance is acceptable, a signal splitter such as the Hustler *Monitor Match 5M5* (Fig. 2-15) may do the trick. This device is intended solely for reception, but it has four outputs for reception of AM–FM, low-band (25–50 MHz), high-band (140–175 MHz), and UHF (300–500 MHz) signals. The *5M5* is designed to use the standard automobile antenna. There are also several combination antennas.

The combination antenna (Fig. 2-16) is usually a better choice than a signal splitter alone, at least in terms of performance.

The choice of how many and what type of antenna is ultimately the customer's. He should be reminded that no antenna is, or can be, all things to all people. As in all things there is no way that cost and space utilization may be cut drastically without sacrifice in other areas—performance and reliability, to name two. You simply can't get something for nothing.

MONITORING ACCESSORIES

Signal splitters might properly be called accessories, as are preamplifiers; but I prefer to consider them part of the antenna system. I must confess some bias against splitters and preamps, whether they are separate or combined.

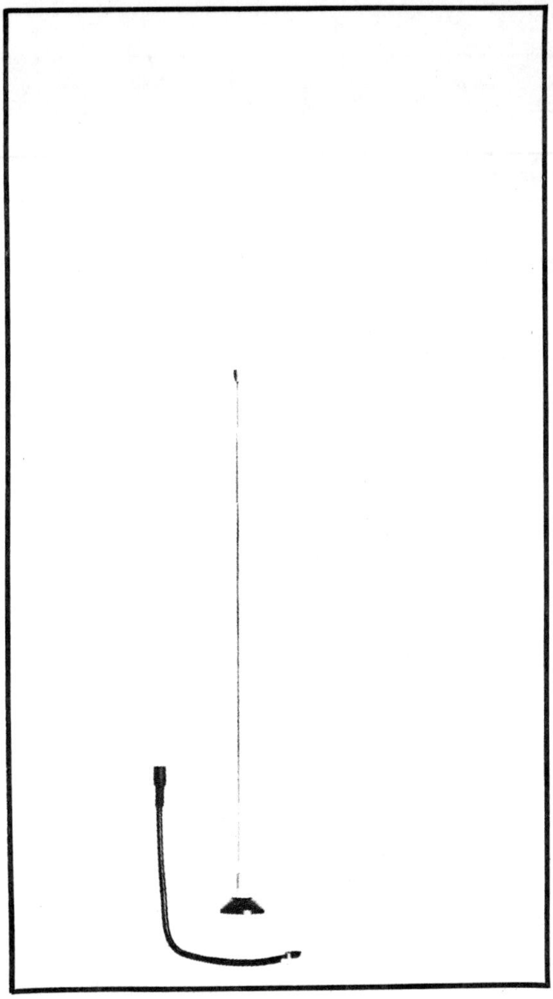

Fig. 2-16. Hustler HFT, a typical high–low multiband mobile antenna.

Preamplifiers really shine when amplifying pulses or on–off signals such as are used in television or RTTY (radio teleprinters). On voice signals preamps may boost the received signal, but they also boost the noise and even add some noise themselves—all of them. This is not to say that preamps are not worth considering in fringe areas. A characteristic of FM is that many noise signals (which are largely AM) may be filtered out and suppressed.

One final justification for using a preamp would be an installation that required reception of signals at a wider spread of frequencies than the design pass band of the monitor.

Most monitors are designed to receive a 6–8 MHz segment of any band, especially in the low band. For example, if a unit were tuned to receive 33–39 MHz and the customer needed reception of both 33.18 MHz and 47.60 MHz, you'd have a problem. No monitor receiver known will give good performance over this 14 MHz spread. If receiver sensitivity is approximately 0.7 μV from 33 to 39 MHz, sensitivity might be as poor as 50 μV, or worse, at 47 MHz. Compromise tuning (for center frequency of 40 MHz) would most likely make both 33 and 47 MHz useless; a good hot preamp would extend the effective pass band somewhat—perhaps enough to cover the desired frequencies.

External speakers are useful accessories, particularly if the monitor is operated in an environment of high ambient noise such as a truck or factory.

A speaker built into a metal or plastic case can be mounted away from the scanner and thus closer to the operator's ear.

Be careful to watch the scanner's audio output for signs of overheating or excess current whenever auxiliary speakers are added. Some scanners are direct coupled at the audio output and can easily be damaged by excessive audio current drawn by too many or mismatched speakers.

Private listening is possible using headphones or earphones—if you don't mind risking ear damage. Most headphones require far less drive than do speakers, and listening volume should be approached slowly from the low side.

A spare power cord for DC operation and a portable antenna can be used with a battery pack (either wet or dry) to provide emergency reception when AC power is out. This could come in quite handy in the aftermath of a flood, tornado, or other natural disaster.

Chapter 3

General Service Information

Before any monitor is sold or put on display, it should be checked for proper operation. This checkout does not require special equipment, and actually the only requirements are power, proper crystals, and an antenna.

PRESALE CHECKOUTS

Follow the manufacturer's directions for installing and programming the crystals for local "hot" frequencies. Then merely listen to see if all channels are receiving properly. If there is doubt as to proper operation, then it's time to go to the bench.

There is no really firm procedure for bench checkouts, but the following items should be checked:

1. Power consumption
2. Receiver sensitivity
3. Scanning and manual operation normal
4. Proper squelch operation

If one or more channels don't receive but scanning operation is normal, the problem may be faulty or wrong crystals. Doublecheck them. Another common problem is defective crystal sockets or switching diodes. A large percentage of problems with any new electronic gear can be traced to faulty solder joints on the printed circuit board (PCB). Always check the simple things first.

Sometimes you may have to change the center frequency of a unit to receive a special frequency. Often merely retuning is sufficient. However many units receiving in the 30–50 MHz range require simple component changes. If enough information is not supplied by the manufacturer (or in this book), a call or letter to the manufacturer will usually get you whatever information you need.

Don't expect a unit to receive 33.0 MHz and 47 MHz with the same RF board at the same time. Compromise tuning to accomplish this kind of frequency spread will lower overall sensitivity to something like 20 or 30 μV, which means that only relatively strong signals may be received.

Explain the problem to your customer and tell him that this problem either requires a unit with two low-band RF boards (available from most manufacturers), or alternatively he may purchase a single-channel unit for one of the frequencies.

INSTALLATION

As a rule monitor installations are not particularly critical. Don't take this to mean that just any old antenna will work as well as any other. Chapter 2 contains information to help you choose the right antenna. Follow the instructions supplied with the antenna unless you are well experienced and have developed your own procedures.

Many antennas used for monitoring (base and mobile) will measure a DC short between the center and shield of the coax. Only very rarely is an antenna loading coil defective. Watch carefully to see that you do not introduce a *real* short at either the antenna or coaxial ca-

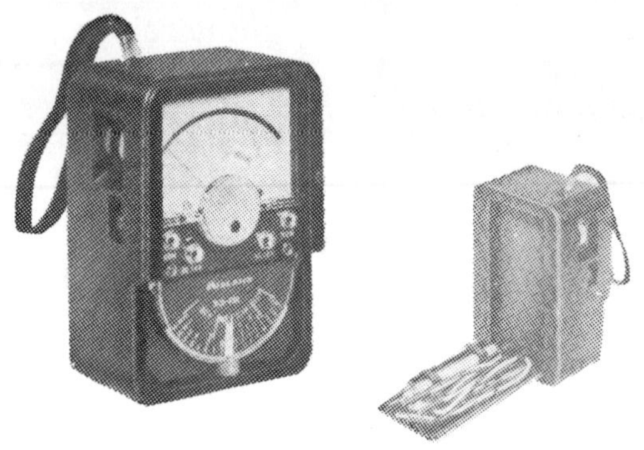

Fig. 3-1. Midland 23–151 FETVOM.

ble connector. One very small wire out of the braid can almost totally block reception. Tuning an antenna to a transmitter is much easier than trying to tune an antenna to an intermittent received signal. Luckily, antennas are less critical from a receiving point of view.

Remember that a CB antenna doesn't do much of a job receiving VHF signals. Be certain that the proper antenna is used.

If reception is noticeably poor after an antenna installation, start over again at the coax connection and check until you find the problem. Broken or shorted coax is the most common problem. A sensitive meter (one that differentiates between 0Ω and 0.5Ω) can be used to localize the short. Figure 3-1 is a typical high-quality FETVOM. An open ground at the antenna end of the coax can cause poor reception; SWR goes up to 3:1 to 5:1 in an otherwise good system.

TOUCHUP TUNING

A good general rule is not to bother a unit that is operating properly. If touchup tuning is deemed necessary, tune for the maximum *clean* signal. Feeding a weak on-channel signal into the antenna input and monitoring the demodulated audio with a scope is the most widely accepted method.

Tuning for maximum noise is almost always wrong. If you have a good rapport with one of the stations in your area—cab company, local fire or police, etc.—you may be able to get a brief on-the-air test. The only proper way is using a highly stable generator such as the Lampkin 107C (Fig. 3-2). Granted, you will have to fix a lot of scanners to pay for the Lampkin; but if you already are servicing other communications equipment, it is well worth the approximately $3000. If you dial 155.25 MHz into the Lampkin, that's just what you get (±0.5 ppm). Either AM or FM from 50 to 6000 Hz is available at the flip of a switch. By the way, the 107C also measures frequency up to 1 GHz and FM deviation. The built-in attenuator lowers output to less than 0.1 μV and you read the output directly from the dial.

ISOLATION OF PROBLEMS TO SPECIFIC SECTIONS

Scanner operation is related to proper squelch operation since opening the squelch stops the scanning process. Don't be too quick to blame the logic board. Failures are much more common in the IF or squelch section. A suspected failure can sometimes be bridged with a 0.001 μF capacitor to restore partial operation. Noise-operated squelch systems do not work if the noise amplifier section fails. Usually voltage checks will pinpoint the trouble spot.

For most of us, complementary-symmetry audio systems are harder to signal-trace than either single-ended or push-pull audio systems. I usually have to sit down and figure out which transistor switches what and then go over voltage readings very carefully. Use a pencil to write them in on a schematic and you won't need to juggle them in your head.

Power supply problems are usually no more complex than a defective diode or capacitor. Any time you find an IC damaged, check back to where its supply voltage originates. Defective regulation accounts for many IC failures.

If you are selling units with little or no service data available, compare the voltage

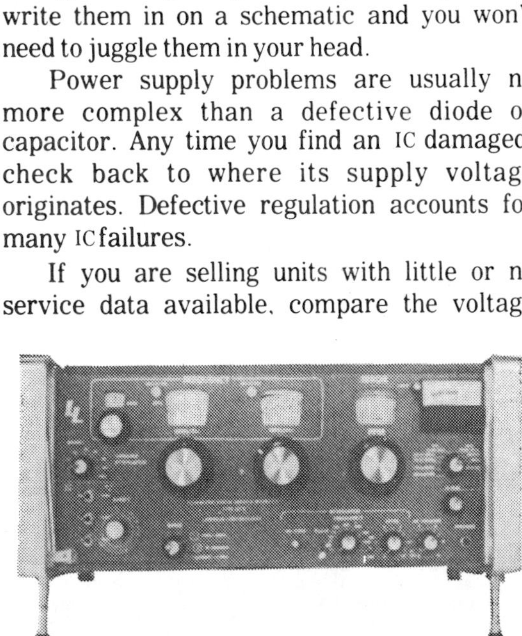

Fig. 3-2. Lampkin 107C communications service monitor.

readings on a good unit with those of the unit being repaired.

Use a sensitive RF probe to look for RF and IF output. Sometimes this can help you pinpoint the defective stage in minutes.

If an RF stage is suspected on a dual-band unit, see if the other RF stage is still doing its thing. If so, you have isolated the problem. If neither RF stage is functioning, look for the problem in the IF section.

Audio output failure is frequently caused by attachment of too many speakers to a unit. Always quiz the customer carefully (and diplomatically) about the complaint. He may give you the answer before you even look inside the unit.

SCANNERS AND THE LAW

Many local governments have ordinances prohibiting the monitoring of police frequencies. Such ordinances are unenforceable and probably pointless. If a criminal wants to monitor police calls while perpetrating a robbery, no threat of a misdemeanor monitoring conviction will deter him. The real problem is not the listening, but to what purpose the information is put.

Federal regulations prohibit making any use of intercepted transmissions (intended for someone else) except distress calls. For example, if one cab company monitors another company's calls and makes use of this information to get there first, that is illegal.

Any private individual who makes a habit of showing up at the scene of police or fire department activities will almost surely find himself with a lot of questions to answer. If he in any way hampers emergency personnel or vehicles, he will probably find himself in jail or court.

It all seems to boil down to a matter of good judgment. Listen, enjoy the excitement, and stay the heck out of the way.

Chapter 4

Regency Monitors

Regency is probably the largest producer of scanning monitors in the world. Regency's strength lies in the fact that it produces a wide range of monitors in both price and frequency capability. Its weakness is the same thing : It can be confusing to dealer and customer alike when you have too many choices. Actually the problem isn't all that bad because careful study of the specs will usually show you which unit is most nearly perfect for the proposed usage.

For example, if a customer wants to monitor UHF only, there is no reason to consider a unit which also incorporates low-band and VHF capability. Also bear in mind that not all the units for which service data is included in this chapter are still in manfuacture. Additionally, Regency is in the process of paring their lines, and new units are constantly being introduced to replace several older units by selecting options of a basic model. The out-of-production models are included here because many of them are still in use.

Since many of the Regency units use the same basic circuit boards, I have endeavored to avoid repeating data excessively. Follow the instructions carefully and make sure that you are using the right data for the right unit and you should have no problem.

UNITS COVERED

Despite many similarities among units, there are differences. Therefore all available schematics are included. While electrical differences exist, many units are identical in layout and tuning data. In these cases redundancy is avoided by including only the layouts with the most detail.

Keeping the above in mind, remember that not every part shown on the layout illustrations is incorporated in every unit using that layout. The same holds true of parts lists. However, where any likelihood of ambiguity exists, additional illustrations and text are provided.

To use the service information in this book, locate the unit being serviced in the listing at the start of the chapter and look to the right of the listing for the page location of the unit in question.

In the alignment data for each unit, you will find that a series of lettered steps (A, B, C, etc.) are suggested. Be careful to follow the instruction under each letter exactly, at least until you are familiar enough with the unit to form your own procedure.

Units are generally grouped by capabilities. Specifications are included for each unit. When referred to a schematic and layout, doublecheck to be certain that you are using the correct data for any particular unit. Read the section on crystal data and general information before proceeding with service.

CRYSTAL DATA FOR ALL UNITS

Miniature plug-in (HC-25/U) crystals are used in the receiver. Shepherd Industries crystals are recommended by Regency. If the crystals are ordered from Regency, it is only necessary to specify part number 301-532 for high-band crystals, 301-542 for low-band crystals, or 301-603 for UHF crystals, and the desired receive frequency.

If desired, the crystals may be purchased from other manufacturers. The following specifications must be included in the order:

High-Band Crystal

1. Crystal frequency f is determined as follows:

$$f = \frac{\text{channel frequency} - 10.7\text{ MHz}}{3}$$

Example

$$f = \frac{155.55 - 10.7\text{ MHz}}{3}$$

$$= \frac{144.85\text{ MHz}}{3} = 48.2833\text{ MHz}$$

2. Frequency tolerance of 0.001%
3. Third overtone
4. Series resonance minus 450 Hz
5. Maximum equivalent series resistance of 35Ω
6. Drive level of 2 mW
7. Holder HC-25/U

Low-Band Crystal

1. Crystal frequency f is determined as follows:

$$f = \text{channel frequency} + 10.7\text{ MHz}$$

Example

$$f = 39.5 + 10.7\text{ MHz} = 50.2\text{ MHz}$$

2. Frequency tolerance of 0.001%
3. Third overtone
4. Series resonance minus 450 Hz
5. Maximum equivalent series resistance of 35Ω
6. Drive level of 2 mW
7. Holder HC-25/U

UHF Band Crystal

1. $f = \frac{\text{receive frequency} - 10.7\text{ MHz}}{9}$

Example

$$f = \frac{458.000 - 10.700\text{ MHz}}{9} = 49.70000\text{ MHz}$$

2. Frequency tolerance of 0.001%
3. Third overtone
4. Parallel resonance, 18 pF load capacity
5. Maximum equivalent series resistance of 35Ω
6. Drive level of 2 mW
7. Holder HC-25/U

REGENCY TMR-1U, -1H

Specifications

Receiver

Frequency range	TMR-1H, 150–174 MHz TMR-1U, 450–470 MHz
Antenna impedance	50Ω
Channels	one, crystal-controlled
Sensitivity (20 dB quieting method)	0.7 μV (max.)
Selectivity (IF)	6 dB down, ±7 kHz (min.) 50 dB down, ±18 kHz (max.)
Selectivity (front end)	
3 dB bandwidth	8 MHz (min.)
6 dB bandwidth	12 MHz (min.)
Spurious rejection (excluding primary image)	50 dB

Table 4-1. Voltage Data—TMR-1 Transistors.

VOLTAGE DATA—TRANSISTORS:				
	Q201	3.1	3.8	7.0
	Q202	1.6	2.3	7.1
	Q203	3.4	4.1	7.0
RF Board	TRANSISTOR	EMITTER (Source)	BASE (Gate)	COLLECTOR (Drain
No. 301-598 (TMR-1U)	Q201 (FET)	0	0	5.5
	Q202	1.5	2.2	6.8
	Q203	2.5	3.1	6.8
	Q204	7.8	7.2	3.0 – 5.0
	Q205	2.9	3.6	7.4
	Q206	2.9	3.6	7.2
	Q207	0	.2	6.8
IF Board No. 500-858	Q101	2.3	3.0	5.8
	Q102	1.0	1.7	4.8
	Q103 (PNP)	8.2	8.2	0 (unsquelched)
		8.2	8.2	1.0 (squelched)
		8.2	8.2	1.5 min. (tight squelch)
	Q104	0	0	1.9 (unsquelched)
		0	.80	.30 (squelched)
		0	.80	.10 (tight squelch)
	Q105	1.4	1.9	5.1 (unsquelched)
		1.1	.10	8.2 (tight squelch)
	Q106	0.7	1.3	12.4
	Q107 (PNP)	13.8	13.1	7.2
	Q108 (PNP)	6.9	6.6	.10
	Q109	6.9	7.2	13.8
	Q110	0	.10	6.9

AFC range	Approx ±6 kHz
Modulation acceptance	±7 kHz (min.)
Intermediate frequencies	First, 10.7 MHz Second, 455 kHz
Squelch system	Noise-operated
Audio output (3.2Ω speaker)	3W at 10% or less distortion, 5W max.

Power

Voltage requirement	117V AC (±10%), 60 Hz, 17W (max.); or 13.8V DC (±10%)
Current requirements at 13.8V DC (receiver squelched)	180 mA
(max. audio output)	800 mA
Fuse size	1.5A, 3AG

General

Front-panel size	$5^5/_8 \times 2^5/_{16}$ in.
Depth (including knobs and rear-panel connectors)	9 in.
Antenna connector	Motorola type
Power connector	4-pin, polarized
Speaker size	4 in. square

Service Data

Note: Since these are single-band, single-channel units, there is no crystal programming.

Voltage data	Tables 4-1, 4-2
Schematics	Schematics 4-1, 4-2
Internal layouts, RF	Figs. 4-2, 4-3 (TMR-1H)
RF	Figs. 4-4, 4-5 (TMR-1U)
IF	Figs. 4-6, 4-7 (all)
Alignment data, steps	(Steps are A, B, C (1H only), D (1U only) E (1H only); F (1U only); G (1H only)
Parts lists	At end of chapter

Table 4-2. Voltage Data TMR-1 IF Board Integrated Circuits.

	Pin Numbers 1 – 7							
	1	2	3	4	5	6	7	
IC101	4.2	0.7	0.7	4.2	7.8	0	4.2	
IC102	4.0	3.5	0	1.3	1.3	1.3	0	
	Pin Numbers 7 – 14							
	7	8	9	10	11	12	13	14
IC101	4.2	7.8	—	—	—	—	—	—
IC102	0	0	0.2	1.4	2.9	3.5	7.6	5.0

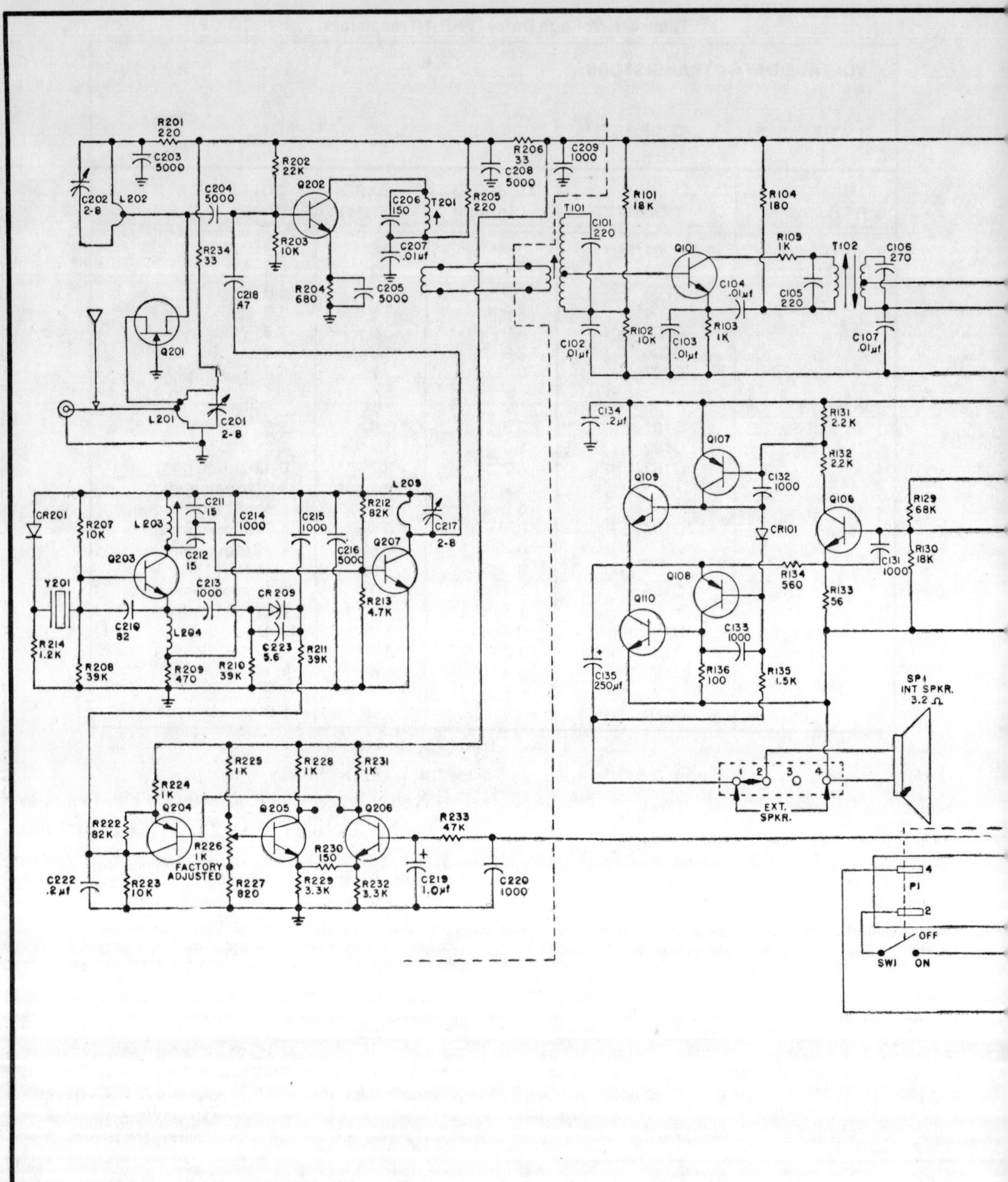
R201
220
C203
5000
C202
2-8
L202
C204
5000
R202
22K
Q202
R203
10K
R234
33
C218
47
Q201
L201
C201
2-8
C206
150
T201
C207
.01µf
R204
680
C205
5000
R205
220
R206
33
C208
5000
C209
1000
T101
C101
220
C102
.01µf
R101
18K
R102
10K
C103
.01µf
Q101
R103
1K
C104
.01µf
R104
180
R105
1K
C105
220
T102
C106
270
C107
.01µf
C134
.2µf
R131
2.2K
R132
2.2K
Q107
Q109
C132
1000
CR101
Q106
R129
68K
R130
18K
C131
1000
R134
560
Q108
Q110
R133
56
C133
1000
C135
250µf
R136
100
R135
1.5K
SP1
INT SPKR.
3.2 Ω
1 2 3 4
EXT.
SPKR.
CR201
R207
10K
L203
C211
15
C212
15
C214
1000
C215
1000
C216
5000
R212
82K
L205
C217
2-8
Q203
Q207
Y201
C210
82
C213
1000
CR209
L204
C223
5.6
R211
39K
R213
4.7K
R214
1.2K
R208
39K
R209
470
R210
39K
R225
1K
R228
1K
R231
1K
R224
1K
Q204
Q205
Q206
R222
82K
R226
1K
FACTORY
ADJUSTED
R230
150
R233
47K
C222
.2µf
R223
10K
R227
820
R229
3.3K
R232
3.3K
C219
1.0µf
C220
1000
P1
4
2
OFF
SW1
ON

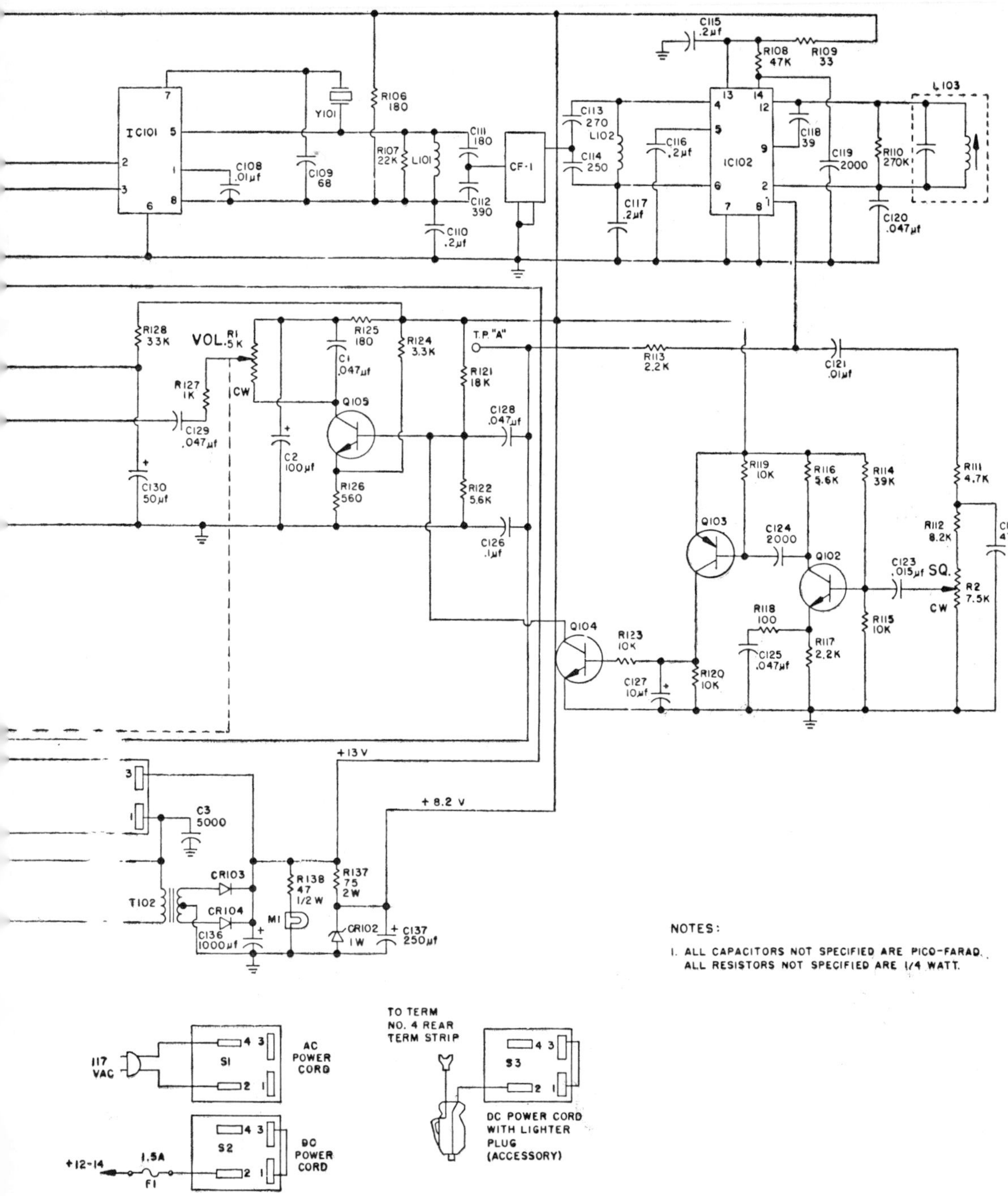

Schematic 4-1. TMR-IU.

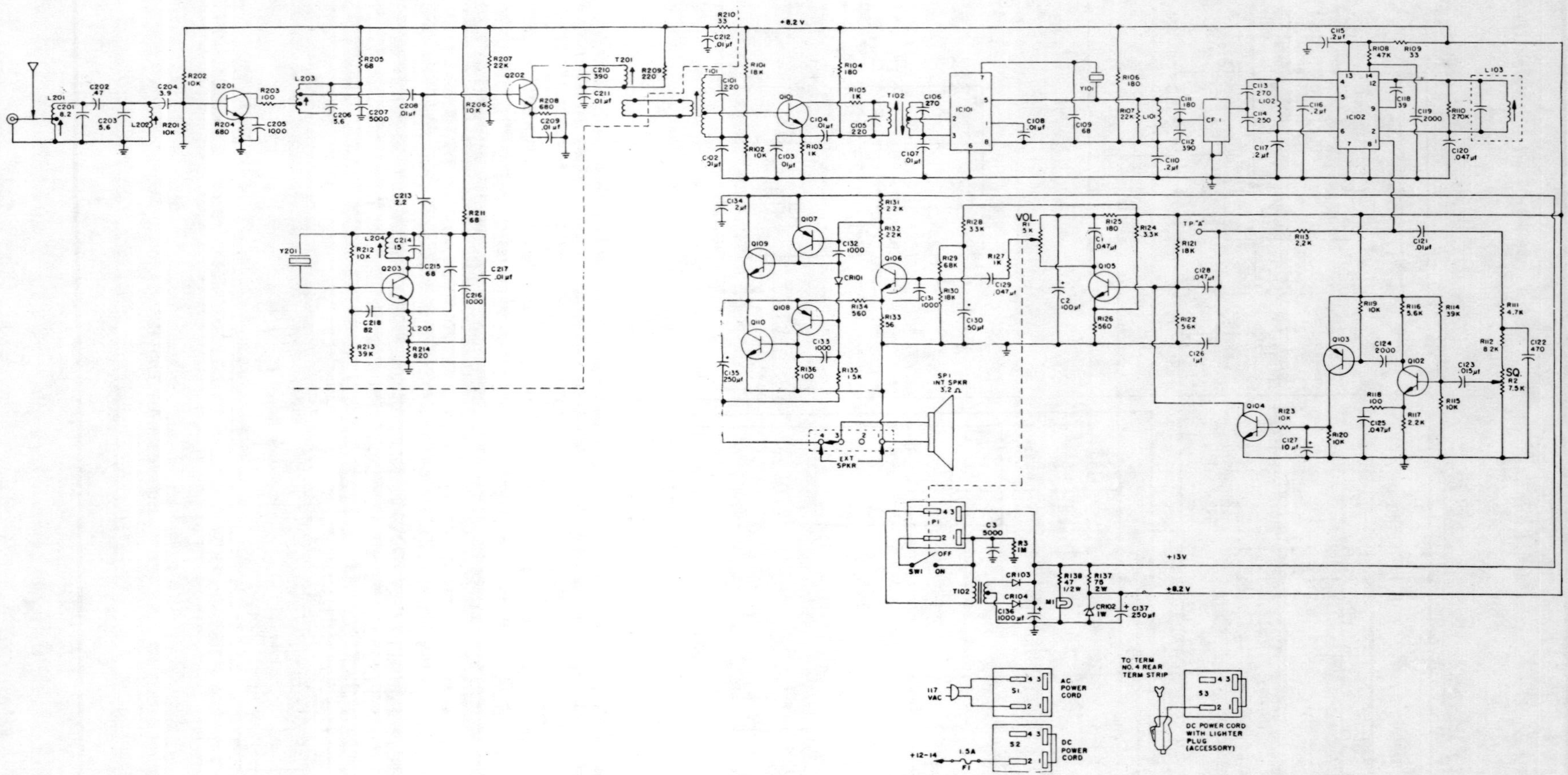

NOTES:
1. ALL CAPACITORS NOT SPECIFIED ARE PICO-FARAD
ALL RESISTORS NOT SPECIFIED ARE 1/4 WATT.

Schematic 4-2. TMR-IH.

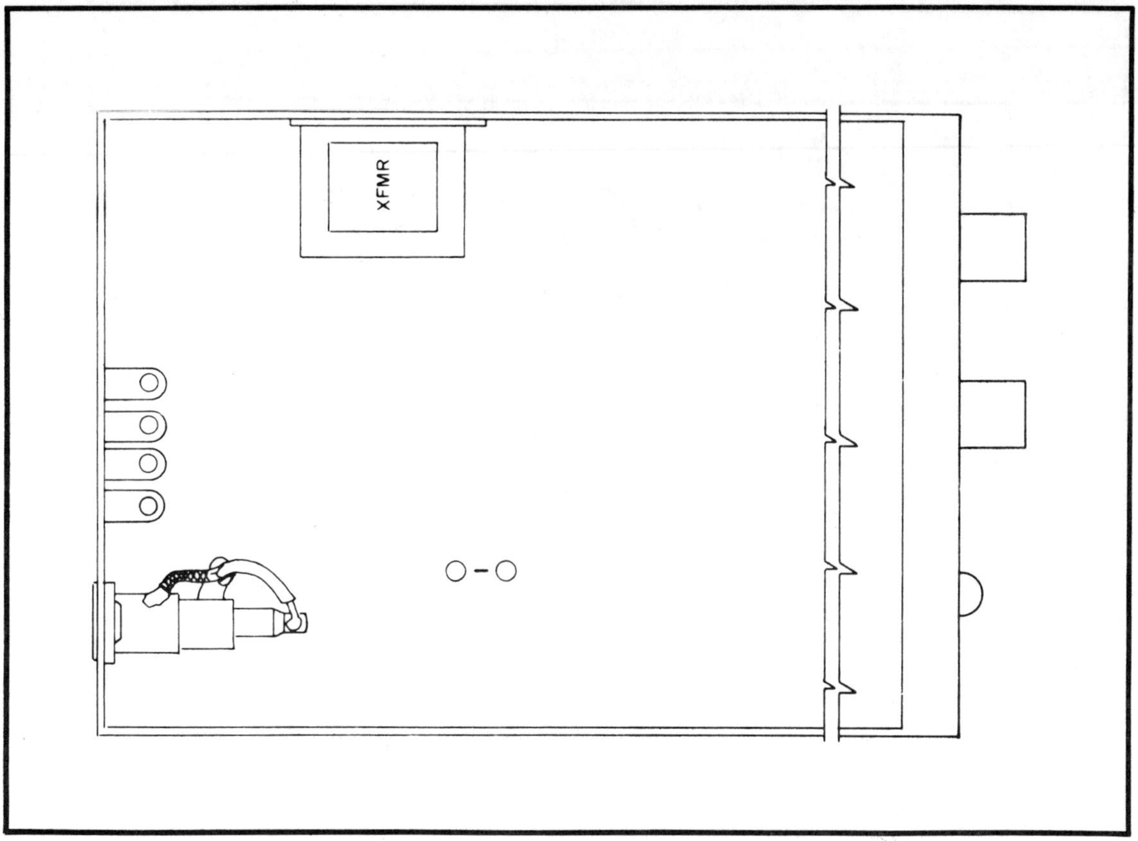

Fig. 4-1. Crystal layout for single-channel TMR series.

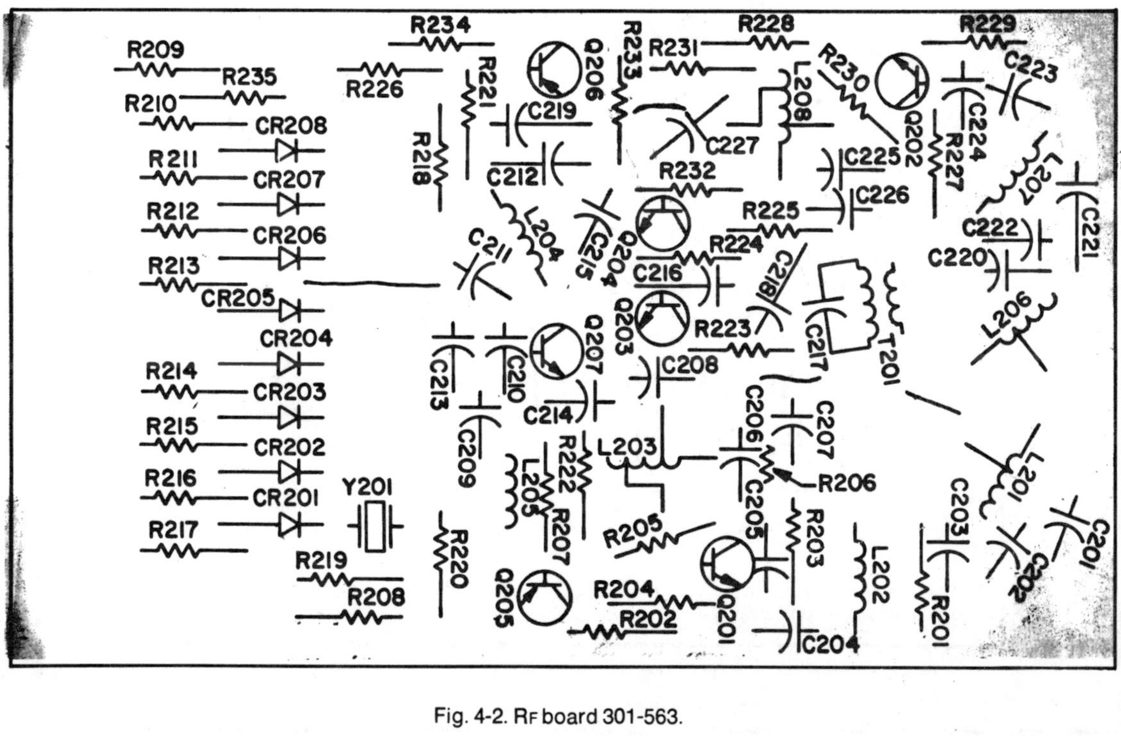

Fig. 4-2. RF board 301-563.

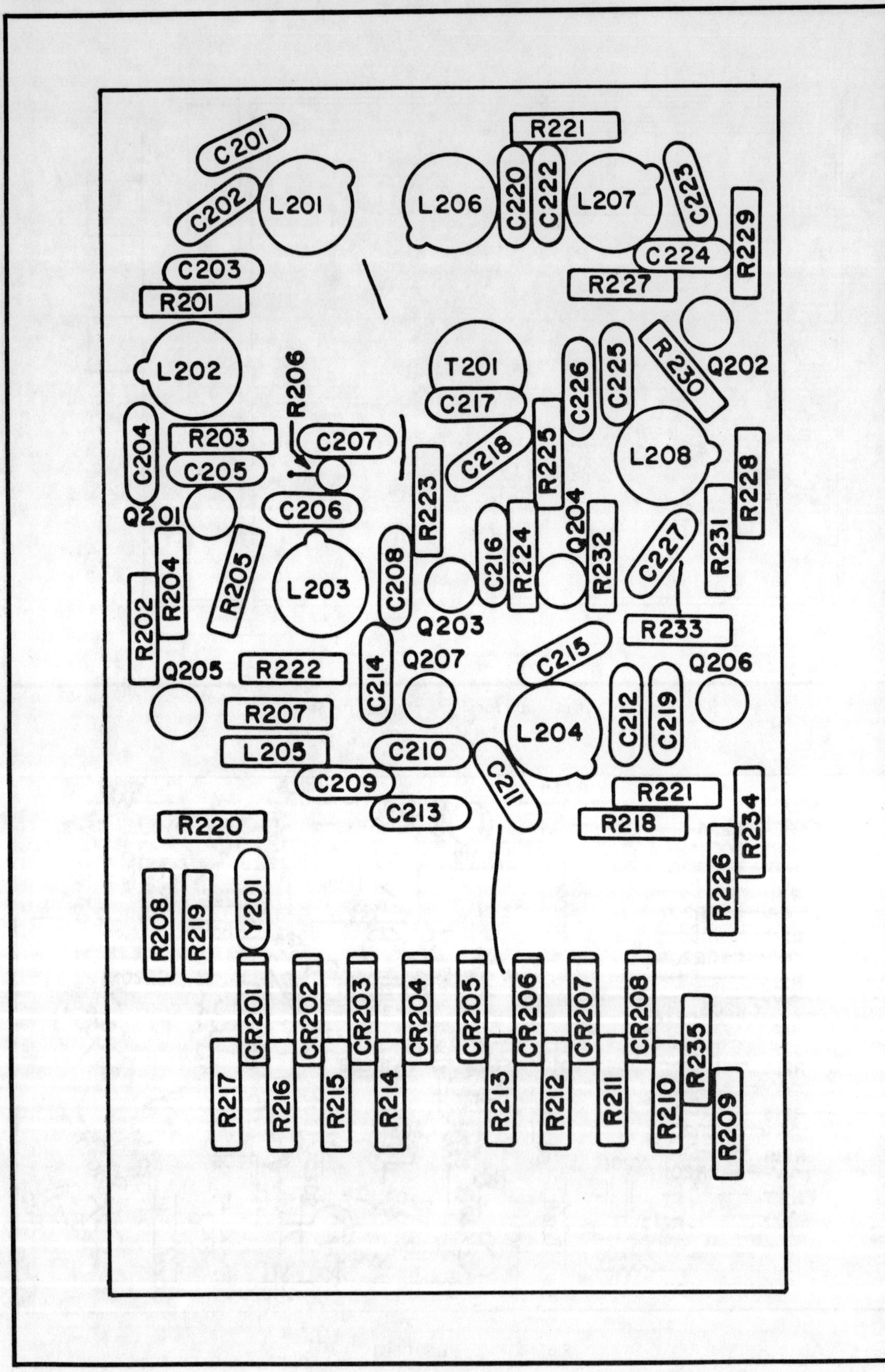

Fig. 4-3. RF board 301-563 parts placement diagram.

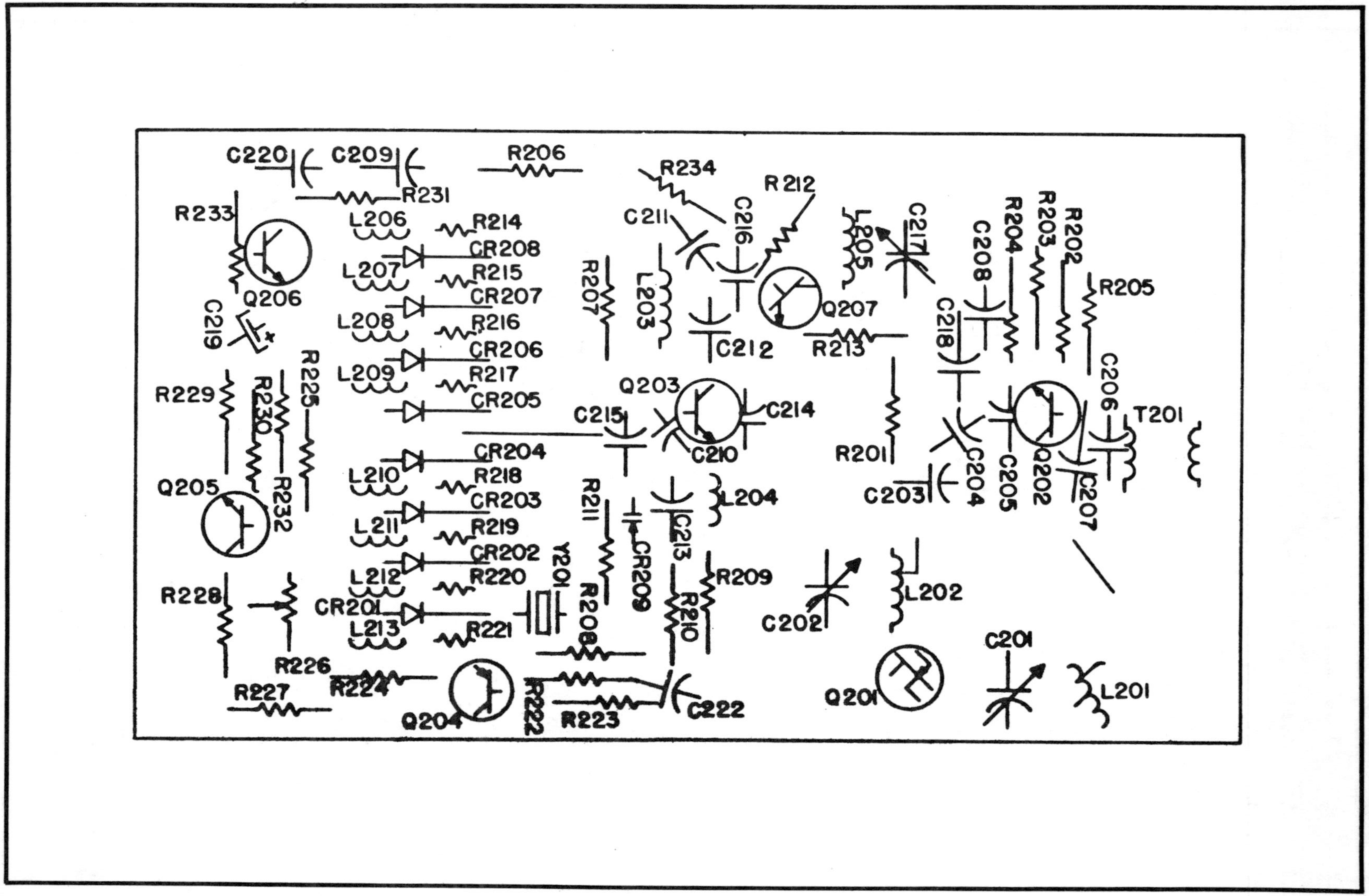

Fig. 4-4. RF board 301-598.

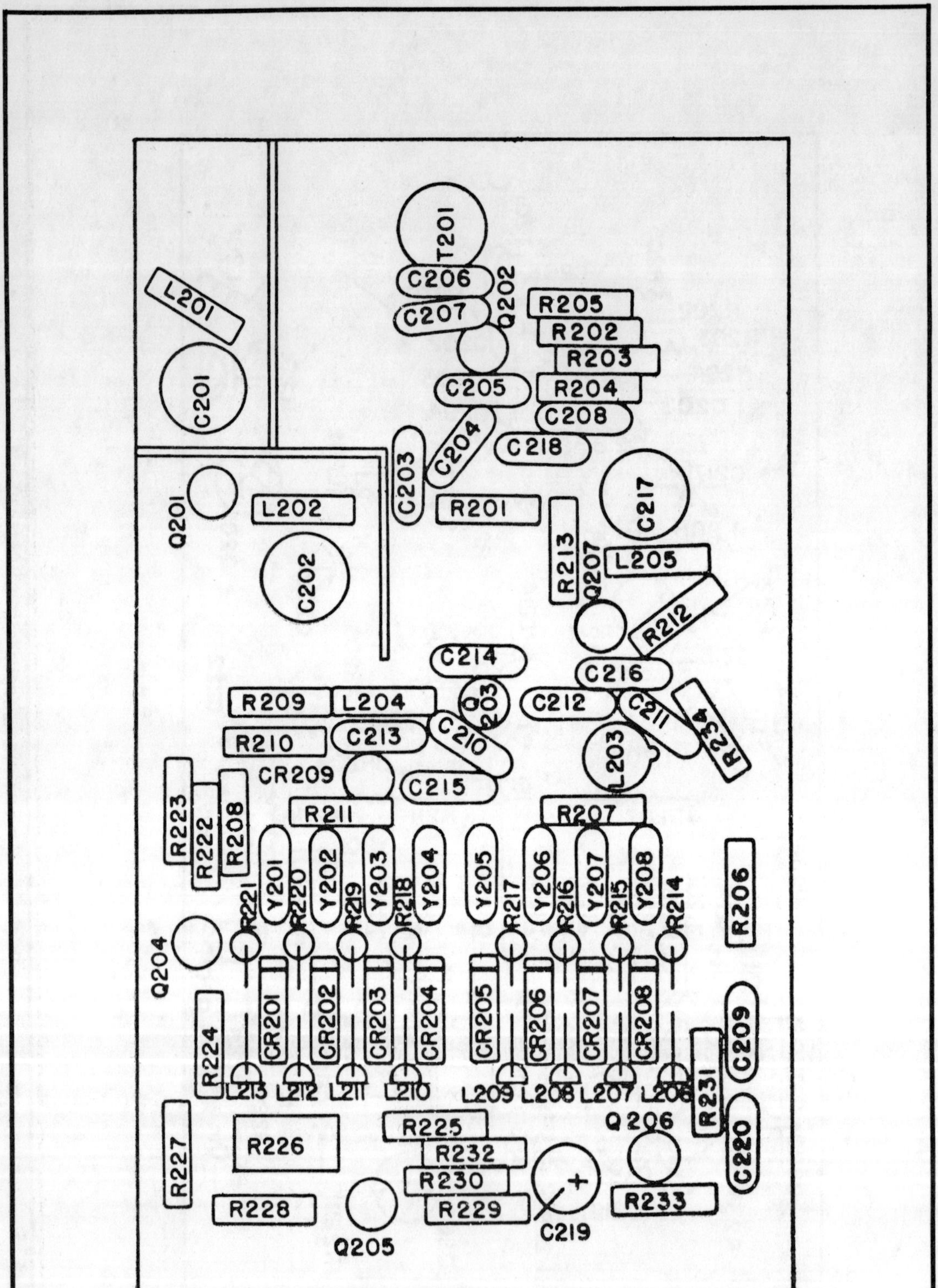

Fig. 4-5. RF board 301-598 parts placement diagram.

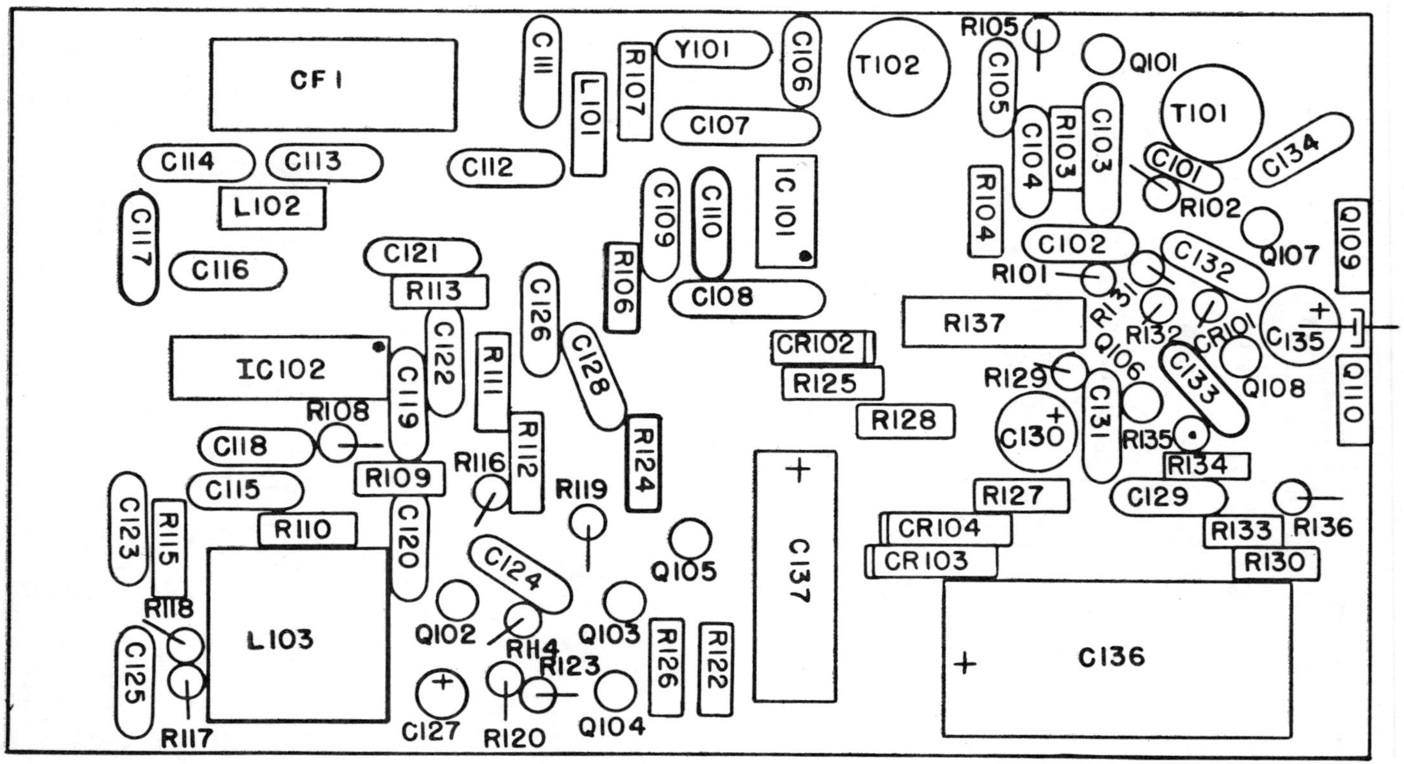

Fig. 4-6. IF—audio board 500-858 parts placement diagram.

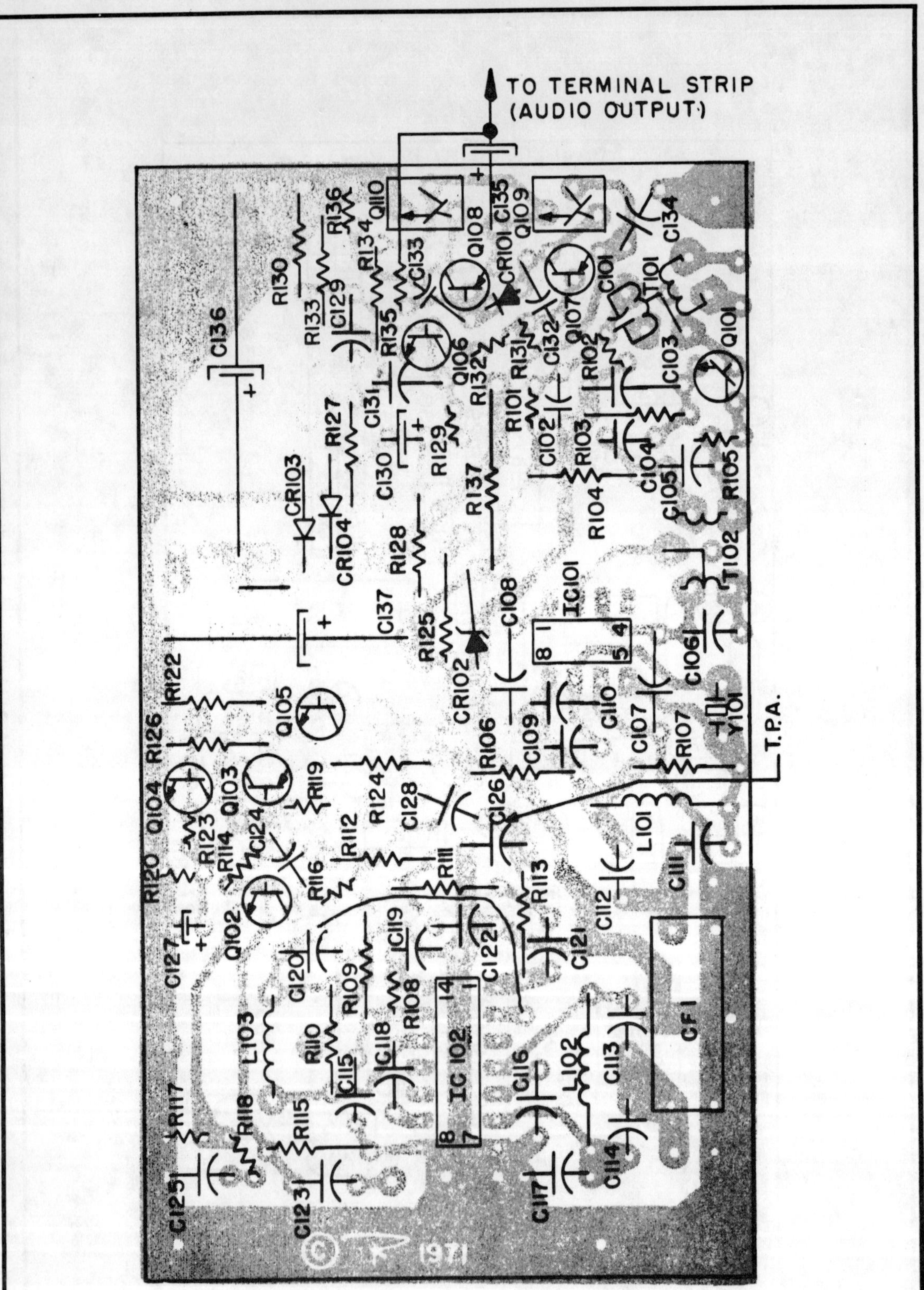

Fig. 4-7. IF board 500-858.

REGENCY TMR-4H, -8H, -8U

Specifications

Receiver

Frequency range	TMR-1H, 150–174 MHz TMR-1U, 450–470 MHz
Antenna impedance	50Ω
Channels	One, crystal-controlled Eight, crystal-controlled
Sensitivity	(20 dB quieting method) 0.7 μV (max.)
Selectivity (IF)	6 dB down, ±7 kHz (min.)
Selectivity (front end)	
3 dB bandwidth	8 MHz (min.)
6 dB bandwidth	12 MHz (min.)
Spurious rejection	(excluding primary image) 50 dB
AFC range	Approx ±6 kHz
Modulation acceptance	±7 kHz (min.)
Intermediate frequencies	First, 10.7 MHz Second, 455 kHz
Squelch system	Noise-operated
Audio output (3.2Ω speaker)	3W at 10% or less distortion, 5W (max.)

Scanner

Scan rate	Approx 15 channels per sec
Scan delay	Approx 0.5 sec

Power

Voltage requirement	117V AC (±10%), 60 Hz, 17W (max.); or 13.8V DC (±10%)
Current requirements at 13.8V DC	
(squelched)	180 mA
(max. audio output)	800 mA
Fuse size	1.5A, 3AG

General

Front-panel size	5⅝ × 2¼ in.
Depth (including knobs and rear-panel connectors)	9 in.
Antenna connector	Motorola type
Power connector	4-pin, polarized
Speaker size	4 in. square

Service Data

Note: All three units are single-band units and no programming is required.

Crystal layout	Fig. 4-8
Voltage data	Tables 4-3, 4-4
Schematics	Schematics 4-3, 4-4, 4-5

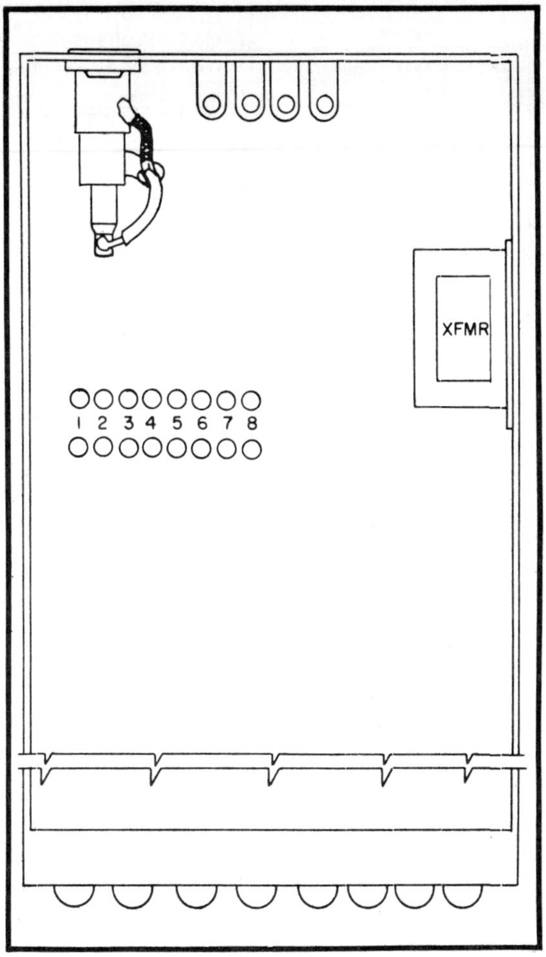

Fig. 4-8. Crystal layout for TMR series (multichannel).

Internal layout	Figs. 4-2, 4-3 (TMR-4H, -8H)
RF	Figs. 4-4, 4-5 (TMR-8U)
IF	Figs. 4-6, 4-7 (all)
Scanner	Figs. 4-9, 4-10 (all)
Alignment data, steps	A, B, C (-4H, -8H only), D (-8U only) E (-4H, -8H only), F (-8U only), G (-4H, -8H only)

REGENCY ACT-P4H

Specifications

Receiver

Frequency range	150–174 MHz
Channels	Four, crystal-controlled
Sensitivity (at tuneup)	0.7 μV (max.)
Frequency separation	10 MHz (max. sensitivity) 14 MHz (usable sensitivity)
Selectivity (IF)	6 dB down, ±7 kHz (min.) 50 dB down, ±20 kHz (max.)

Table 4-3. Voltage Data—TMR-4H, -8H, -8U Transistors.

	TRANSISTOR	EMITTER (Source)	BASE (Gate)	COLLECTOR (Drain)
RF Board 301-598 (TMR-8U)	Q201 (FET)	0	0	5.5
	Q202	1.5	2.2	6.8
	Q203	2.5	3.1	6.8
	Q204	7.8	7.2	3.0 – 5.0
	Q205	2.9	3.6	7.4
	Q206	2.9	3.6	7.2
	Q207	0	0.2	6.8
RF Board No. 301-563 (TMR-8H, -4H)	Q201	3.1	3.8	7.0
	Q202	1.6	2.3	7.1
	Q203	3.4	4.1	7.0
IF Board No. 500-858	Q101	2.3	3.0	5.8
	Q102	1.0	1.7	4.8
	Q103 (PNP)	8.2	8.2	0 (unsquelched)
		8.2	8.2	1.0 (squelched)
		8.2	8.2	1.5 Min. (tight squelch)
	Q104	0	0	7.2 (unsquelched) (TMR-8H, 4H)
		0	.80	.30 (squelched)
		0	.80	.10 (tight squelch)
	Q105	1.4	1.9	5.1 (unsquelched)
		1.1	.10	8.2 (tight squelch)
	Q106	0.7	1.3	12.4
	Q107 (PNP)	13.8	13.1	7.2
	Q108 (PNP)	6.9	6.6	.10
	Q109	6.9	7.2	13.8
	Q110	0	.10	6.9
Scan Board No. 500-841	Q301	.2	.2	6.0 Manual
		3P	3P	6.0 Scan
	Q302	5.2	5.9	6.0 Manual
		3P	3P	6.0 Scan
	Q303	0	.7	.1 Manual
		0	.1	1.6 Scan

Note: All voltages are nominal and are measured with a VTVM. **Scan** indicates the unit is scanning. **Man** indicates the unit is not scanning and is stopped at channel 1. A **P** beside a voltage indicates that the meter reading is pulsating (fluctuating) because the scanner section of the unit is operating.

Table 4-4. Voltage Data TMR-4H, -8H, -8U Integrated Circuits.

	IC No.	1	2	3	4	5	6	7	8	9	10	11	12	13	14	
IF Board	IC 101	4.2	0.7	0.7	4.2	7.8	0	4.2	7.8	–	–	–	–		–	
	IC 102	4.0	3.5	0	1.3	1.3	1.3	0	0	0.2	1.4	2.9	3.5	7.6	5.0	
Scan Board	IC 301	2.0P	2.0P	9.P	9.P	2.0P	2.0P	0	2.0P	2.0P	9P	9P	2P	2P	3P	(SCAN)
		.2	3.6	11	.7	3.6	3.6	0	3.6	.2	11	11	.2	.2	5.2	(MAN)
	IC 302	2.0P	2.0P	9P	9P	2.0P	2.0P	0	2.0P	2.0P	9P	9P	2P	2P	3P	(SCAN)
		.2	3.6	11	11	3.6	3.6	0	3.6	.2	11	11	.2	.2	0	(MAN)
	IC 303	2.7P	2.0P	2.0P	2.0P	2.7P	2.0P	0	3P	4P	3P	3P	3P	4P	4.8	(SCAN)
		1.5	3.6	.2	.2	3.6	3.6	0	.2	1.5	5.9	5.9	.2	5.9	4.8	(MAN)
	IC 304	1.6	1.3P	1.8P	1.3P	1.3P	1.3P	0	2P	2.7P	2.0P	2.0P	2.0P	2.0	4.8	(SCAN)
		.1	.2	3.8	3.8	1.5	1.5	0	.2	1.5	3.6	3.6	.2	3.6	4.8	(MAN)

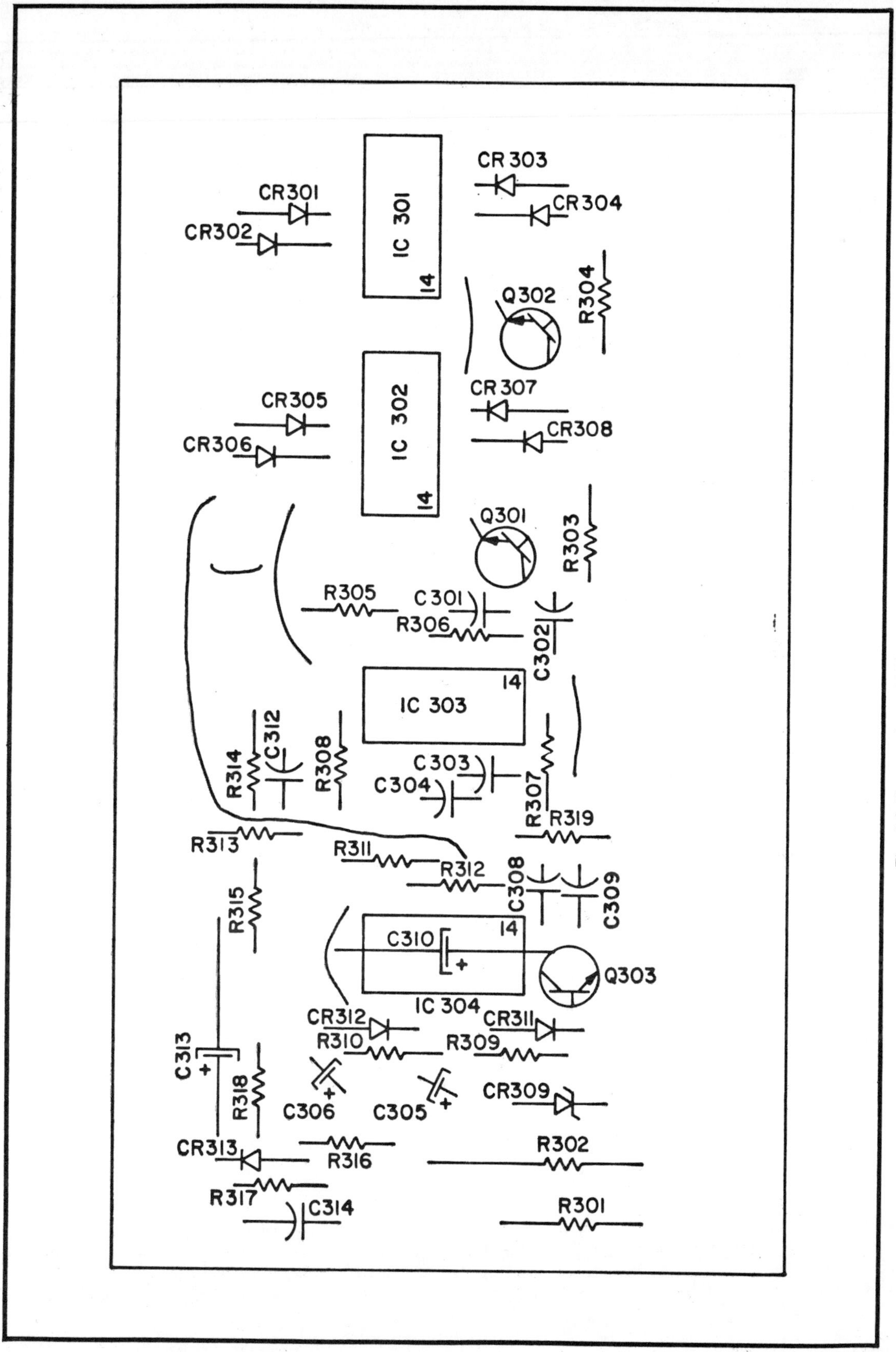

Fig. 4-9. Scanner board 500-841 (Bottom view).

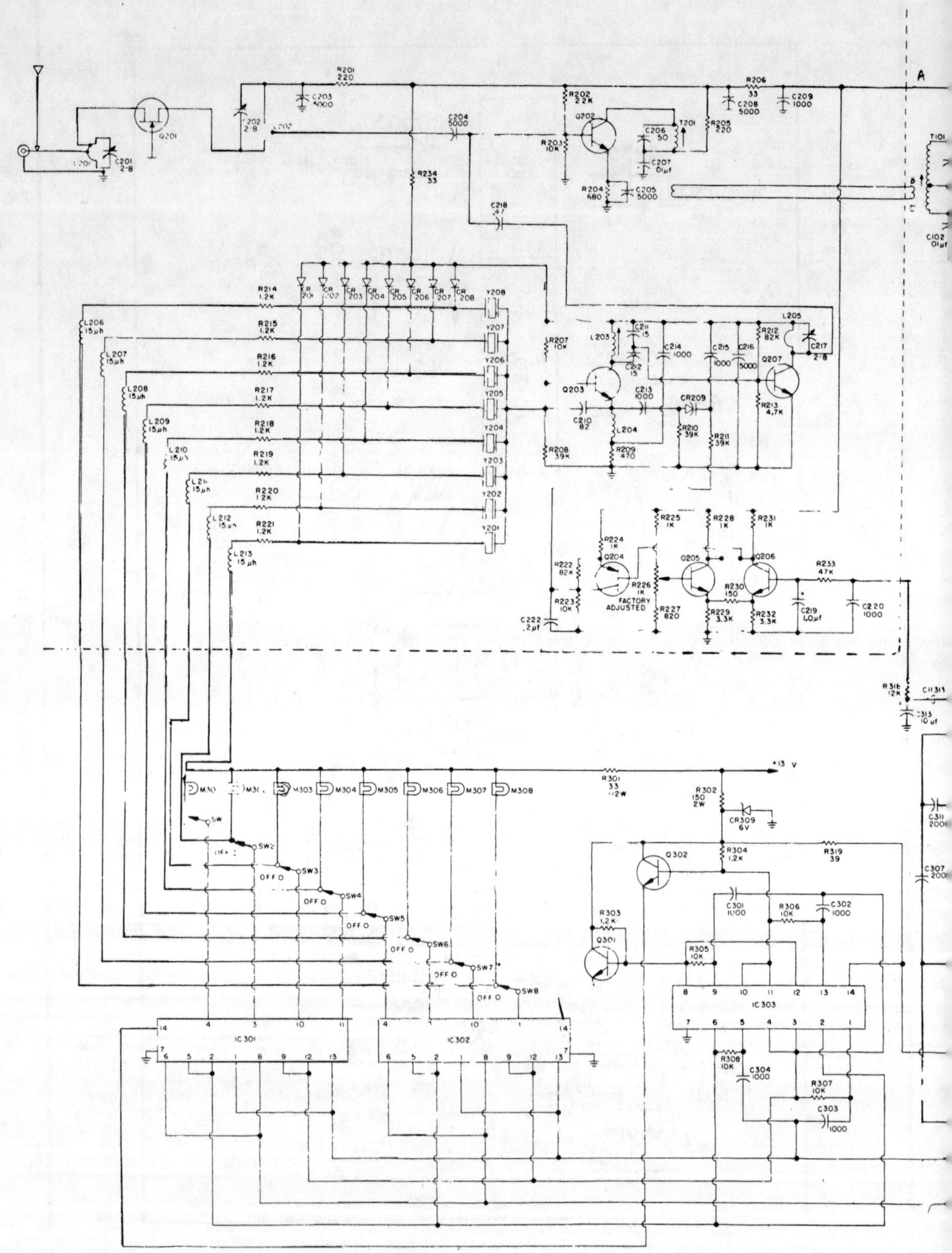

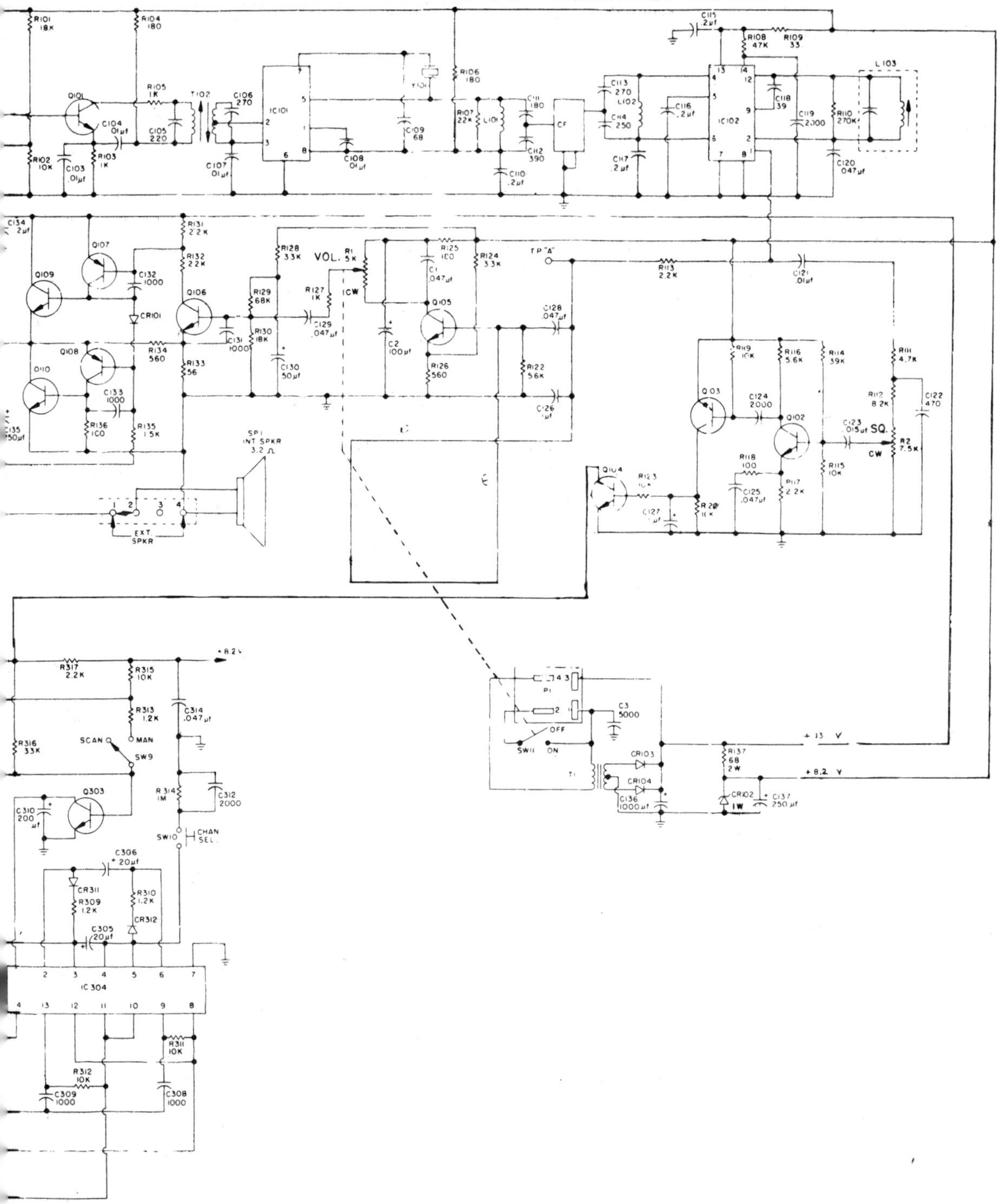

Schematic 4-3. TMR-8U.

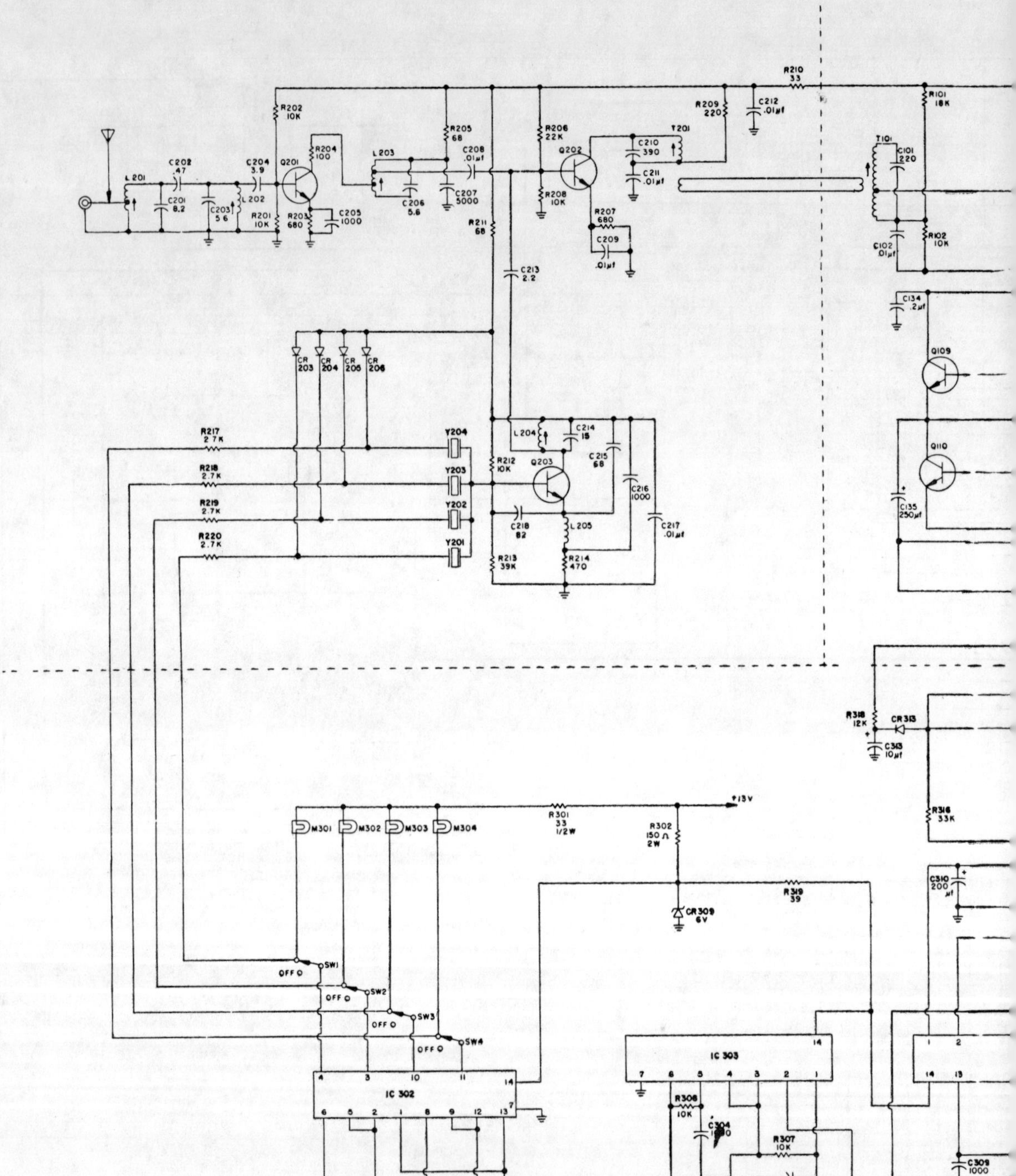
R210
33
R202
10K
R209
220
C212
.01μf
R101
18K
R205
68
R206
22K
T201
T101
C101
220
C202
.47
C204
3.9
Q201
R204
100
L203
C208
.01μf
Q202
C210
390
C211
.01μf
L201
C201
8.2
C203
5.6
L202
R201
10K
R203
680
C205
1000
C206
5.6
C207
5000
R208
10K
R207
680
C209
.01μf
R211
68
R102
10K
C102
.01μf
C213
2.2
C134
.2μf
Q109
CR
203
CR
204
CR
205
CR
206
Q110
R217
2.7K
Y204
L204
C214
15
R218
2.7K
Y203
R212
10K
Q203
C215
68
C216
1000
R219
2.7K
Y202
C218
82
L205
R220
2.7K
Y201
R213
39K
R214
470
C217
.01μf
C135
250μf
R318
12K
CR313
C313
10μf
R316
33K
+13V
M301
M302
M303
M304
R301
33
1/2W
R302
150 Ω
2W
R319
39
C310
200
μf
CR309
6V
SW1
OFF
SW2
OFF
SW3
OFF
SW4
OFF
IC 302
IC 303
R306
10K
C304
1000
R307
10K
C303
1000
C309
1000

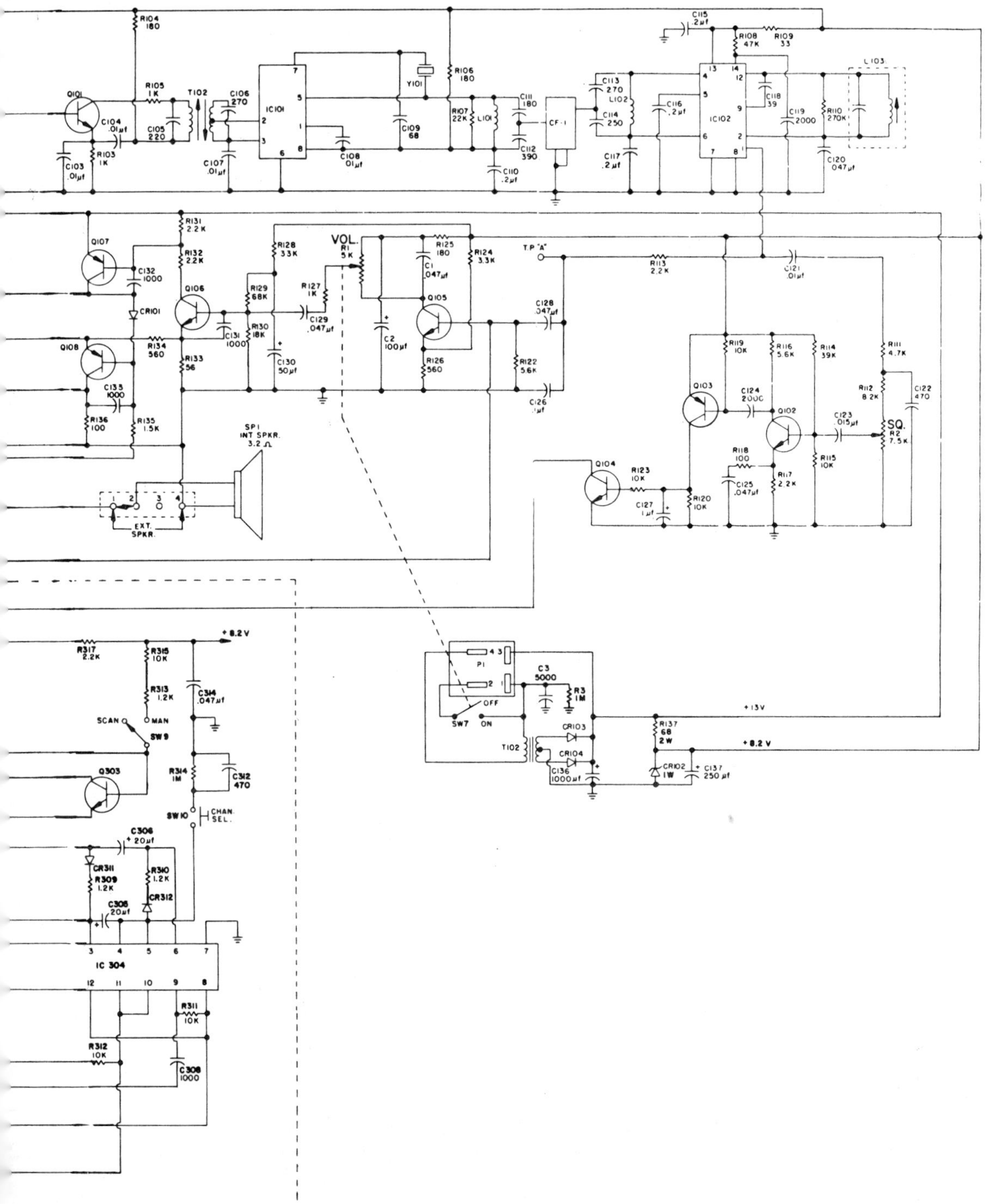

Schematic 4-4. TMR-4H.

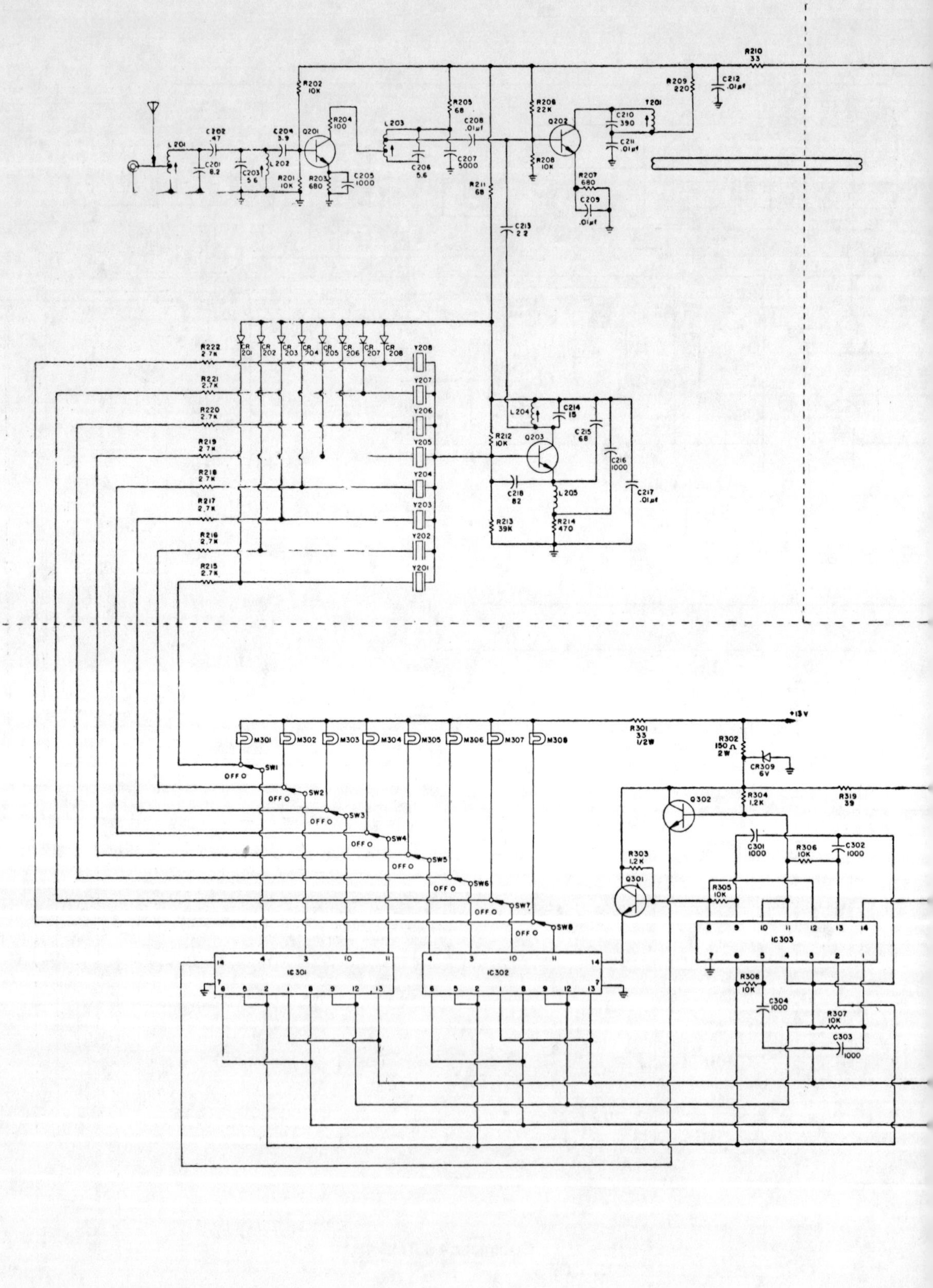

R210 33
R202 10K
R205 68
R206 22K
R209 220
C212 .01µf
T201
C210 390
C211 .01µf
Q202
C202 47
C204 3.9
Q201
R204 100
L203
C208 .01µf
L201
C201 8.2
C203 5.6
L202
R201 10K
R203 680
C205 1000
C206 5.6
C207 5000
R208 10K
R207 680
C209 .01µf
R211 68
C213 22
CR201
CR202
CR203
CR204
CR205
CR206
CR207
CR208
R222 2.7K
R221 2.7K
R220 2.7K
R219 2.7K
R218 2.7K
R217 2.7K
R216 2.7K
R215 2.7K
Y208
Y207
Y206
Y205
Y204
Y203
Y202
Y201
L204
C214 15
C215 68
R212 10K
Q203
C216 1000
C218 82
L205
C217 .01µf
R213 39K
R214 470
+13 V
R301 33 1/2W
M301
M302
M303
M304
M305
M306
M307
M308
R302 150Ω 2W
CR309 6V
OFF O
SW1
SW2
SW3
SW4
SW5
SW6
SW7
SW8
Q302
R304 1.2K
R319 39
C301 1000
R306 10K
C302 1000
R303 1.2K
Q301
R305 10K
IC303
R308 10K
C304 1000
R307 10K
C303 1000
IC301
IC302

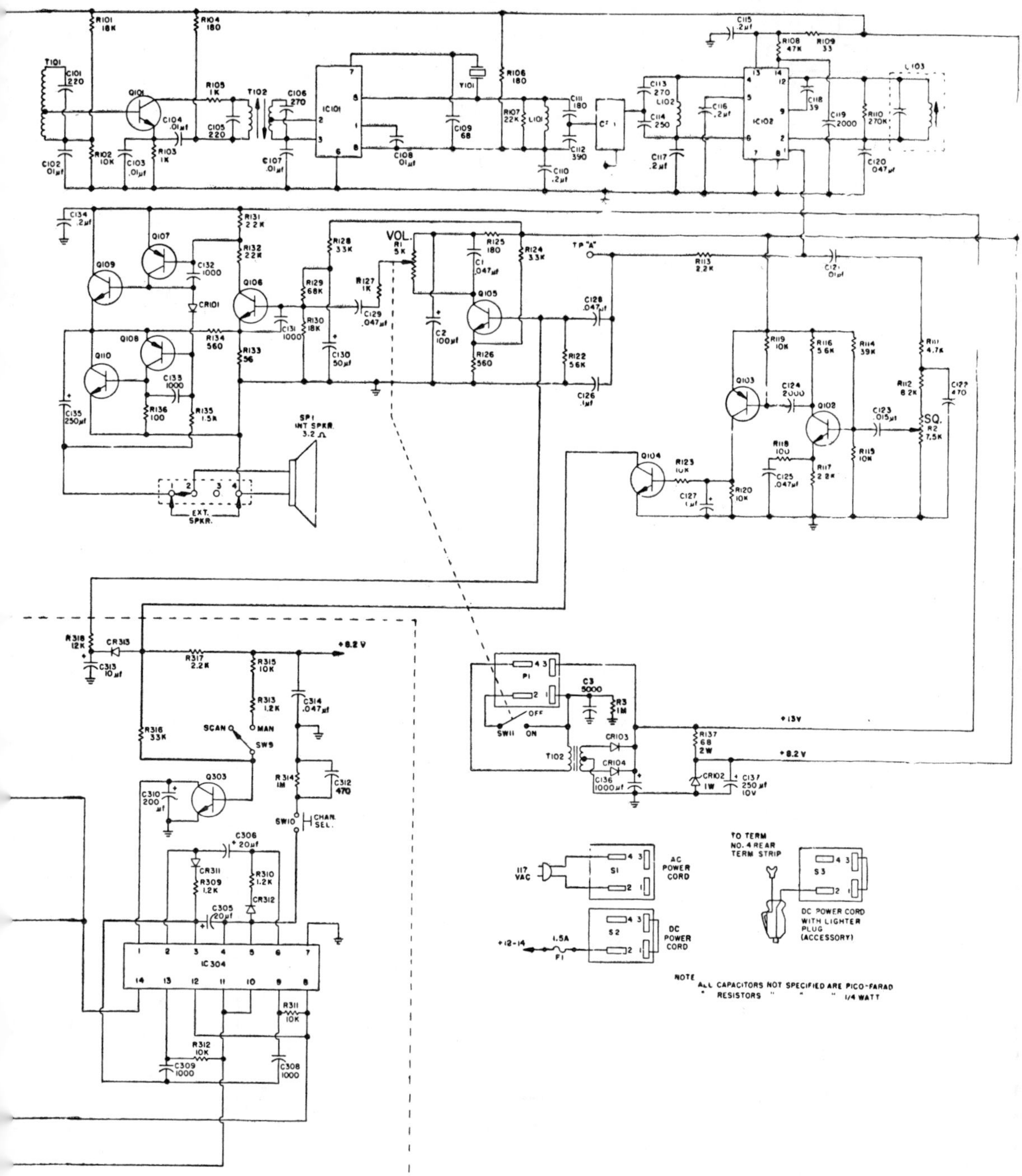

Schematic 4-5. TMR-8H.

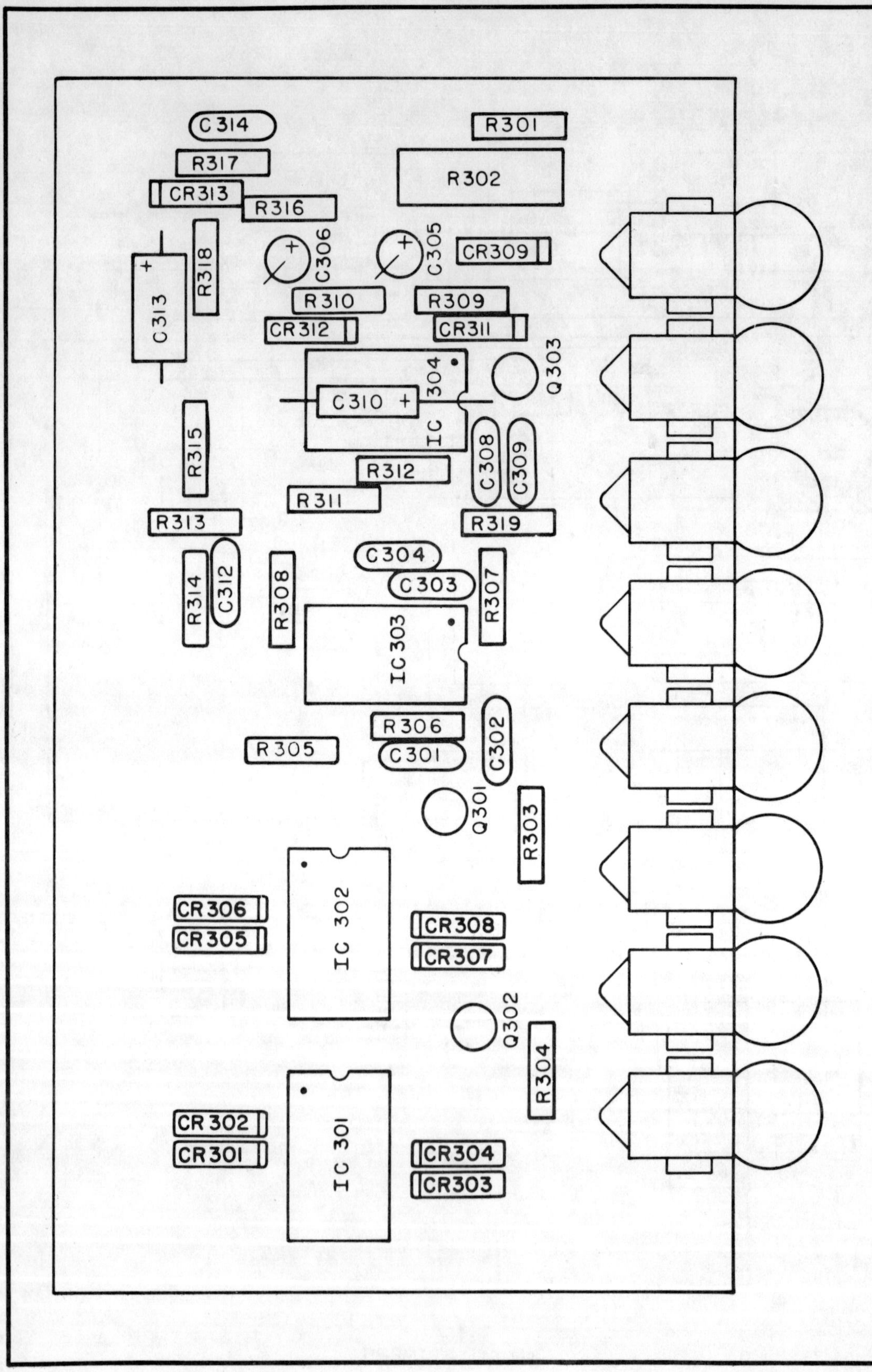

Fig. 4-10. Scanner board parts placement diagram. All parts not on some models.

Spurious rejection (excluding primary image)	50 dB
Modulation acceptance	±7 kHz (min.)
Intermediate frequencies	First, 10.7 MHz Second, 455 kHz
Squelch system	Noise-operated
Audio output (16Ω speaker)	0.4W peak

Scanner

Scan rate	Approx 10 channels per sec
Scan delay	Approx 0.4 sec

Power

Batteries	Two 9V transistor radio batteries (Eveready No. 216 or equivalent), or 8.75V 225 mA-h rechargeable ni-cad battery pack (Regency MA-72)
Current requirements at 8.7V DC	
(squelched and scanning)	20–25 mA (avg)
(max. audio output)	100 mA

General

Physical dimensions	$3\frac{3}{4} \times 6\frac{3}{4} \times 1\frac{3}{8}$ in.
Speaker size	$2\frac{1}{4}$ in., round
Earphone jack	2.5 mm

Service Data

Note: The ACT-P4H is a 4-channel, single-band hand-held unit requiring no crystal programming.

Crystal layout	Fig. 4-11
Voltage data	Tables 4-5, 4-6, 4-7
Schematic	Schematic 4-6
Internal layouts, receiver	Figs. 4-12, 4-13, 4-14
Scanner	Figs. 4-15, 4-16, 4-17
Alignment data, steps	A-2, H

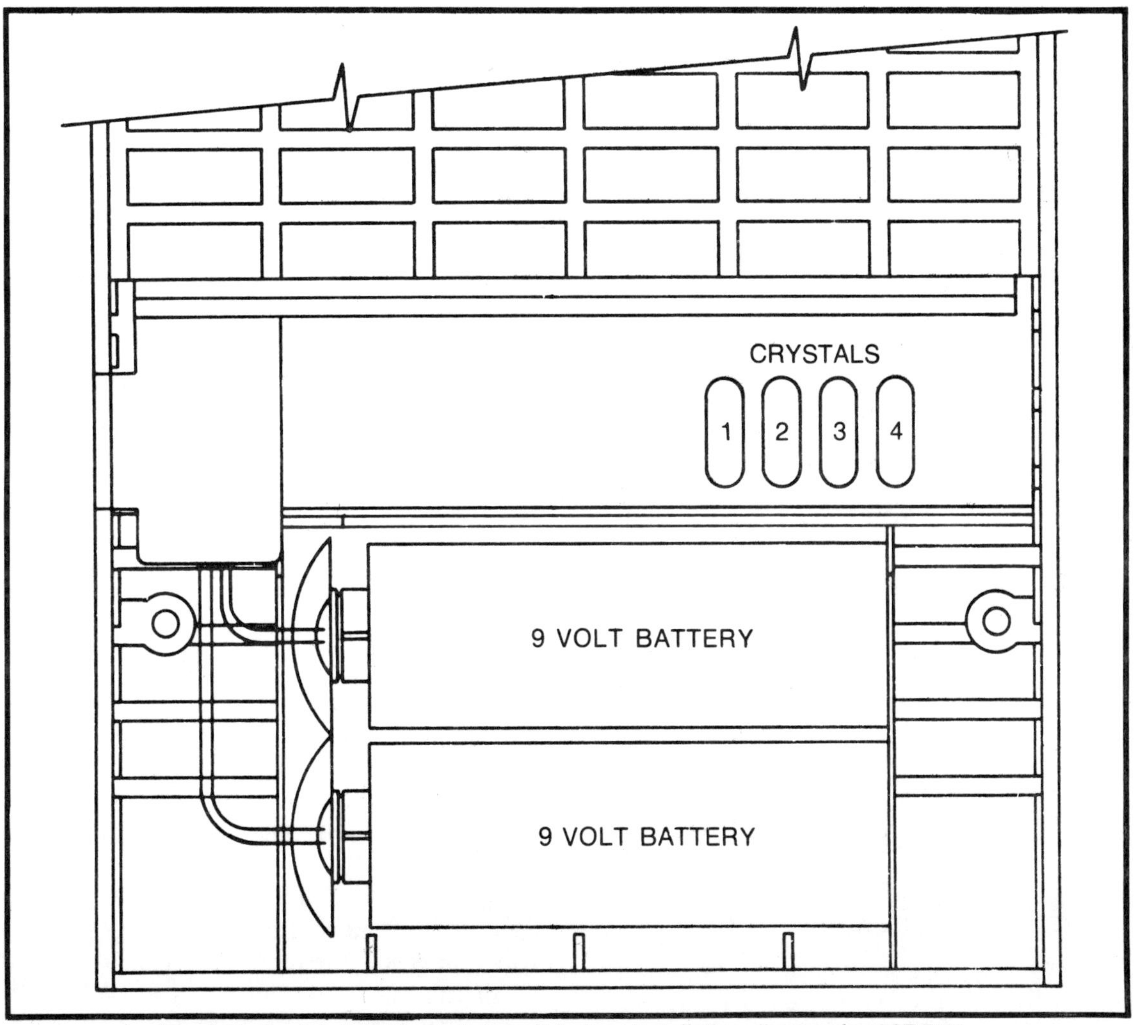

Fig. 4-11. Crystal location and dry cell battery installation diagram for ACT-P4H.

REGENCY TMR-8H/L, -8H/LL, -8H/LM, -8H/LH

Specifications

Receiver

Frequency range	Low band, 30–50 MHz High band, 150–174 MHz
Antenna impedance	50Ω
Channels	Eight, crystal-controlled
Sensitivity (at tuneup)	Low band, 0.5 μV (max.) High band, 0.7 μV (max.)
Frequency separation	
Low-band	
Model 8 H/LL	33–39 MHz
Model 8 H/LM	37–43 MHz
Model 8 H/LH	41–47 MHz
High-band model	6 dB bandwidth, 8 MHz 9 dB bandwidth, 12 MHz
Selectivity (IF)	6 dB down, ±7 kHz (min.) 50 dB Down, ±20 kHz (max.)
Spurious rejection (excluding primary image)	50 dB
Modulation acceptance	±7 kHz (min.)
Intermediate frequencies	First, 10.7 MHz Second, 455 kHz
Squelch system	Noise-operated
Audio output (3.2Ω speaker)	3W at 10% or less distortion, 5W max.

Scanner

Scan rate	Approx 15 channels per sec
Scan delay	Approx 0.5 sec

Power

Voltage requirement	117V AC (±10%), 60 Hz 17W max.; or 13.8V DC (±10%)
Current requirements at 13.8V DC	
(squelched)	180 mA
(max. audio output)	800 mA
Fuse size	1.5A, 3AG

General

Front-panel size	5⅝ × 2¼ in.
Depth (including knobs and rear-panel connectors)	9 in.
Antenna connector	Motorola type
Power connector	4-pin, polarized
Speaker size	4 in. square

Table 4-5. Voltage Data—ACT-P4H Main Board 501-055.

Scan Bd. No. 501-056	Q201			
	Scan (Slow)	8.4	8.3	0-8 (P)
	Stop (CH 4)	8.2	7.2	8.1
	Q202			
	Scan (Slow)	8.4	8.3	0-8 (P)
	Stop (CH 3)	8.2	7.2	8.1
	Q203			
	Scan (Slow)	8.4	8.3	0-8 (P)
	Stop (CH 2)	8.2	7.2	8.1
	Q204			
	Scan (Slow)	8.4	8.3	0-8 (P)
	Stop (CH 1)	8.2	7.2	8.1
	Q205			
	Scan (Slow)	0	0-0.7 P	0-8.5 P
	Stop (CH 4)	0	0.7	0
	Q206			
	Scan (Slow)	0	0-0.7 P	0-8.5 P
	Stop (CH 3)	0	0.7	0
	Q207			
	Scan (Slow)	0	0-0.7 P	0-8.5 P
	Stop (CH 2)	0	0.7	0
	Q208			
	Scan (Slow)	0	0-0.7 P	0-8.5 P
	Stop (CH 1)	0	0.7	0
	Q209			
	Stop (Any Channel)	5.8	6.4	6.8

Service Data

Insert the crystal (or crystals) in the proper socket pins as indicated on the crystal location drawing, Fig. 4-18A. The number by each pair of sockets matches the channel number on the front panel.

If the crystal inserted is for the high-band (148–174 MHz), place the proper color-coded wire and socket onto the proper high-band pin; if the crystal is for the low-band (30–50 MHz), place the proper wire and socket onto the proper low-band pin. Pictorial B illustrates how the band-selection wires are properly connected. Pictorial C shows an example of a partially programmed board. (See Fig. 4-18.)

If a particular channel is not used (in other words, if there is no crystal installed for that channel), the band-selection wire must still be connected to either a high-band pin or a low-band pin. Thus, for proper scanner operation, all of the band-selection wires *must* be connected, even though not all channels are used.

NOTES: 1. All voltages are nominal and are measured with a VTVM.
2. For transistors Q101 through Q113 "SCAN" indicates that the receiver is scanning all four channels in the "FAST" scan mode.
3. "STOP" indicates that the receiver is stopped on oen of the channels 1-4.
4. For transistors Q201 through Q209 "SCAN" indicates that the receiver is scanning in the "SLOW" scan mode. "STOP" indicates that the scanner is stopped ON THE INDICATED CHANNEL.

	Transistor	Emitter (Source)	Base (Gate)	Collector (Drain)
Main Bd.	Q101			
No. 501-055	Scan	4.2P	3.6P	1.5P
	Stop	4.8	4.1	2.0
	Q102			
	Scan	1.5P	2.2P	4.9P
	Stop	2.0	2.7	5.8
	Q103			
	Scan	0.9P	1.6P	4.8P
	Stop	1.2	1.9	5.8
	Q104			
	Scan			
	Stop	3.8	3.8	5.8
	Q105			
	Scan	0.8	1.4	3.8
	Stop	1.2	1.8	4.5
	Q106	8.3	8.3	5.5
	Q107	7.8	7.8	3.9
	Q108			
	Squelch Tight	8.2	8.0	0
	Squelch Open	6.3	5.8	0.6
	Q109			
	Squelch Tight	0	0	8.0
	Squelch Open	0	0.6	3.9
	Q110			
	Squelch Tight	8.3	8.7	9.3
	Squelch Open	4.0	4.3	8.6
	Q111	9.0	9.0	
	Squelch Tight	9.0	9.0	8.2
	Squelch Open	9.0	8.9	4.0
	Q112			
	Squelch Tight	8.2	8.0	0
	Squelch Open	4.0	3.8	0
	Q113			
	Squelch Tight	8.3	8.3	7.5
	Squelch Threshold	8.3	8.3	7.0
	Squelch Open	7.8	7.8	0.3

After the crystals are installed and any necessary band programming changes are completed, reinstall the speaker assembly. Then carefully reinstall the cover and its hardware.

Crystal layout	Figs. 4-8, 4-18
Voltage data	Tables 4-3, 4-4 (8H/LL, -8H/LM, -8H/LH only), 4-8, 4-9, 4-10 (TMR-8/H/L only)
Schematics	Schematics 4-7, 4-8
Internal layout, RF	Figs. 4-2, 4-3
IF	Figs. 4-6, 4-7 (not available for TMR-8H/L)
Scanner	Figs. 4-9, 4-10 (not available for TMR-8H/L)
Alignment data, are given as last text of the chapter, just before the illustrations.)	Steps A, B, I, E

Table 4-7. Voltage Data—ACT-P4H Integrated Circuits.

NOTES:

1. All voltages are nominal and are measured with a VTVM
2. For IC's 101, 102 and 103, "SCAN" indicates that the receiver is in the "FAST" scan mode.
3. For IC201, "SCAN" indicates that the receiver is in the "SLOW" scan mode.
4. A "P" beside a reading indicates that the meter is pulsating (fluctuating) because the scanner section is operating.
5. For IC201, "STOP" indicates that the receiver is stopped *ON THE INDICATED CHANNEL.*

	IC101								
Main Bd.	pin	1	2	3	4	5	6	7	8
No. 501-055	Scan	1.8P	0.8 P	0.8 P	1.8 P	4.7 P	0	1.8 P	2.8 P
	Stop	2.0	0.8	0.8	1.9	5.8	0	1.9	3.3
	IC102								
	Scan	2.3	0	0	0	2.4 P	0	6.3 P	7.8
	Stop	2.9	0	0	0	2.9	0	6.0	7.7
	IC103								
	Scan (slow)	4.2 P	4.5 P	4.5 P	4.0 P	0.3	8.2	0	4.5 P
	Stop (squelched)	0	8.1	8.0	0	0.3	8.0	0	4.5
	Unsquelched	0	7.8	7.6	0	4.3	5.9	0	4.4
Scan Bd.	IC201								
No. 501-056	Scan (slow)	0/9 P	0/7.5 P	0/9 P	0/9 P	0/7.5 P	0	0/9 P	0
	Stop (squelched)	0	0/7.5 (CH 1)	0/7.5 (CH 2)	0	0/7.5 (CH 4)	0	0	0
	Unsquelched	0	0/7	0/7	0	0/7	0	0	0

	IC101								
Main Bd.	pin	9	10	11	12	13	14	15	16
No. 501-055	Scan								
	Stop								
	IC102								
	Scan	0	2.4 P	4.5P	0	0	2.4 P		
	Stop	0	3.0	5.7	0	0	3.0		
	IC103								
	Scan (slow)	4.0	0	7.5	8.3	0	8.3		
	Stop (squelched)	4.0	0	7.2	8.0	0	8.0		
	Unsquelched	4.0	7.8	0.2	0	7.8	7.8		
Scan Bd.	IC201								
No. 501-056	Scan (slow)	0	0/9 P	0/7.5 P	0/9 P	0	0/9 P	0	9
	Stop (squelched)	0	0	0/7.5 (CH 3)	0/9	0	0/9	0	9
	Unsquelched	0	0	0/7	7.0	7.0	0	0	8.4

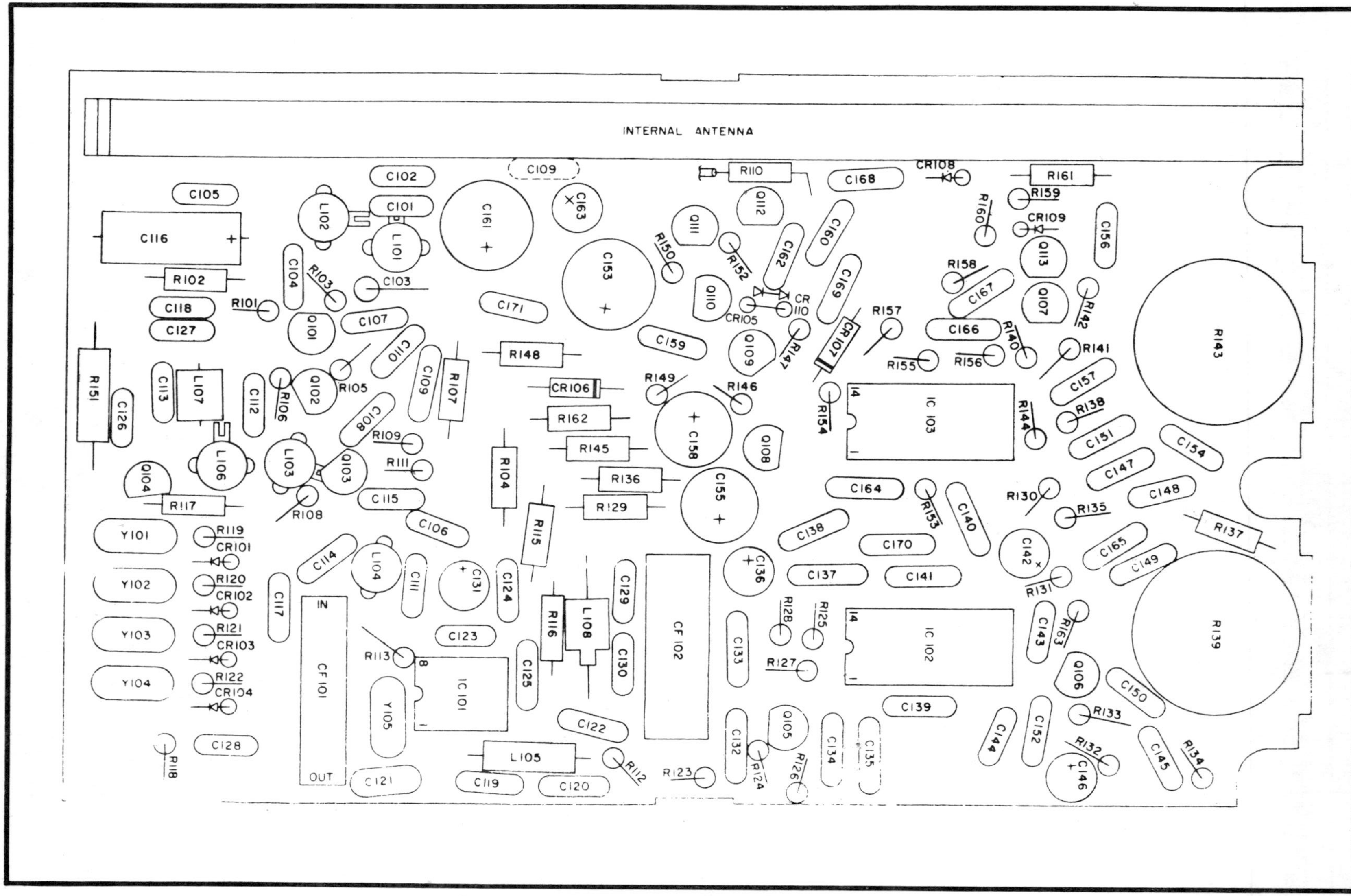

Fig. 4-12. Receiver board 501-055 parts placement diagram.

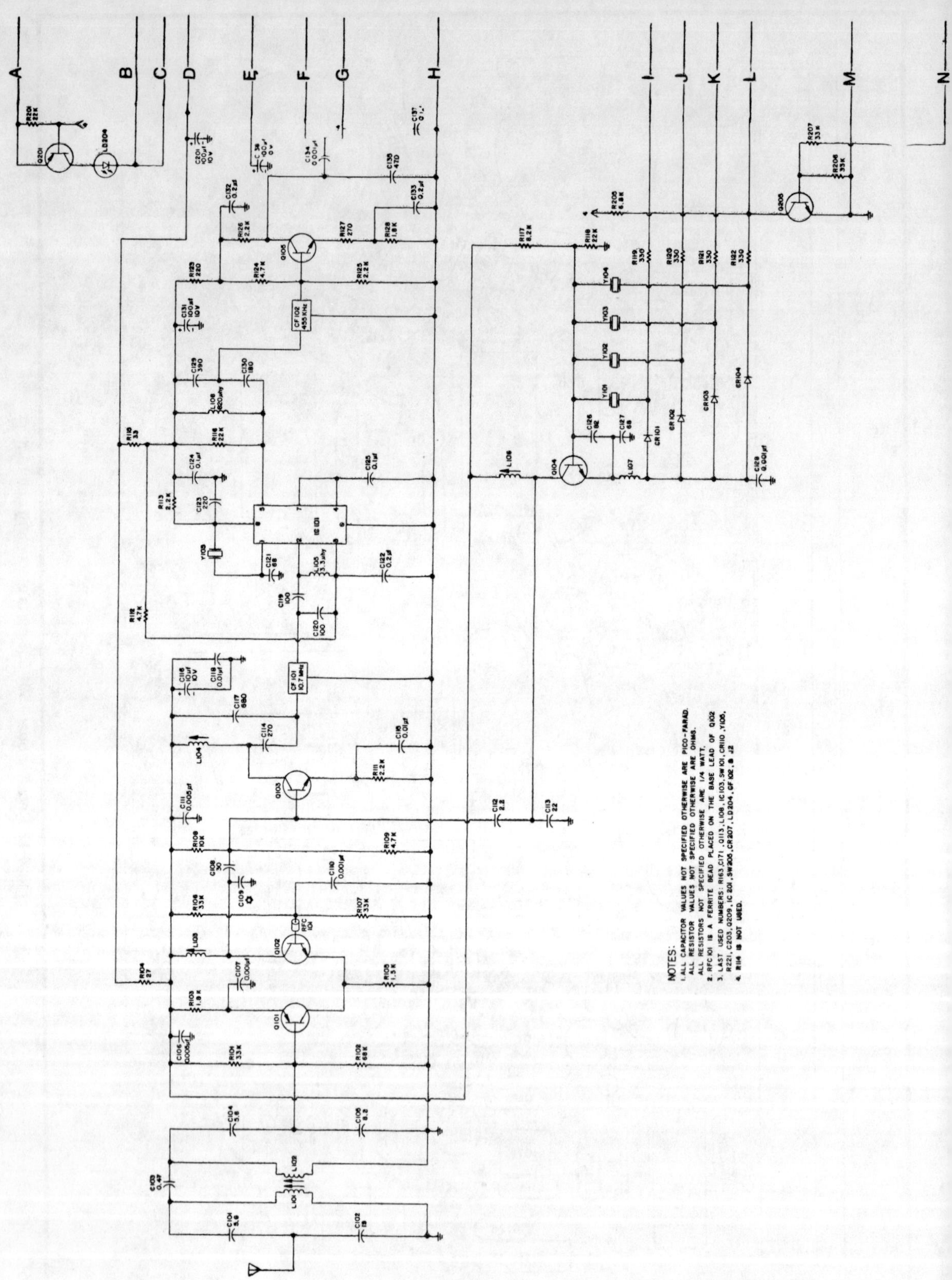
NOTES:
1. ALL CAPACITOR VALUES NOT SPECIFIED OTHERWISE ARE PICO-FARAD.
ALL RESISTOR VALUES NOT SPECIFIED OTHERWISE ARE OHMS.
ALL RESISTORS NOT SPECIFIED OTHERWISE ARE 1/4 WATT.
2. RFC101 IS A FERRITE BEAD PLACED ON THE BASE LEAD OF Q102
3. LAST USED NUMBERS: R163, C171, Q113, L108, IC103, SW101, CR110, Y106, R221, C203, Q209, IC201, SW205, CR207, LD204, CF102, & J2
4. R114 IS NOT USED.

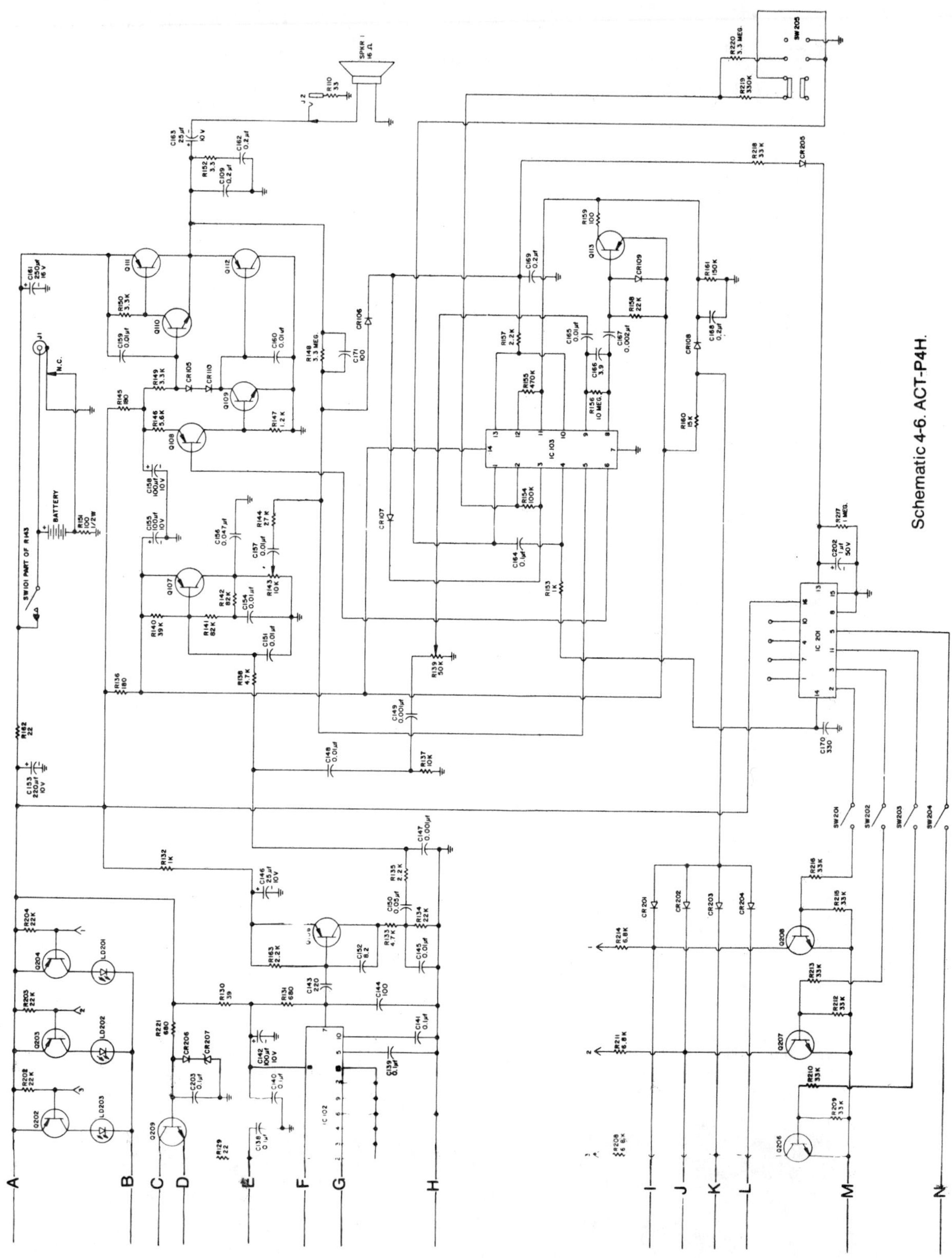

Schematic 4-6. ACT-P4H.

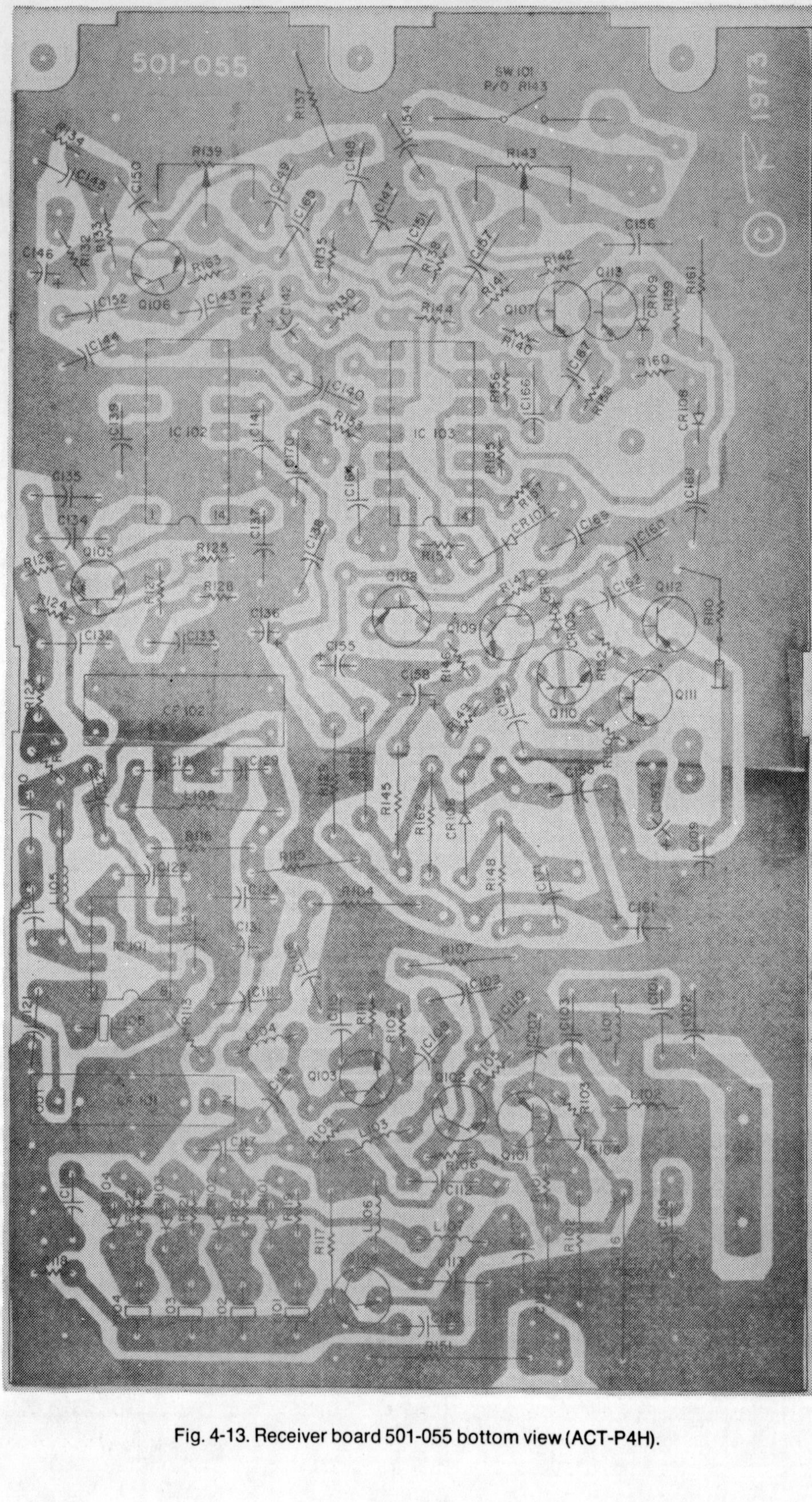

Fig. 4-13. Receiver board 501-055 bottom view (ACT-P4H).

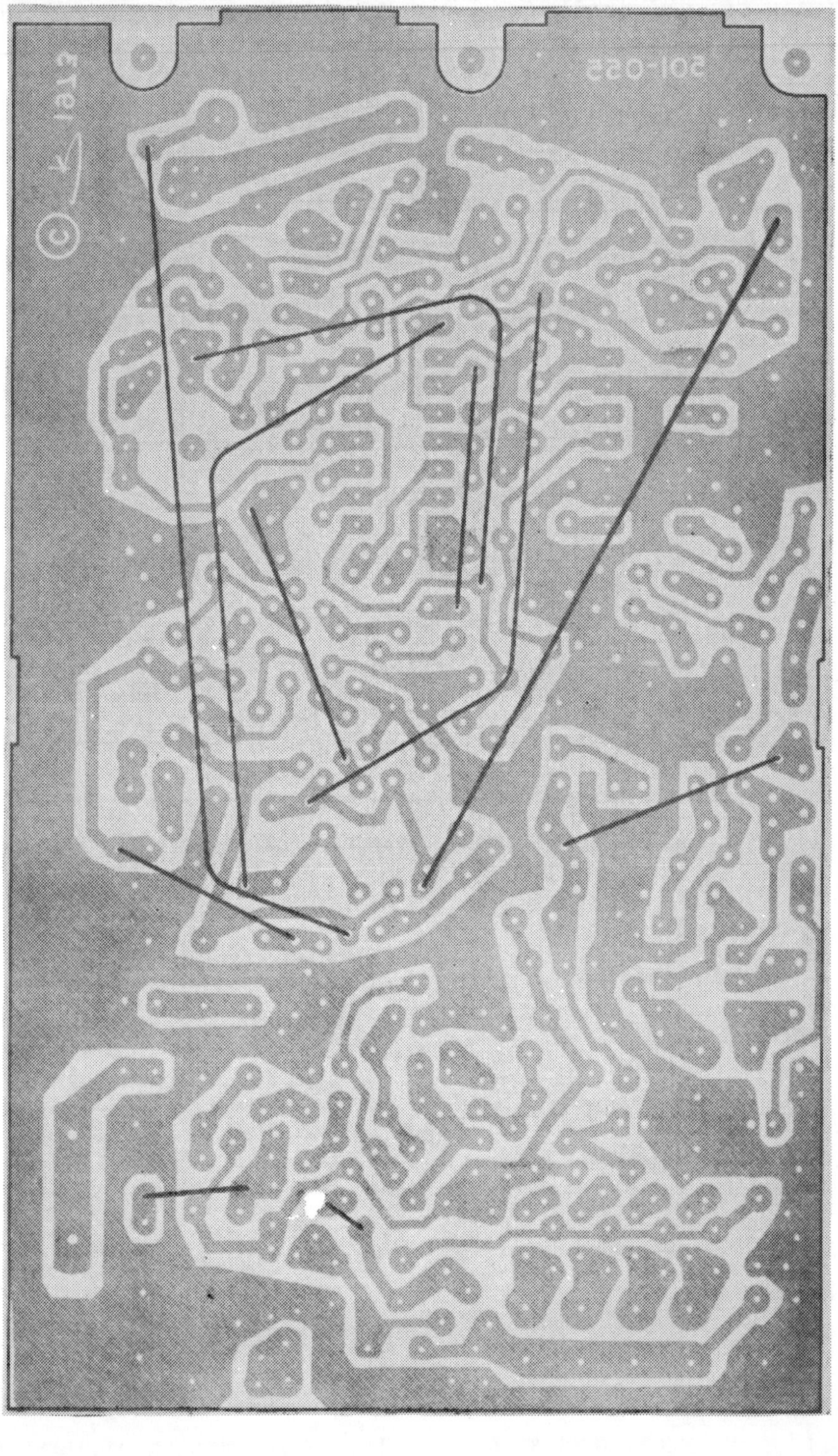

Fig. 4-14. Receiver board 501-055 jumper overlay diagram (ACT-P4H). (Note: jumpers show point-to-point connections. Not to be used as actual routing patterns.)

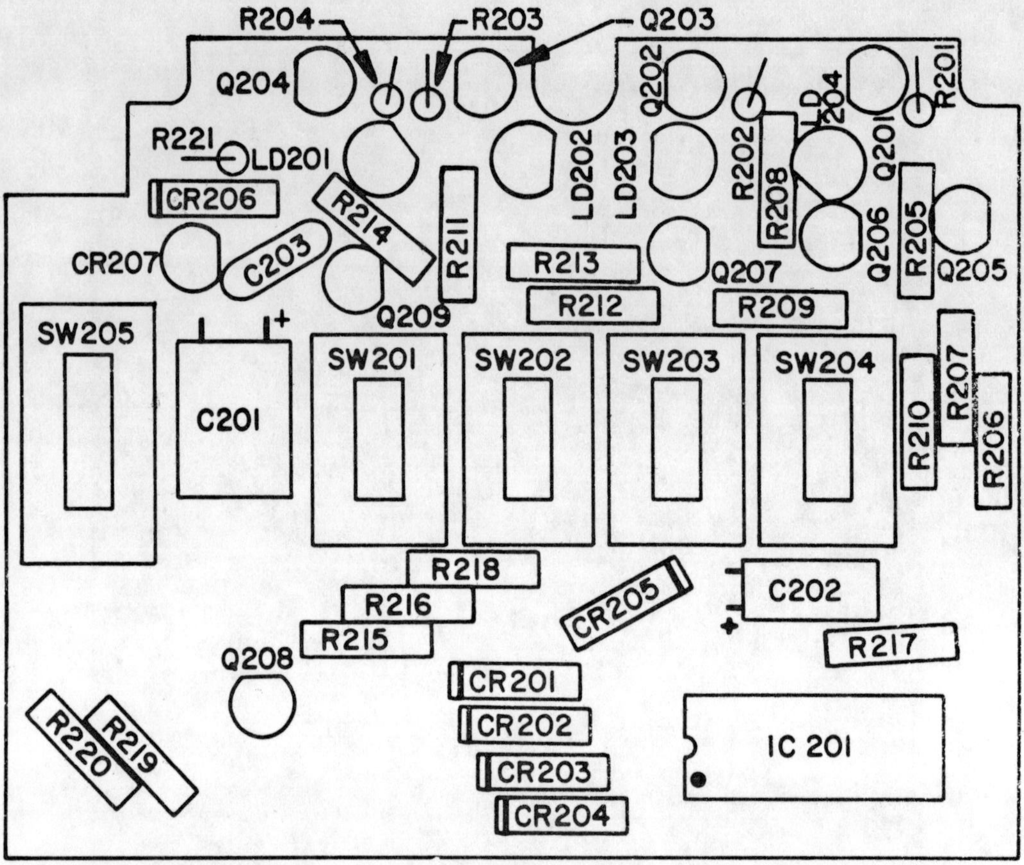

Fig. 4-15. Scanner board 501-056 parts placement diagram (ACT-P4H).

REGENCY TME-8H, -8H/L, -8H/LL, -8H/LM, -8H/LH

Specifications

Receiver (Model TME-8H/L)

Frequency range	Low band. 30–50 MHz High band: 150–174 MHz
Antenna impedance	50Ω
Channels	Eight: crystal-controlled
Sensitivity (at tuneup)	Low band: 0.5 μV (max.) High band: 0.7 μV (max.)
Frequency separation	
Low-band model	
Model 8H/LL	33–39 MHz
Model 8H/LM	37–43 MHz
Model 8H/LH	41–47 MHz
High-band model	6 dB bandwidth: 8 MHz 9 dB bandwidth: 12 MHz
Selectivity (IF)	6 dB down: ±7 kHz (min.) 50 dB down: ±20 kHz (max.)
Spurious rejection (excluding primary image)	50 dB
Modulation acceptance	±7 kHz (min.)
Intermediate frequencies	First. 10.7 MHz Second. 455 kHz
Squelch system	Noise-operated
Audio output (3.2Ω speaker)	3W at 10% or less distortion: 5W max.

Receiver (Model TME-8H)

Frequency range	150–174 MHz
Antenna impedance	50Ω
Channels	Eight. crystal-controlled
Sensitivity (at tuneup)	0.7 μV (max.)
Frequency separation	6 dB bandwidth. 8 MHz 9 dB bandwidth. 12 MHz
Selectivity (IF)	6 dB down. ±7 kHz (min.) 50 dB down. ±20 kHz (max.)

Spurious rejection (excluding primary image)	50 dB
Modulation acceptance	±7 kHz (min.)
Intermediate frequencies	First, 10.7 MHz Second 455 kHz
Squelch system	Noise-operated
Audio output (3.2Ω speaker)	3W at 10% or less distortion, 5W max.

Scanner

Scan rate	Approx 15 channels per sec
Scan delay	Approx 0.5 sec

Power

Voltage requirement	117V AC (±10%), 60 Hz, 17W max.; or 13.8V DC (±10%)
Current requirements at 13.8V DC	
(squelched)	180 mA
(max. audio output)	800 mA
Fuse size	1.5A, 3AG

General

Front-panel size	$3\frac{7}{8} \times 10\frac{1}{8}$ in.
Depth (including knobs and rear-panel connectors)	$9\frac{1}{2}$ in.
Antenna connector	Motorola type
Power connector	4-pin, polarized
Speaker size	$3\frac{1}{2}$ in., square

Service Data

Prior to installing a crystal, the receiver will have to be partially pulled out of its cabinet. First, remove the telescopic antenna if it is installed. Second, remove the two knobs (VOLUME and SQUELCH). Third, remove the rear panel (cover) by removing the four mounting screws. Fourth, remove the four rubber feet by unscrewing each one. The receiver may then be slid rearward from the cabinet until the crystal socket pins are accessible.

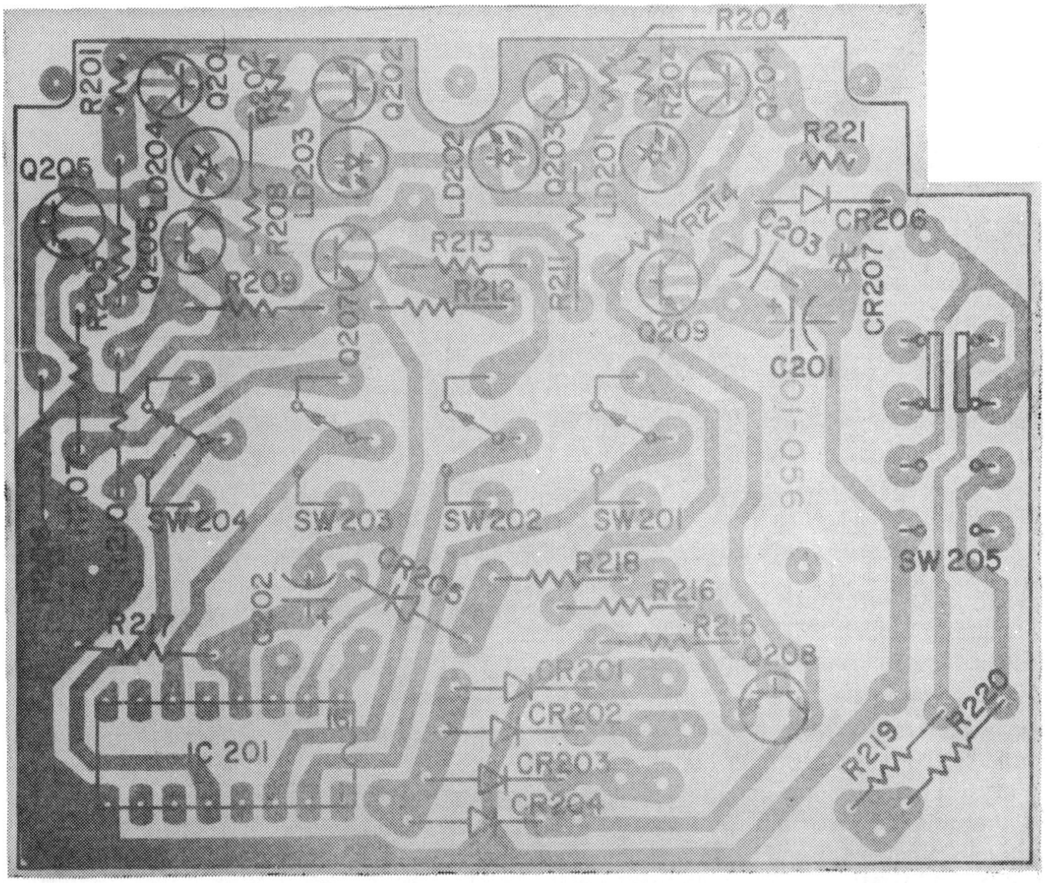

Fig. 4-16. Scanner board 501-056, bottom view (ACT-P4H).

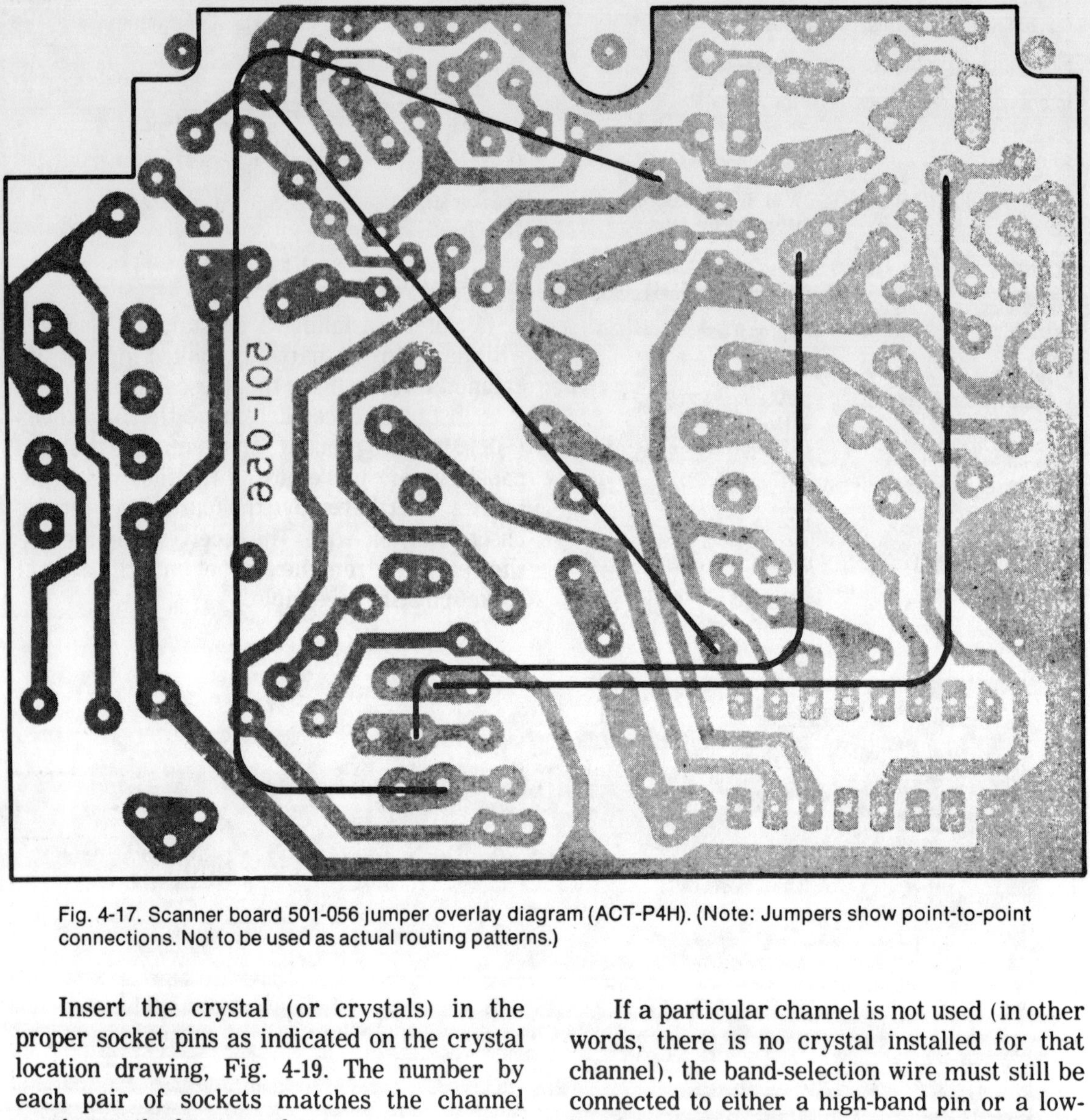

Fig. 4-17. Scanner board 501-056 jumper overlay diagram (ACT-P4H). (Note: Jumpers show point-to-point connections. Not to be used as actual routing patterns.)

Insert the crystal (or crystals) in the proper socket pins as indicated on the crystal location drawing, Fig. 4-19. The number by each pair of sockets matches the channel number on the front panel.

Band Programming

Band programming will have to be performed on Model TME-8H/L only. This is described in the following paragraph.

If the crystal inserted is for the high band (148–174 MHz), place the proper color-coded wire and socket onto the proper high-band pin; if the crystal is for the low band (30–50 MHz), place the proper wire and socket onto the proper low-band pin. Figure 4-18B illustrates how the band-selection wires are properly connected. Pictorial C shows an example of a partially programmed board. (See Fig. 4-18.)

If a particular channel is not used (in other words, there is no crystal installed for that channel), the band-selection wire must still be connected to either a high-band pin or a low-band pin. Thus, for proper scanner operation, all of the band-selection wires *must* be connected, even though not all channels are used.

Crystal layout	Figs. 4-18, 4-19
Voltage data	Tables 4-11, 4-12, 4-13
Schematics	Schematics 4-9, 4-10
Internal layout, RF	Figs. 4-2, 4-3
IF	Figs. 4-6, 4-7
Scanner	Figs. 4-20, 4-21
Alignment data (Steps are given as last text of chapter, just before the illustrations.)	A, G, B, I (-H/LL, -8H/LM, -8H/LH only), C (TME-8H only), E (all)

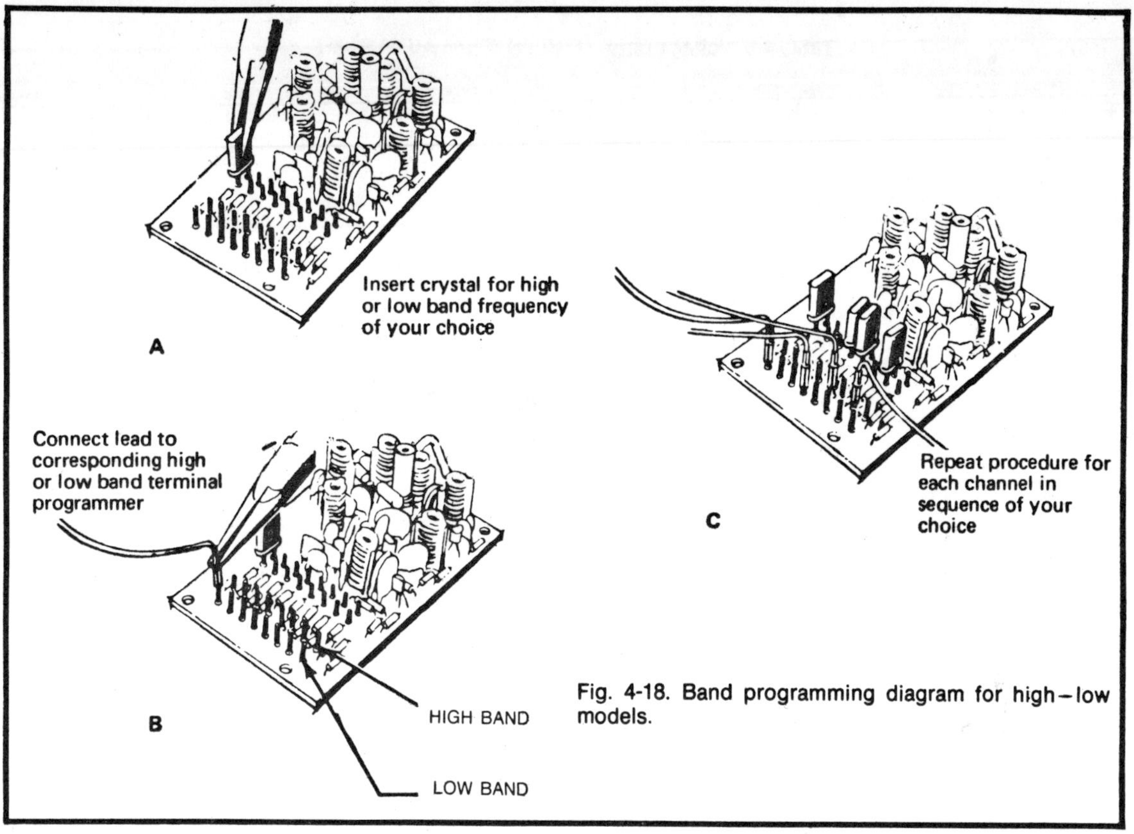

Fig. 4-18. Band programming diagram for high–low models.

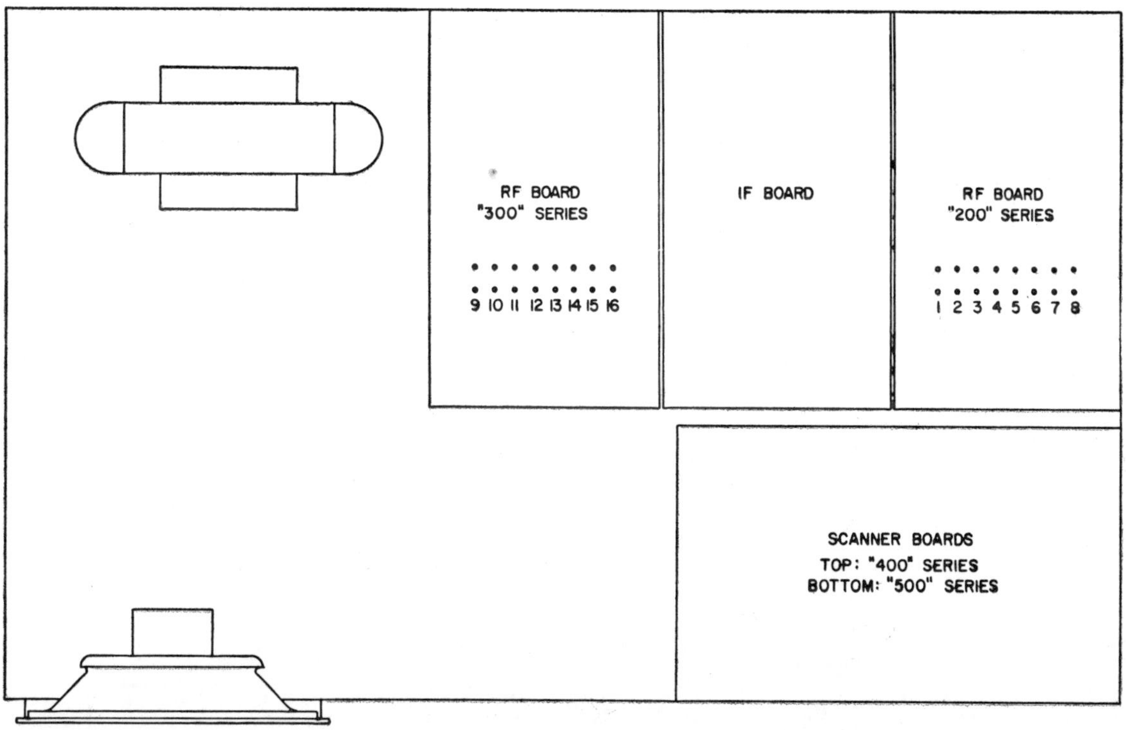

Fig. 4-19. Crystal location diagram for TME series (only 200 board used in TME-8 units).

Table 4-8. Voltage Data—TMR-8H IF and Audio Board.

TRANSISTOR	EMITTER	BASE	COLLECTOR
Q101	3.0V	3.3V	7.5V
Q102	3.8V	4.4V	6.3V
Q103	8.5V 8.5V 8.5V	8.5V 8.5V 8.5V	OV Unsquelched .8V Squelched 1.5V Tight Squelch
Q104	OV OV OV	OV Unsquelched .8V Squelched 1.5V Tight Squelch	6.8V Unsquelched .10V Squelched- Scanning
Q105	1.6V Unsquelched 1.1V Squelched- Scanning	1.9V .10V	5.1V 8.3V
Q106	5.8V	6.6V	12.5V
Q107	13V DC Operation 14V AC Operation	12.5V	7.2V
Q108	6.5V	7.2V	.10V
Q109	6.5V 6.5V	7.2V 7.2V	13V DC Operation 14V AC Operation
Q110	OV	.10V	6.5V

	EMITTER	BASE	COLLECTOR	
Q-201	3.1 0V	3.8 0V	7.2 8.0V	Low Band Activated High Band "
Q-202	0 3.1	0 3.8	8.0 7.2	Low Band " High Band "
Q-203	1.6 1.6	2.2 0	7.4 7.4	Low Band " High Band "
Q-204	1.6 1.6	0 2.2	7.4 7.4	Low Band " High Band "
Q-205	8.0 8.0	7.4 11.0 +	7.8 0V	Low Band " High Band "
Q-206	8.0 8.0	11.0 + 7.4	0V 7.8	Low Band " High Band "
Q-207	3.4	4.7	7.1	

Table 4-9. IC Voltages—TMR-8H only.

IC NO.		1	2	3	4	5	6	7	8	9	10	11	12	13	14
IC 101		.70V	0V	0V	.70V	7.5V	0V	4.2V	7.5V	3.32V	3.3V				
IC 102		4.0V	3.5V		1.3V	1.3V	1.3V						3.5V	7.6V	5.0V
IC 301	Activated	5.8V	5.8V	5.8V	5.8V	5.8V	5.8V	0V	5.8V	5.8V	5.8V	5.8V	5.8V	5.8V	6.0V
	Deactivated	.1V	.1V	1.6V	.1V	1.6V	.1V	0V	.1V	1.6V	.1V	1.6V	.1V	1.6V	6.0V
IC 302 IC 303	Activated	5.8V	5.8V	1.0V	1.0	5.8V	5.8V	0V	5.8V	5.8V	1.0V	1.0V	5.8V	5.8V	5.1V
	Deactivated	.1V	.1V	11V	11V	.1V	.1V	0V	.1V	.1V	11V	11V	.1V	.1V	0V

REGENCY ACT-R10H/L/U

Specifications

Receiver (Model ACT-R10H/L/U)

Frequency range	Low band, 30–50 MHz High band, 150–174 MHz UHF band, 450–470 MHz
Antenna impedance	50Ω
Channels	Ten, crystal-controlled

Table 4-10. Voltage Data—TME-8H Scanner Board.

TRANSISTOR	EMITTER	BASE	COLLECTOR	
Q-301	0V	.60V	.20V	Deactivated
	0V	.20V	6.2V	Activated
Q-302	4.0V	3.9V	6.3V	Activated
	0.6V	0.2V	8.3V	Deactivated
Q-303	6.8	6.3	8.1	Activated
	7.8	8.3	7.9	Deactivated
Q-304	0V	.6V	.1V	
Q-305	0V	.1V	8.5V	
Q-306	5.1	5.8V	6.0	Activated
	0	0	6.0	Deactivated
Q-307	5.1V	5.8V	6.0V	Activated
	0V	0V	6.0V	Deactivated

Sensitivity (at tuneup)	Low band, 0.5 μV (max.) High band, 0.6 μV (max.) UHF band, 0.7 μV (max.)
Frequency separation	
Low-band model	6 dB bandwidth, 33–46 MHz 10 dB bandwidth, 30–50 MHz
High-band model	6 dB bandwidth, 8 MHz 10 dB bandwidth, 10 MHz
UHF-band model	6 dB bandwidth, 8 MHz 10 dB bandwidth, 12 MHz
Selectivity (IF)	6 dB down, ±7 kHz (min.) 50 dB down, ±18 kHz (max.)
Spurious rejection (excluding primary image)	50 dB
AFC range (UHF band only)	Approx ±5 kHz
Modulation acceptance	±7 kHz (min.)
Intermediate frequencies	First, 10.7 MHz Second, 455 kHz
Squelch system	Noise-operated
Audio output (3.2Ω speaker)	1W at 10% or less distortion, 2W max.

Scanner

Scan rate	Approx 15 channels per sec
Scan delay	Approx 0.5 sec

Power

Voltage requirement	117V AC (±10%), 60 Hz, 13W max.; or 13.8V DC (±10%)
Current requirements at 13.8V DC	
(squelched)	180 mA
(max. audio output)	600 mA
Fuse size	1.5A, 3AG

Table 4-11. Voltage Data—TME-8H, 8HLL, -8HLM, -8HLH—IF and RF Boards.

NOTE: All voltages are nominal and are measured with a VTVM. SCAN indicates the unit is scanning. MAN indicates the unit is not scanning and is stopped at channel 1. A "P" beside a voltage indicates that the meter reading is pulsating (fluctuating) because the scanner section of the unit is operating.

	TRANSISTOR	EMITTER (Source)	BASE (Gate)	COLLECTOR (Drain)	
RF Board No. 301-563 (TME-8H/L only)	Q201	3.1	3.8	7.0	Low Band Activated
		0	0	7.6	High Band Activated
	Q202	0	0	7.6	Low Band Activated
		3.1	3.8	7.0	High Band Activated
	Q203	1.6	2.3	7.1	Low Band Activated
		1.6	0	7.1	High Band Activated
	Q204	1.6	0	7.1	Low Band Activated
		1.6	2.3	7.1	High Band Activated
	Q205	7.8	7.4	7.6	Low Band Activated
		7.8	11.0	0	High Band Activated
	Q206	7.8	11.0	0	Low Band Activated
		7.8	7.0	7.6	High Band Activated
	Q207	3.4	4.1	7.0	
RF Board No. 301-563 (TME-8H only)	Q201	3.1	3.8	7.0	
	Q202	1.6	2.3	7.1	
	Q203	3.4	4.1	7.0	
IF Board No. 500-858	Q101	2.3	3.0	5.8	
	Q102	1.0	1.7	4.8	
	Q103 (PNP)	8.2	8.2	0	(unsquelched)
		8.2	8.2	1.0	(squelched)
		8.2	8.2	1.5	Min. (tight squelch)
	Q104	0	0	7.2	(unsquelched)
		0	.80	.30	(squelched)
		0	.80	.10	(tight squelch)
	Q105	1.4	1.9	5.1	(unsquelched)
		1.1	.10	8.2	(tight squelch)
	Q106	0.7	1.3	12.4	
	Q107 (PNP)	13.8	13.1	7.2	
	Q108 (PNP)	6.9	6.6	.10	
	Q109	6.9	7.2	13.8	
	Q110	0	.10	6.9	

Table 4-12. Voltage Data—TME-8H, -8HLL, -8HLM, -8HLH Scanner Board.

Scan Board No. 500-850	Q301	0	.70	.10 (SCAN)
		0	.70	0 (MAN)
	Q302	0	.70	.10
	Q303	0	.10	3.6 (SCAN)
		0	.10	1.2 (MAN)
	Q304	0	.10	2.8 (SCAN)
		0	.70	0.2 (MAN)
	Q305	3.1	3.6	3.2 (SCAN)
		0.7	1.2	0.7 (MAN)
	Q307 (PNP)	5.1	4.5	2.0 (SCAN)
		5.1	4.6	.10 (MAN)
	Q308	0	.10	1.5
		Base 1	**Emitter**	**Base 2**
	Q306 (unijunction)	.20	3.2	4.6 (SCAN)
		.20	0.7	4.6 (MAN)

General

Front-panel size	$6\frac{1}{2} \times 2\frac{5}{8}$ in.
Depth (including knobs and rear-panel connectors)	$9\frac{3}{4}$ in.
Antenna connectors	Motorola type
Speaker size	2 × 6 in.. 8Ω

Service Data

Prior to installing a crystal, the receiver's cover will have to be removed.

- To remove the cover, first remove telescopic antennas if they are in-

Table 4-13. Voltage Data—TME-8H, -8HLL, -8HLM, -8HLH Integrated Circuits.

Voltage Data   Integrated Circuits

NOTE: A "P" beside a voltage indicates that the meter reading is pulsating (fluctuating) because the scanner section of the unit is operating. MAN indicates the unit is not scanning and is at channel 1 (M301 is lighted).

	IC No.	1	2	3	4	5	6	7	8	9	10	11	12	13	14
IF Board 500-858	IC 101	4.2	0.7	0.7	4.2	7.8	0	4.2	7.8	–	–	–	–	–	-
	IC 102	4.0	3.5	0	1.3	1.3	1.3	0	0	0.2	1.4	2.9	3.5	7.6	5.0
Scan Board	IC 301	.7P	.7P	9P	9P	.7P	.7P	0	.7P	.7P	9P	9P	.7P	.7P	5.1 (Scan)
		3.3	3.3	0.5	10.8	0.2	0.2	0	0.2	0.2	10.8	10.8	0.2	0.2	5.1 (Man)
	IC 302	.7P	.7P	9P	9P	.7P	.7P	0	.7P	.7P	9P	9P	.7P	.7P	5.1 (Scan)
		0.2	0.2	10.8	10.8	0.2	0.2	0	0.2	0.2	10.8	10.8	0.2	0.2	5.1 (Man)
	IC 303	0.1	0	0	0	0	0	0	1.5	1.5	.7P	.7P	.7P	.7P	5.1 (Scan)
		0	0	0	0	0	0	0	1.5	1.5	0.2	0.2	0.2	3.3	5.1 (Man)
	IC 304	.7P	0	0	0	0	0	0	1.5	1.5	.7P	.7P	.7P	.7P	5.1 (Scan)
		0.2	0	0	0	0	0	0	1.5	1.5	0.2	0.2	0.2	0.2	5.1 (Man)

stalled. Second, unscrew the two large bolts located at the sides of the unit. The cover may then be slipped off by sliding it toward the rear of the unit.

- Insert the crystal in the proper socket pins as indicated on the crystal location diagram (Fig. 4-22). The number by each pair of sockets matches the channel number on the front panel.
- If the crystal inserted is for the low VHF band (30–50 MHz), place the proper color-coded wire and socket onto the proper low-band pin (row of pins nearest front of unit); if the crystal is for the high VHF band (148–174 MHz), place the proper color-coded wire onto the proper high-band pin (middle row of pins); if the crystal is for the UHF band (450–470 MHz), place the proper color-coded wire onto the proper UHF-band pin (row of pins near the rear of set). (See crystal location, Fig. 4-22.) Be sure that each channel has its color-coded wire programmed properly with respect to crystal installed. Reading from left to right (channels 1 through 10), the color-coded wires should be in the following order: brown, red, orange, yellow, green, blue, purple (or violet), pink, white, and black.
- *Note*: If a particular channel is not used (in other words, there is no crystal installed for that channel), the band-selection wire must still be connected to either a high-band, low-band, or UHF-band pin. Thus, for proper scanner operation, all of the band-selection wires *must* be connected, even though not all channels are used.
- As shipped from the factory, the first three channels are programmed for low band, the next four (channels 4 through 7) for high band, and the last three (channels 8 through 10) for the UHF band. If desired, this arrangement can be changed to any combination of high, low, or UHF channels.

470–500 MHz Operation

This unit can be retuned to cover an 8 MHz segment of the 470–500 MHz band. The major difference is that the crystal frequency is determined by taking the channel frequency minus 10.7 MHz and dividing by 10. This allows for L119 (injection coil for the high VHF band) and drive coil for the UHF tripler (Q108) not to be retuned and, therefore, not to affect VHF high-band sensitivity. Capacitor C139, the UHF-band injection tuning capacitor, is tuned to the tenth harmonic of the crystal. The tuning procedure for 470–500 MHz operation is the same as normal UHF alignment.

UHF Band Crystal (470–500 MHz)

1. Crystal frequency

$$= \frac{\text{receiver frequency} - 10.7\ \text{MHz}}{10}$$

Example .

Crystal frequency

$$= \frac{485.10\ \text{MHz} - 10.7\ \text{MHz}}{10}$$

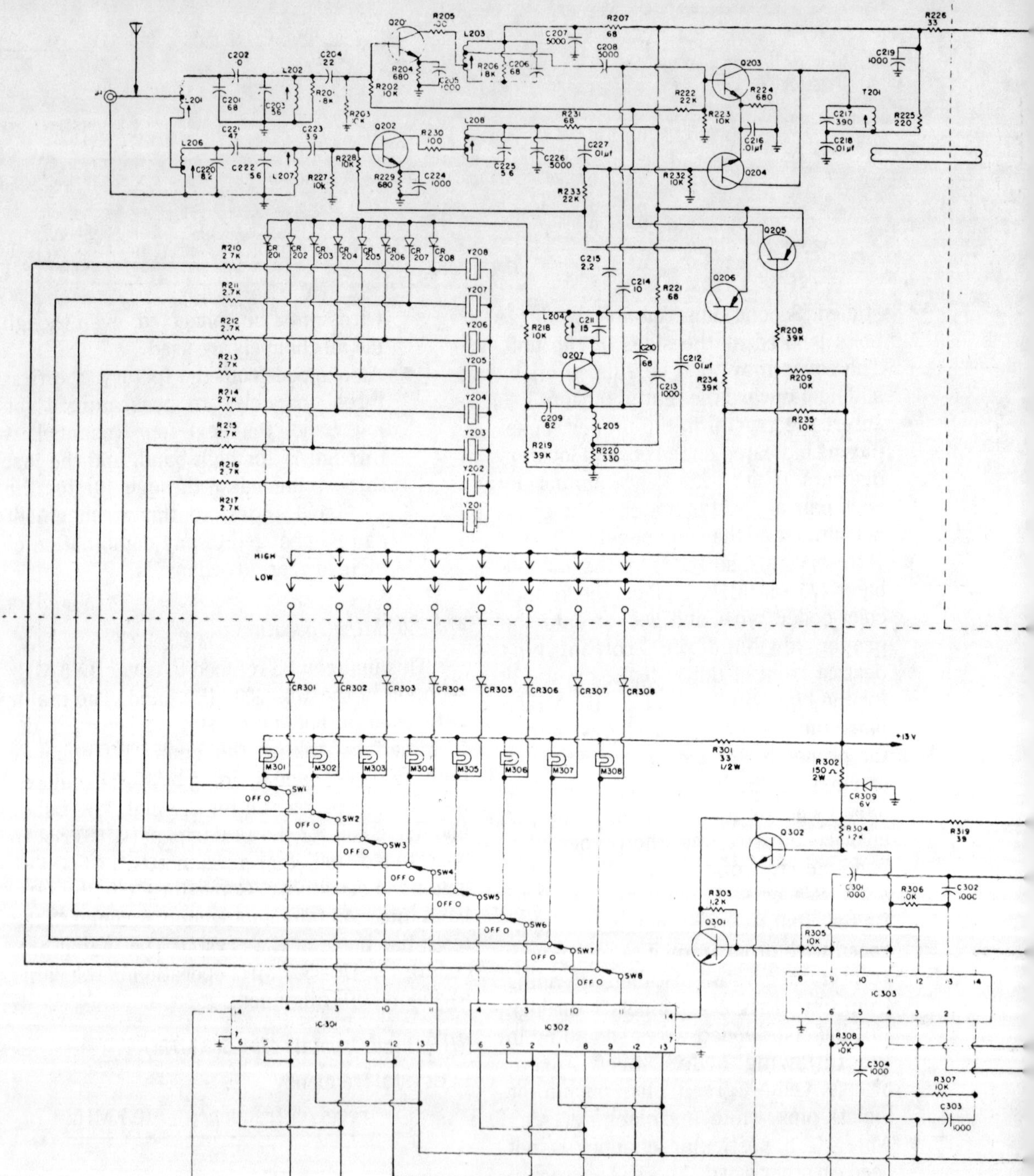

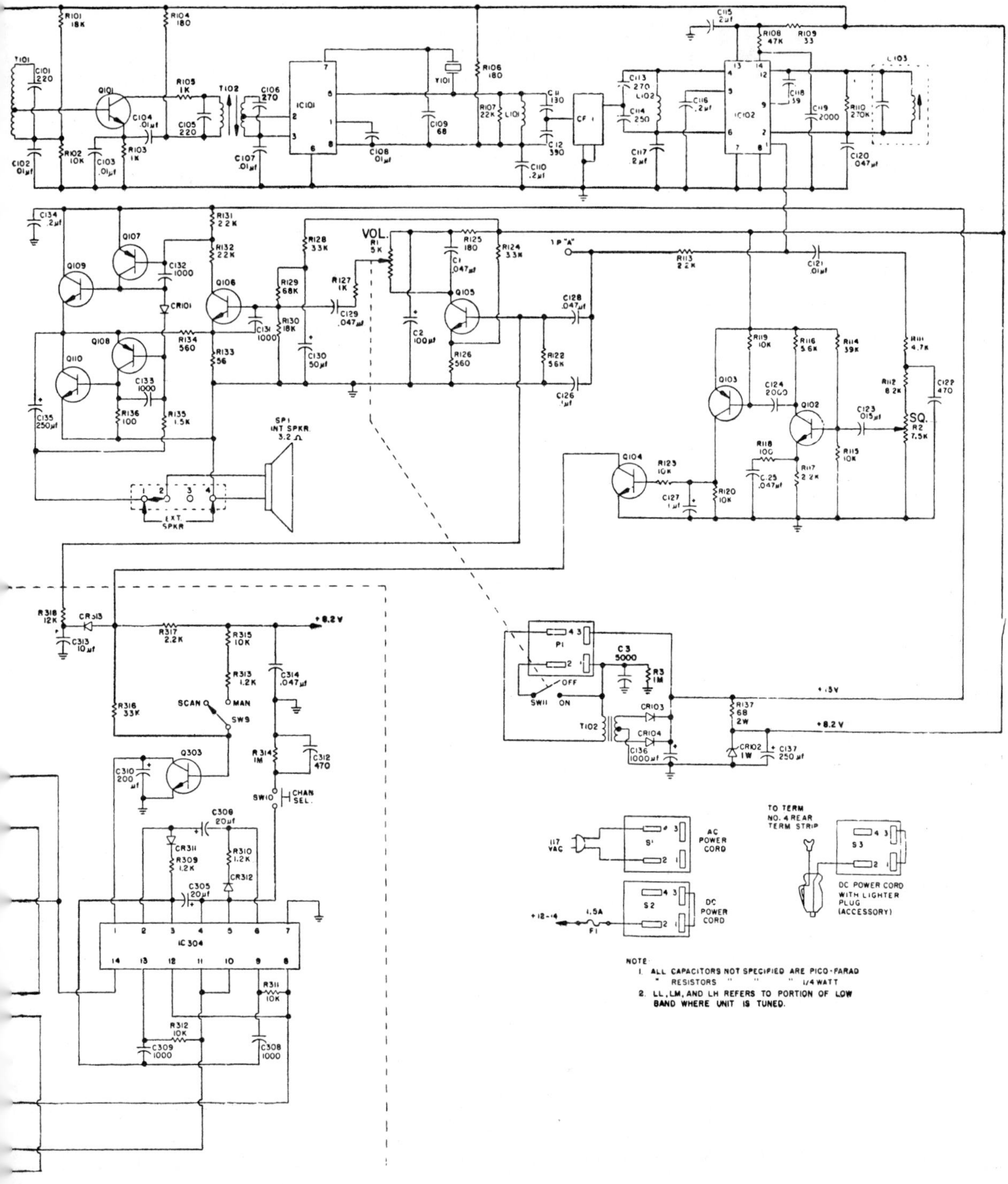

Schematic 4-7. TMR-8H/LL, TMR-8H/LM, TMR-8H/LH.

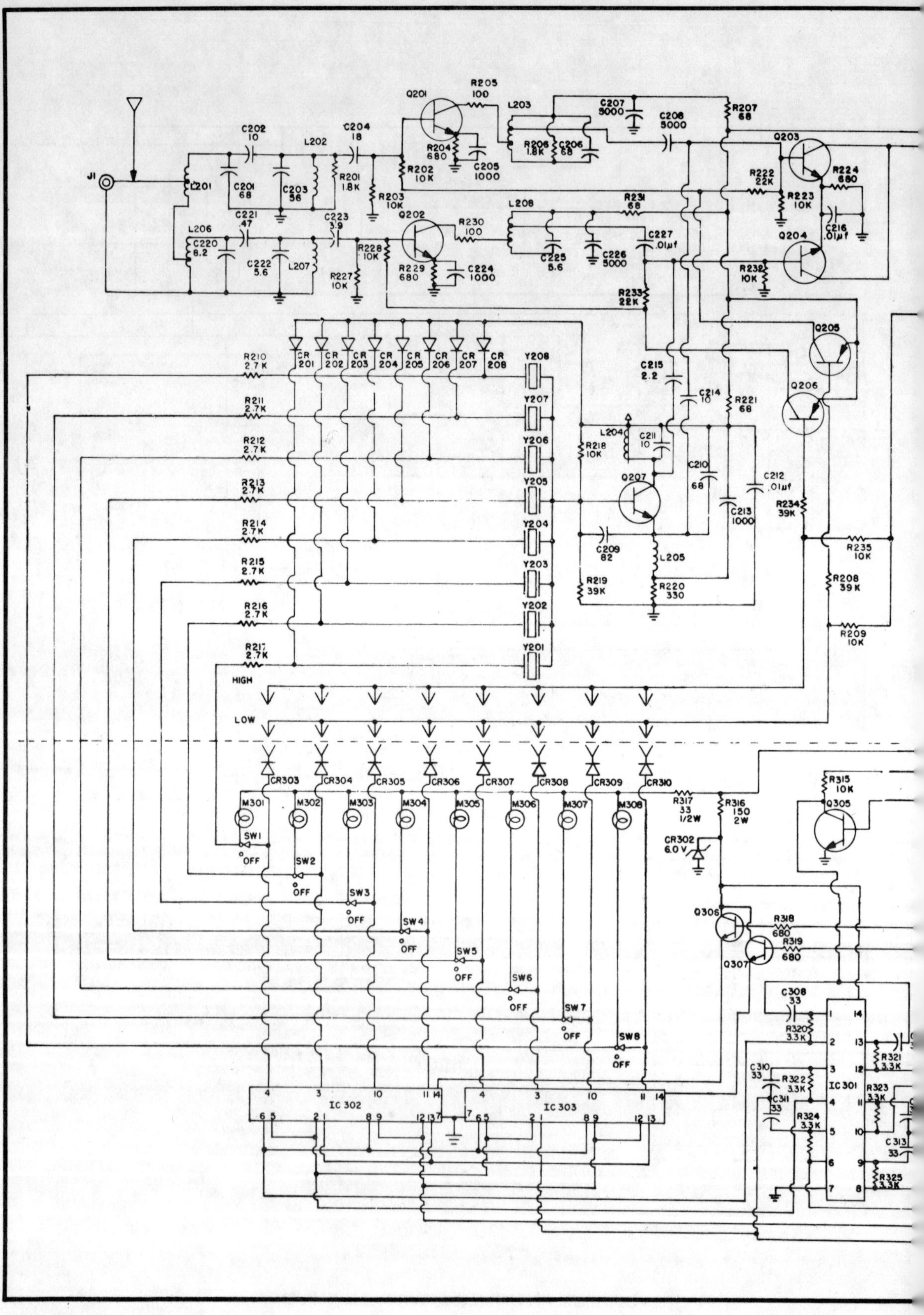
J1
L201
C202 10
C201 68
C203 56
L202
R201 1.8K
C204 18
R202 10K
R203 10K
Q201
R205 100
R204 680
C205 1000
L203
R206 1.8K
C206 68
C207 5000
C208 5000
R207 68
Q203
R222 22K
R223 10K
R224 680
C216 .01µf
Q204
R232 10K
L206
C220 8.2
C221 .47
C222 5.6
L207
C223 3.9
R227 10K
R228 10K
Q202
R229 680
R230 100
C224 1000
L208
C225 5.6
C226 5000
R231 68
C227 .01µf
R233 22K
CR 201
CR 202
CR 203
CR 204
CR 205
CR 206
CR 207
CR 208
R210 2.7K
R211 2.7K
R212 2.7K
R213 2.7K
R214 2.7K
R215 2.7K
R216 2.7K
R217 2.7K
Y208
Y207
Y206
Y205
Y204
Y203
Y202
Y201
Q205
Q206
C215 2.2
C214 10
R221 68
L204
C211 10
R218 10K
Q207
C210 68
C212 .01µf
C213 1000
R234 39K
C209 82
L205
R219 39K
R220 330
R235 10K
R208 39K
R209 10K
HIGH
LOW
CR303
CR304
CR305
CR306
CR307
CR308
CR309
CR310
M301
M302
M303
M304
M305
M306
M307
M308
SW1
SW2
SW3
SW4
SW5
SW6
SW7
SW8
OFF
R317 33 1/2W
R316 150 2W
CR302 6.0V
R315 10K
Q305
Q306
Q307
R318 680
R319 680
C308 33
R320 3.3K
C310 33
C311 33
R322 3.3K
R324 3.3K
IC301
R321 3.3K
R323 3.3K
C313 33
R325 3.3K
IC 302
IC 303

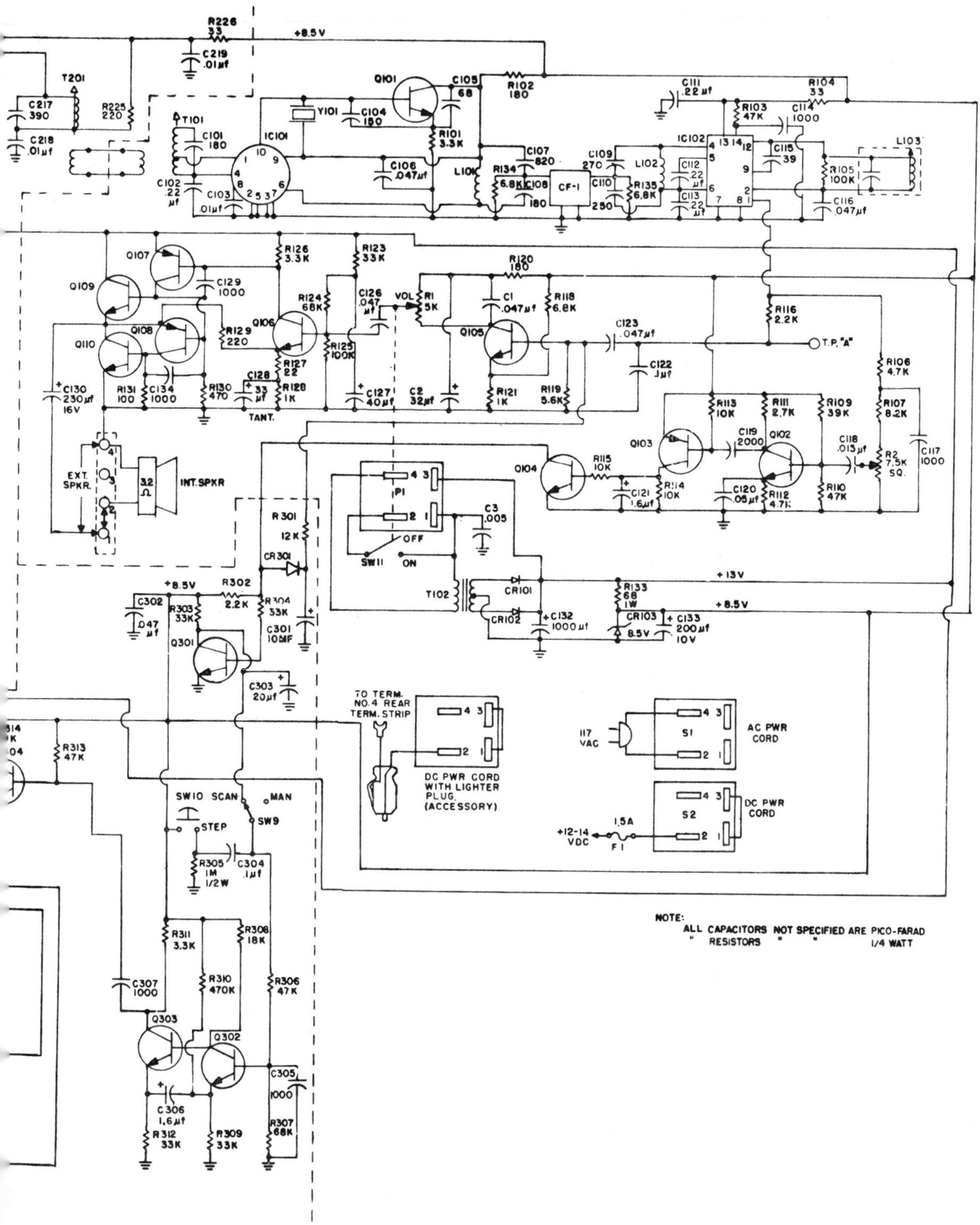

Schematic 4-8. TMR-8H/L.

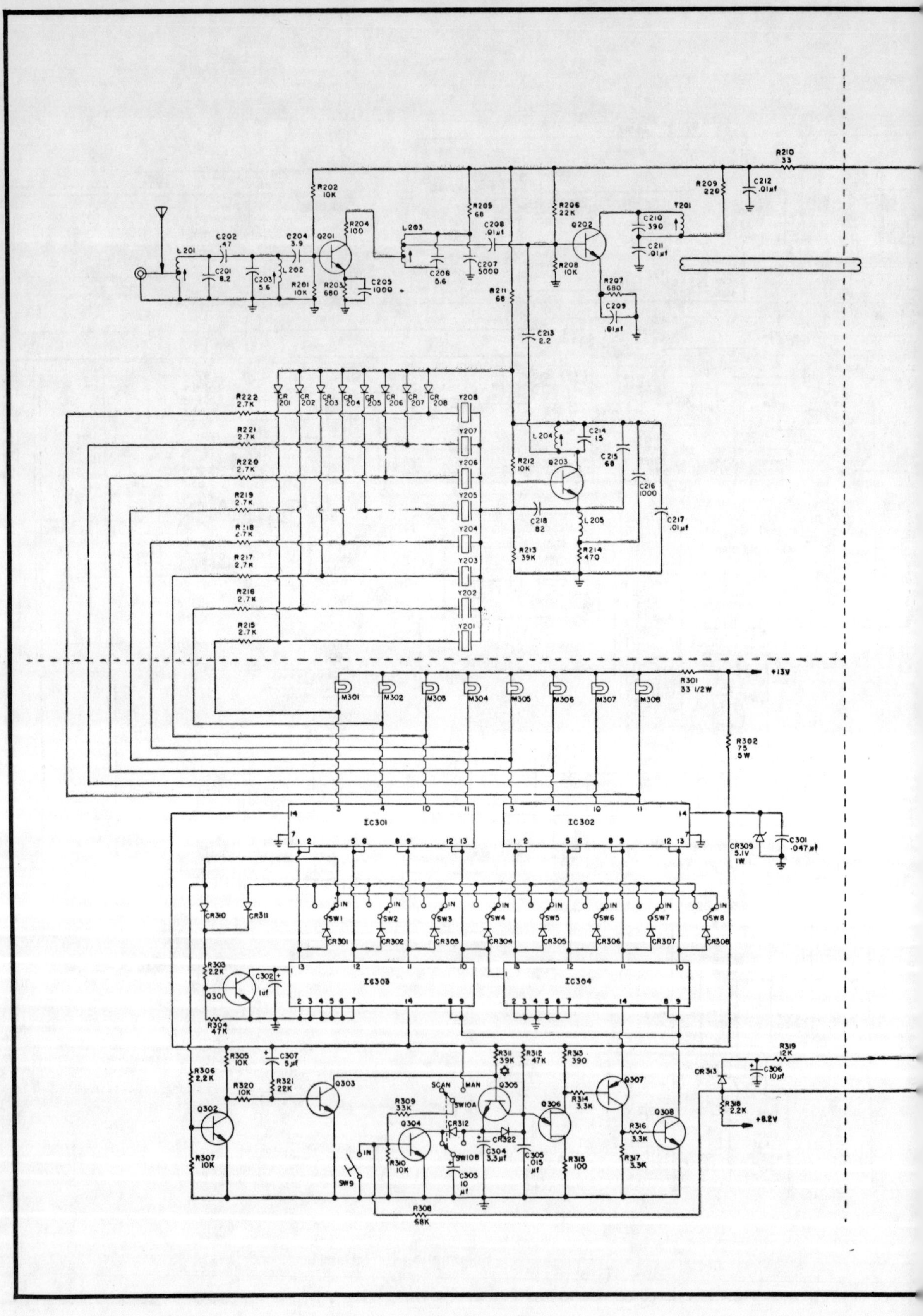
R210 33
R209 220
C212 .01µf
R202 10K
R205 68
R206 22K
T201
C202 .47
C204 3.9
R204 100
L203
C208 .01µf
Q202
C210 390
C211 .01µf
L201
C201 8.2
C203 5.6
L202
Q201
R201 10K
R203 680
C205 1000
C206 5.6
C207 5000
R208 10K
R207 680
C209 .01µf
R211 68
C213 2.2
R222 2.7K
R221 2.7K
R220 2.7K
R219 2.7K
R218 2.7K
R217 2.7K
R216 2.7K
R215 2.7K
CR 201
CR 202
CR 203
CR 204
CR 205
CR 206
CR 207
CR 208
Y208
Y207
Y206
Y205
Y204
Y203
Y202
Y201
L204
C214 15
C215 68
R212 10K
Q203
C216 1000
C218 82
L205
C217 .01µf
R213 39K
R214 470
M301
M302
M303
M304
M305
M306
M307
M308
R301 33 1/2W
+13V
R302 75 .5W
IC301
IC302
CR309 5.1V 1W
C301 .047µf
CR310
CR311
SW1
SW2
SW3
SW4
SW5
SW6
SW7
SW8
IN
CR301
CR302
CR303
CR304
CR305
CR306
CR307
CR308
R303 2.2K
C302 1µf
Q301
IC303
IC304
R304 47K
R305 10K
C307 5µf
R306 2.2K
R320 10K
R321 22K
Q303
R311 39K
R312 47K
R313 390
R319 12K
SCAN
MAN
SW10A
Q305
Q307
CR313
C306 10µf
R309 33K
Q304
CR312
CR322
Q306
R314 3.3K
R318 2.2K
+8.2V
Q302
R316 3.3K
Q308
C304 3.3µf
C305 .015 µf
SW10B
R307 10K
R310 10K
C303 10 µf
R315 100
R317 3.3K
SW9
R308 68K

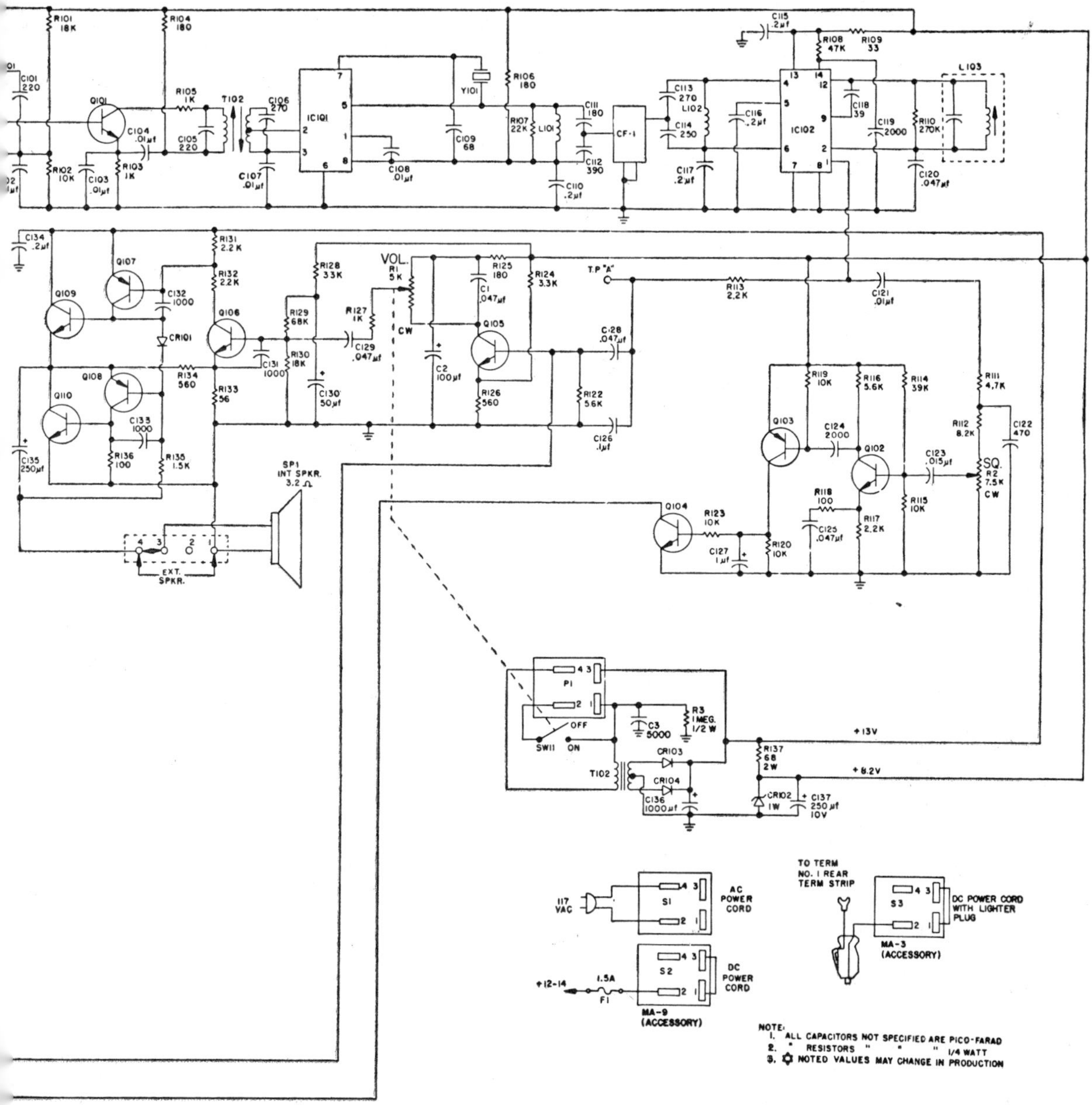

Schematic 4-9. TME-8H.

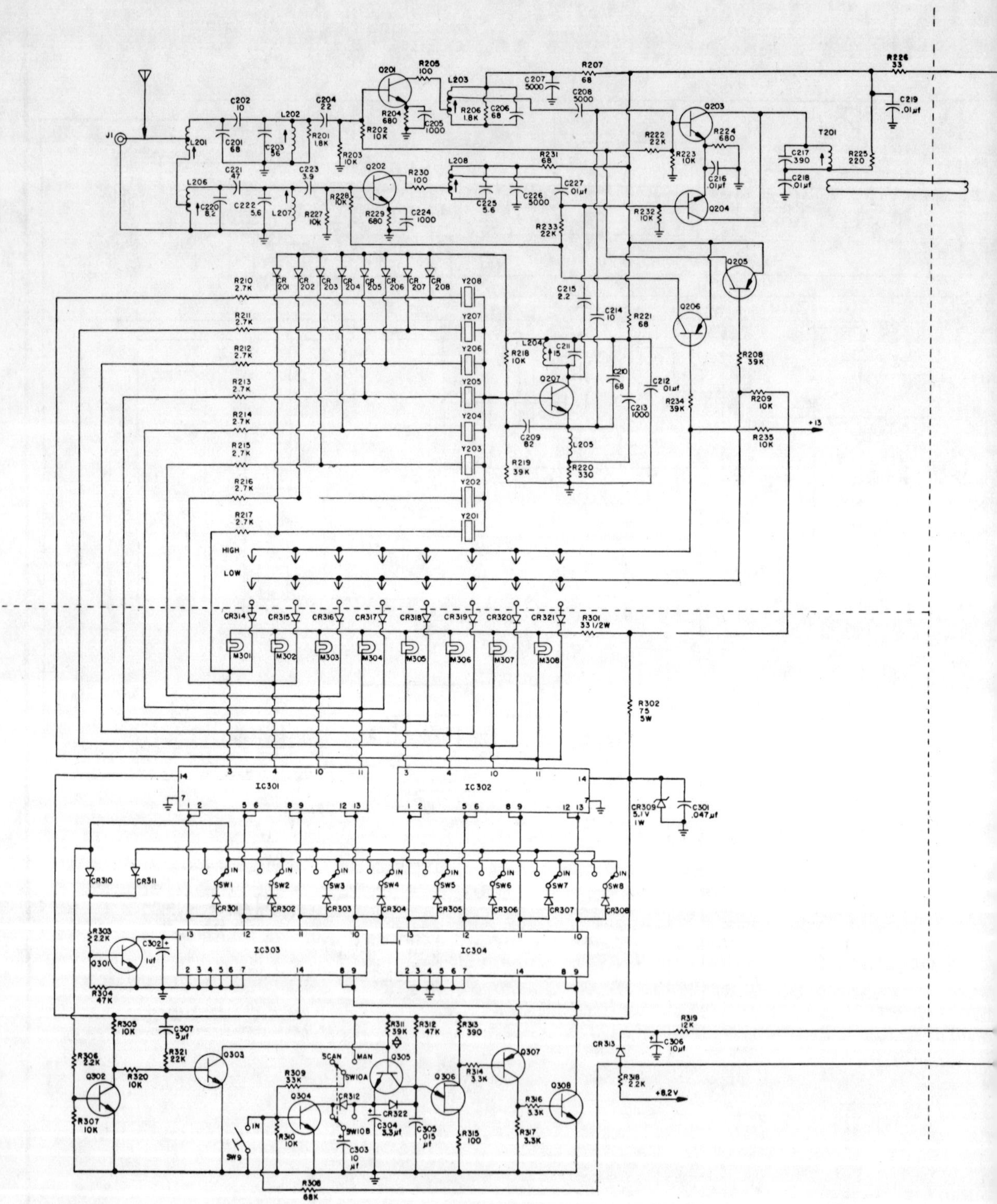
J1
L201
C202 10
C201 68
C203 56
L202
R201 1.8K
C204 22
R203 10K
R202 10K
Q201
R205 100
R204 680
C205 1000
L203
R206 1.8K
C206 68
C207 5000
R207 68
C208 5000
Q203
R222 22K
R223 10K
R224 680
C216 .01µf
C217 390
C218 .01µf
T201
R225 220
R226 33
C219 .01µf
L206
C221 47
C220 8.2
C222 5.6
L207
C223 3.9
R227 10k
R228 10k
Q202
R230 100
R229 680
C224 1000
L208
C225 5.6
C226 5000
R231 68
C227 .01µf
R232 10K
Q204
R233 22K
CR 201
CR 202
CR 203
CR 204
CR 205
CR 206
CR 207
CR 208
Y208
Y207
Y206
Y205
Y204
Y203
Y202
Y201
R210 2.7K
R211 2.7K
R212 2.7K
R213 2.7K
R214 2.7K
R215 2.7K
R216 2.7K
R217 2.7K
Q205
C215 2.2
C214 10
R221 68
Q206
L204
C211 15
R218 10K
Q207
C210 68
C213 1000
C212 .01µf
R234 39K
R208 39K
R209 10K
R235 10K
+13
C209 82
L205
R219 39K
R220 330
HIGH
LOW
CR314
CR315
CR316
CR317
CR318
CR319
CR320
CR321
R301 33 1/2W
M301
M302
M303
M304
M305
M306
M307
M308
R302 75 5W
IC301
IC302
CR309 5.1V 1W
C301 .047µf
CR310
CR311
SW1
SW2
SW3
SW4
SW5
SW6
SW7
SW8
IN
CR301
CR302
CR303
CR304
CR305
CR306
CR307
CR308
R303 2.2K
Q301
C302 1µf
IC303
IC304
R304 47K
R305 10K
C307 5µf
R321 22K
R306 2.2K
Q302
R320 10K
Q303
R307 10K
R311 39K
R312 47K
R313 390
SCAN
MAN
SW10A
Q305
R309 33K
Q304
CR312
CR322
Q306
R314 3.3K
Q307
R310 10K
SW10B
C304 3.3µf
C305 .015 µf
R315 100
R316 3.3K
R317 3.3K
Q308
SW9
C303 10 µf
R308 68K
CR313
C306 10µf
R319 12K
R318 2.2K
+8.2V

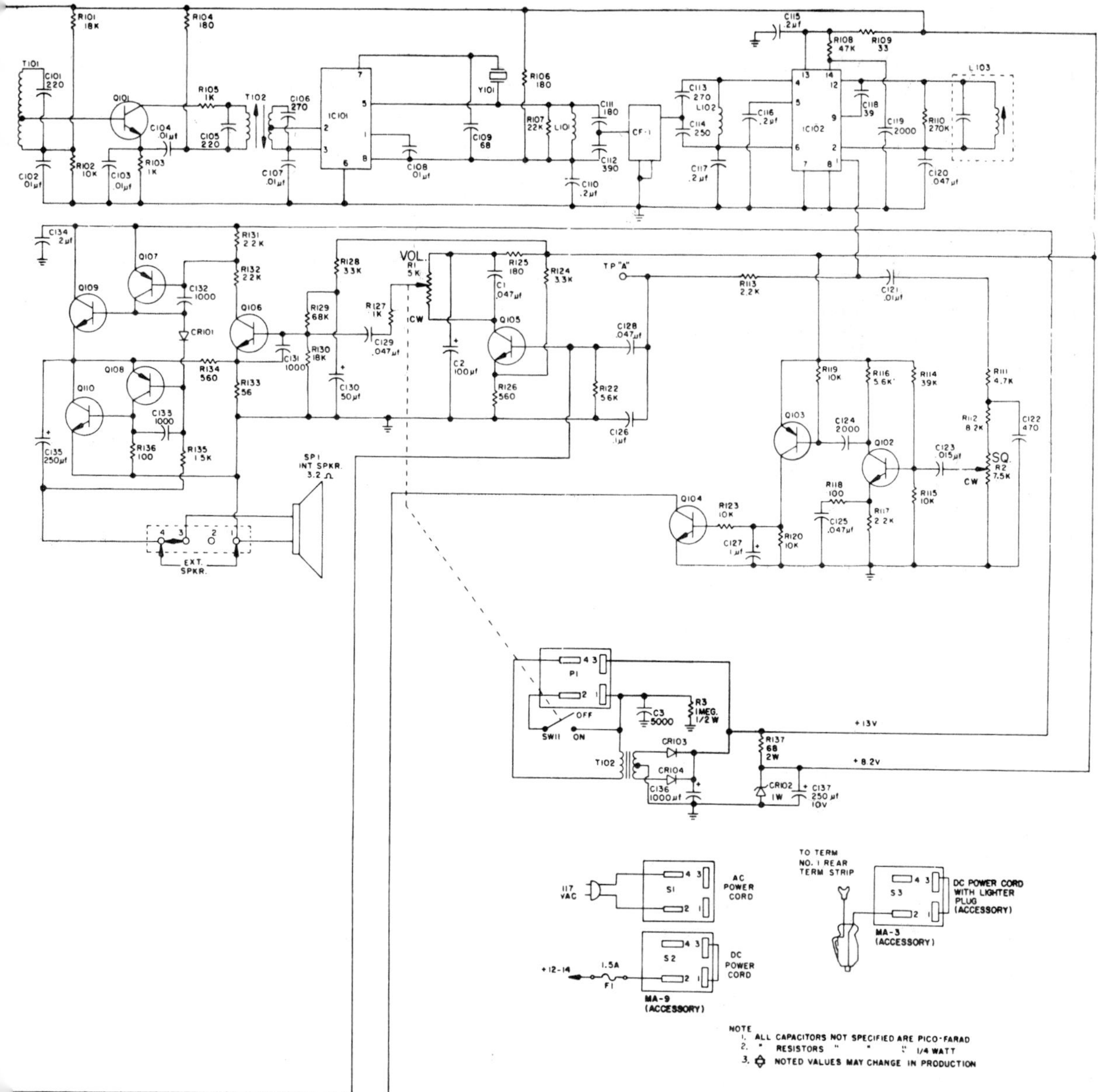

Schematic 4-10. TME-8H/LL, TME-8H/LM, TME-8H/LH.

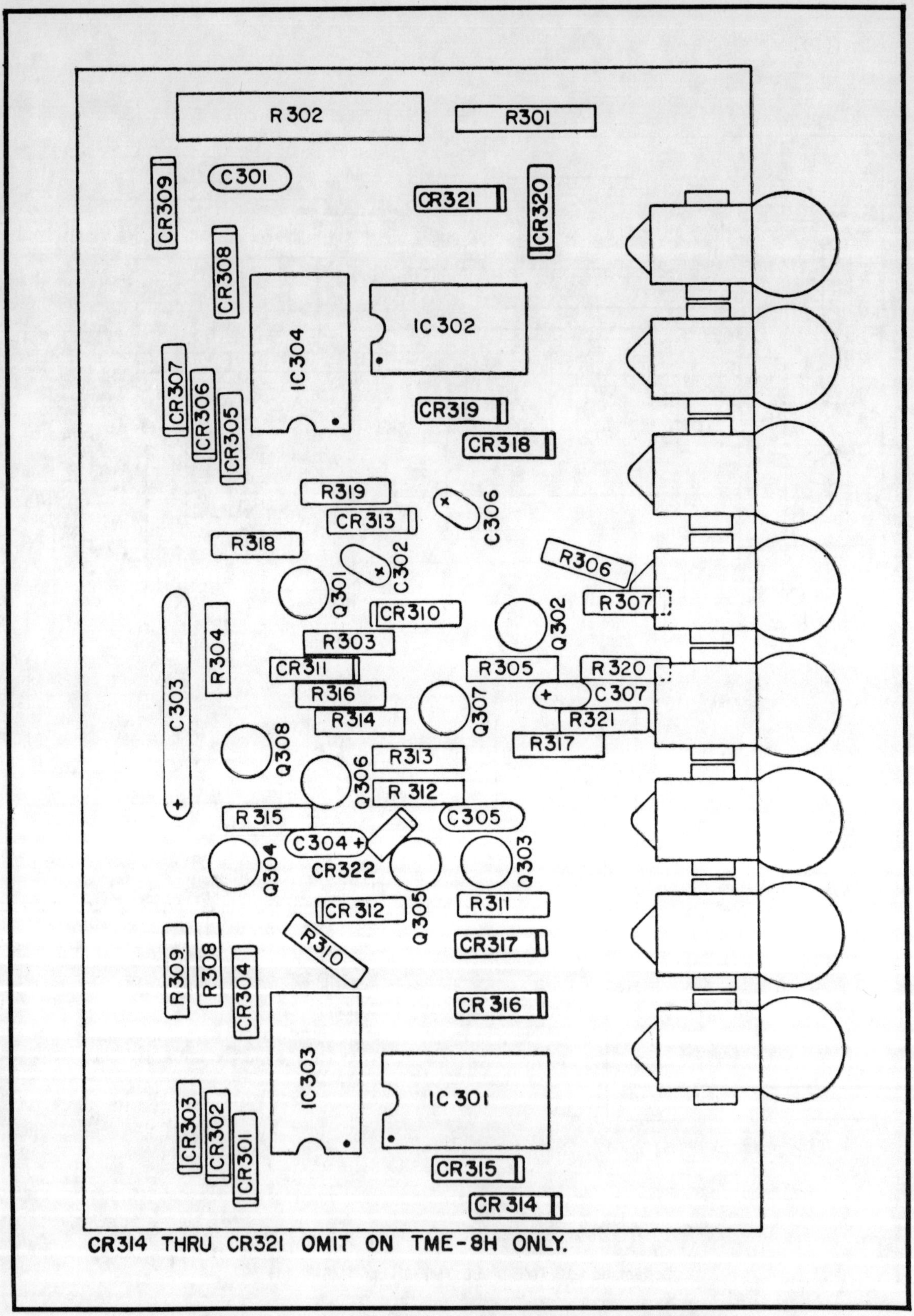

Fig. 4-20. Scanner board 500-850 parts placement diagram, TME-8 series.

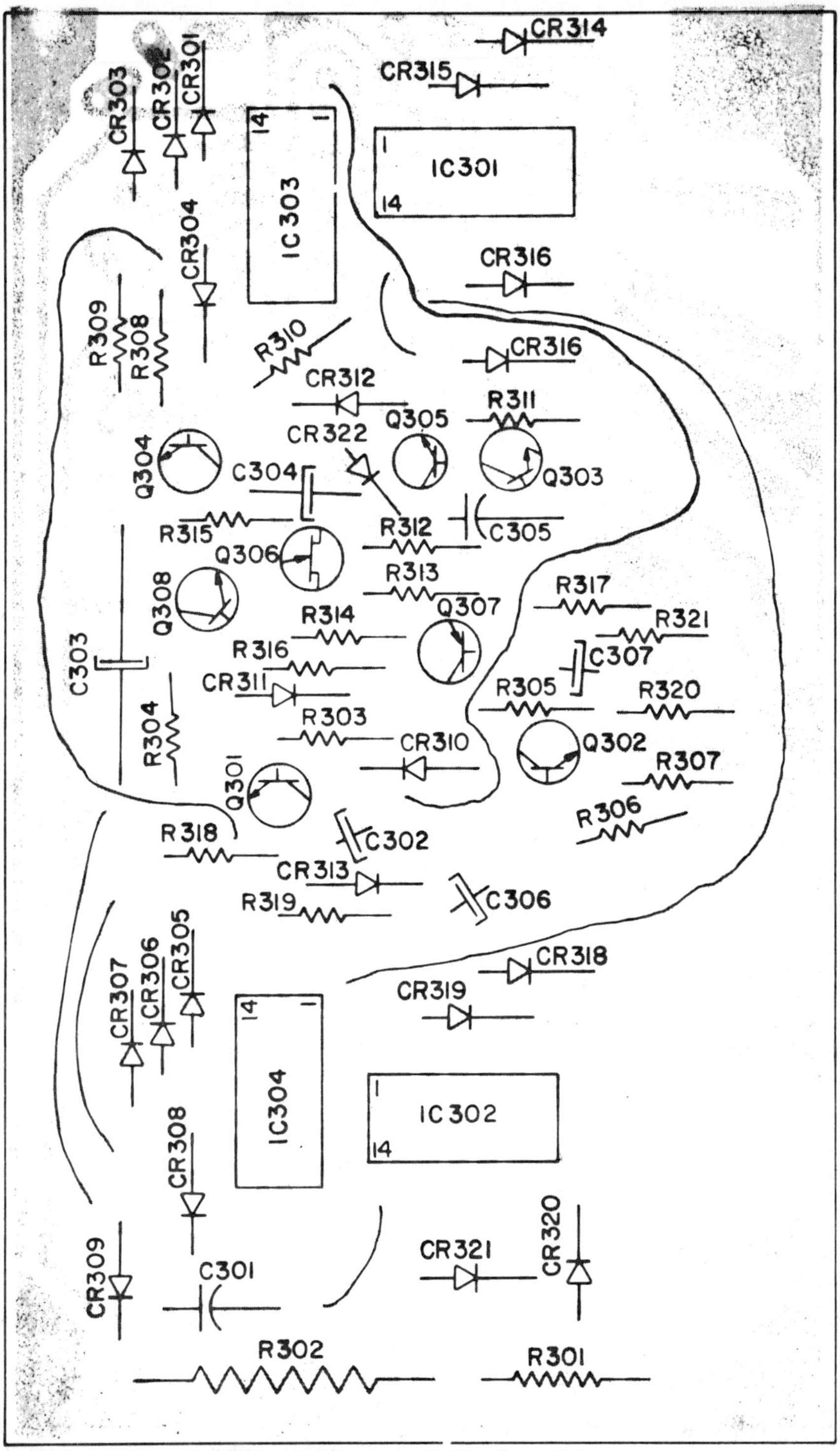

CR314 THRU CR321 OMIT ON TME-8H ONLY.

Fig. 4-21. Scanner board bottom view, TME-8 series.

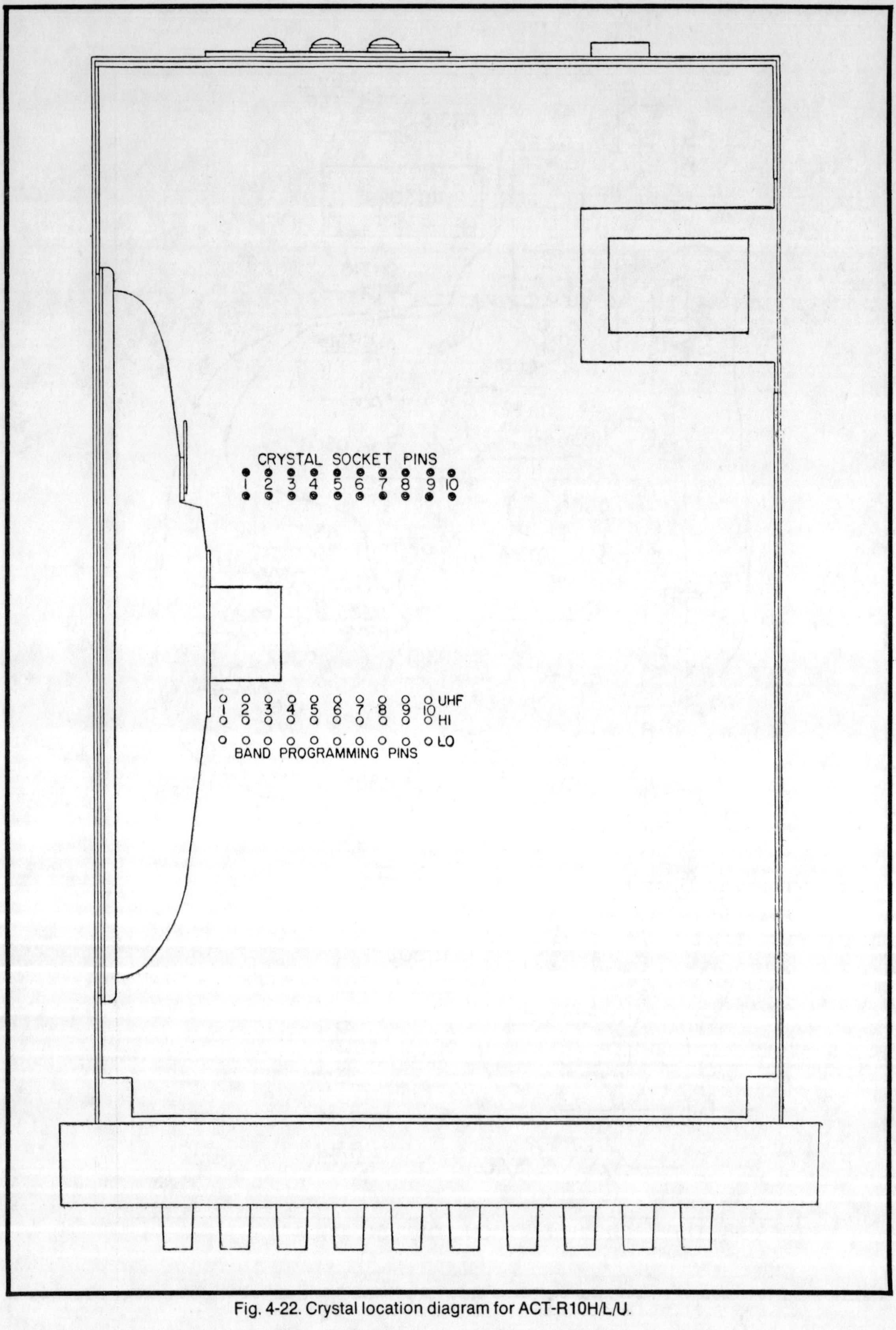

Fig. 4-22. Crystal location diagram for ACT-R10H/L/U.

2. Frequency tolerance of 0.001%
3. Third overtone
4. Parallel resonance, 18 pF load capacity
5. Maximum equivalent series resistance of 35Ω
6. Drive level of 2 mW
7. Holder, HC-25/U

Low-Band Crystal

Crystal frequency = channel frequency + 10.7 MHz

Crystal layout		Fig. 4-22
Voltage data		Tables 4-14, 4-15, 4-16
Schematic		Schematic 4-11
Internal layout		
	Main	Figs. 4-23, 4-24, 4-25
	Light board	Fig. 4-26
Alignment data		A-2, J, K, L, M

Table 4-14. Voltage Data—ACT-R10H/L/U Transistors.

NOTE: All voltages are nominal and are measured with a VTVM. SCAN indicates the unit is scanning. MAN indicates the unit is not scanning and is stopped at channel 1. A "P" beside a voltage indicates that the meter reading is pulsating (fluctuating) because the scanner section of the unit is operating.

VOLTAGE DATA – SEMICONDUCTORS:

TRANSISTOR	EMITTER (Source)	BASE (Gate)	COLLECTOR (Drain)	
Q101	3.1	3.8	7.0	Low Band Activated
	0	0	0	High Band Activated
	0	0	0	UHF Band Activated
Q102	1.6	2.3	7.7	Low Band Activated
	1.6	0	7.7	High Band Activated
	1.4	0	7.7	UHF Band Activated
Q103	0	0	8.2	Low Band Activated
	3.1	3.8	7.9	High Band Activated
	0	0	8.2	UHF Band Activated
Q104	1.6	0	7.7	Low Band Activated
	1.6	2.3	7.7	High Band Activated
	1.4	0	7.7	UHF Band Activated
Q105 (FET)	0	0	6.0	
Q106	1.6	0	7.7	Low Band Activated
	1.6	0	7.7	High Band Activated
	1.6	2.3	7.7	UHF Band Activated
Q107	3.7	4.4	7.4	No Crystal
	3.4	4.0	7.4	With Crystal
Q108	0	0	0	Low High Band Activated
	0	.4	8.1	UHF Band Activated (No Crystal)
	0	.1	4.0	UHF Band Activated (With Crystal)
Q109 (PNP)	8.1	8.1	.8	Low Band Activated
	8.1	8.1	.8	High Band Activated
	7.9	7.2	3-5	UHF Band Activated
Q110	8.2	8.2	8.2	Low Band Activated
	8.2	8.2	8.2	High Band Activated
	2.9	3.6	7.2	UHF Band Activated
Q111	8.2	8.2	4.0	Low Band Activated
	8.2	8.2	4.0	High Band Activated
	2.9	3.6	7.4	UHF Band Activated
Q112	0	0	8.2	Low Band Activated
	0	0	8.2	High Band Activated
	0	.7	.1	UHF Band Activated

Table 4-15. Voltage Data—ACT-R10HLU Semiconductors.

TRANSISTOR	EMITTER (Source)	BASE (Gate)	COLLECTOR (Drain)	
Q113(PNP)	8.2	7.5	8.1	Low Band Activated
	8.2	8.2	0	High Band Activated
	8.2	8.2	0	UHF Band Activated
Q114(PNP)	8.2	8.2	0	Low Band Activated
	8.2	7.5	8.1	High Band Activated
	8.2	8.2	0	UHF Band Activated
Q115(PNP)	8.2	8.2	0	Low Band Activated
	8.2	8.2	0	High Band Activated
	8.2	7.5	8.1	UHF Band Activated
Q116	0.4	1.1	4.5	
Q117	1.0	1.7	5.0	
Q118	8.2	8.2	0	(Unsquelched)
	8.2	8.2	1.0	(Squelched)
	8.2	8.2	1.5	Min. (Tight Squelched)
Q119	0	0	7.2	(Unsquelched)
	0	0.8	0.2	(Squelched)
	0	0.8	0.1	(Tight Squelched)
Q120	13.8	13.1	13.6	(SCAN)
	13.8	13.1	13.6	(MAN)
	13.8	13.8	0	(No Lights)
Q121	0	0	4.0	(SCAN)
	0	0.8	0.1	(MAN)
Q122	0	0.8	0.1	
Q124	0	.2	4.9	(SCAN)
	0	.2	5.1	(MAN)
	BASE 1	EMITTER	BASE 2	
Q123	0.2	3.8	5.1	(SCAN)
(Unijunction)	0.2	0.5	5.1	(MAN)
	CATHODE	ANODE		
CR113	1.8	2.4		(Unsquelched)
	1.0	0		(Squelched)

Table 4-16. Voltage Data—ACT-R10HLU Integrated Circuits.

NOTE: A "P" beside a voltage indicates that the meter reading is pulsating (fluctuating) because the scanner section of the unit is operating. MAN indicates the unit is not scanning and is at channel 1 (M301 is lighted).

IC No.	1	2	3	4	5	6	7	8	9	10	11	12	13	14	15	16	
IC 101	4.2	0.7	0.7	4.2	7.8	0	4.2	7.8	–	–	–	–	–	–	–	–	
IC102	4.0	3.5	0	1.3	1.3	1.3	0	0	0.2	1.4	2.9	3.5	7.6	5.0	–	–	
IC103	7.1	.01	0	0	0	.01	0	6.9	0	0	0	0	0	13.8	–	–	
IC104	5.1	0	0	0	5.1	5.1	5.1	.1	.1	0	.1	.1	0	.1	–	–	Manual
	4.9	0	0	0	5.1	5.1	5.1	1P	2P	0	2P	1P	0	2P	–	–	SCAN
IC105	.5	11.2	11.2	11.2	11.2	11.2	11.2	0	11.2	11.2	11.2	.1	.1	.1	.1	5.1	Manual
	9P	9P	9P	9P	9P	9P	9P	0	9P	9P	9P	1P	2P	1P	2P	5.1	SCAN

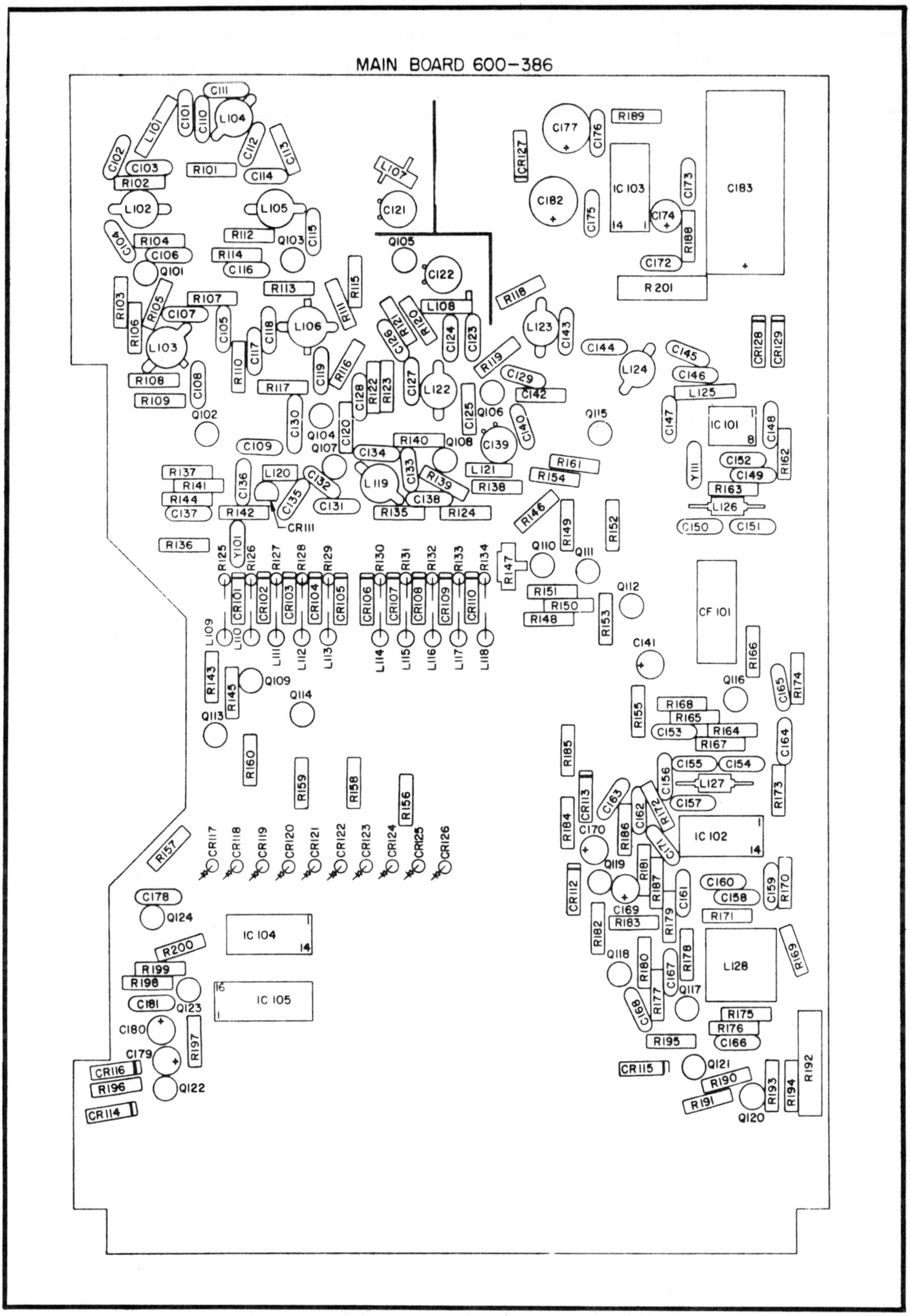

Fig. 4-23. Main-board parts placement diagram for ACT-R10H/L/U.

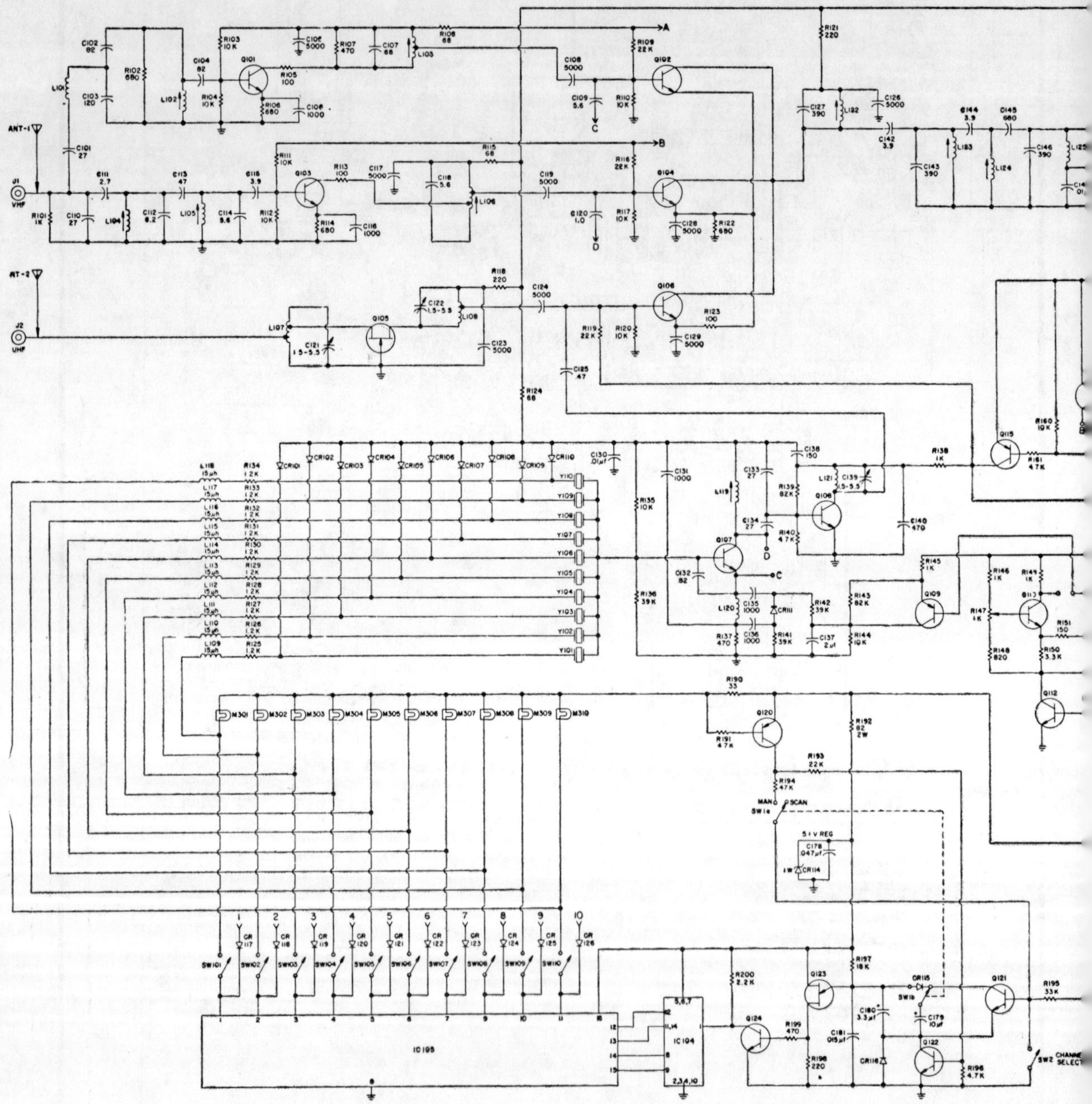
ANT-1
J1 VHF
ANT-2
J2 UHF
Q101
Q102
Q103
Q104
Q105
Q106
Q107
Q108
Q109
Q112
Q113
Q115
Q120
Q121
Q122
Q123
Q124
IC194
IC195
5.1 V REG
MAN
SCAN
SW1a
SW1b
SW2 CHANNEL SELECT
M301
M302
M303
M304
M305
M306
M307
M308
M309
M310
SW101
SW102
SW103
SW104
SW105
SW106
SW107
SW108
SW109
SW110
Y101
Y102
Y103
Y104
Y105
Y106
Y107
Y108
Y109
Y110
CR101
CR102
CR103
CR104
CR105
CR106
CR107
CR108
CR109
CR110
CR111
CR114
CR115
CR116
R121 220
R190 33
R192 82 2W
R195 33K
R196 4.7K
R197 18K
R200 2.2K
C178 .047µf
C179 10µf
C180 3.3µf
C181 .015µf

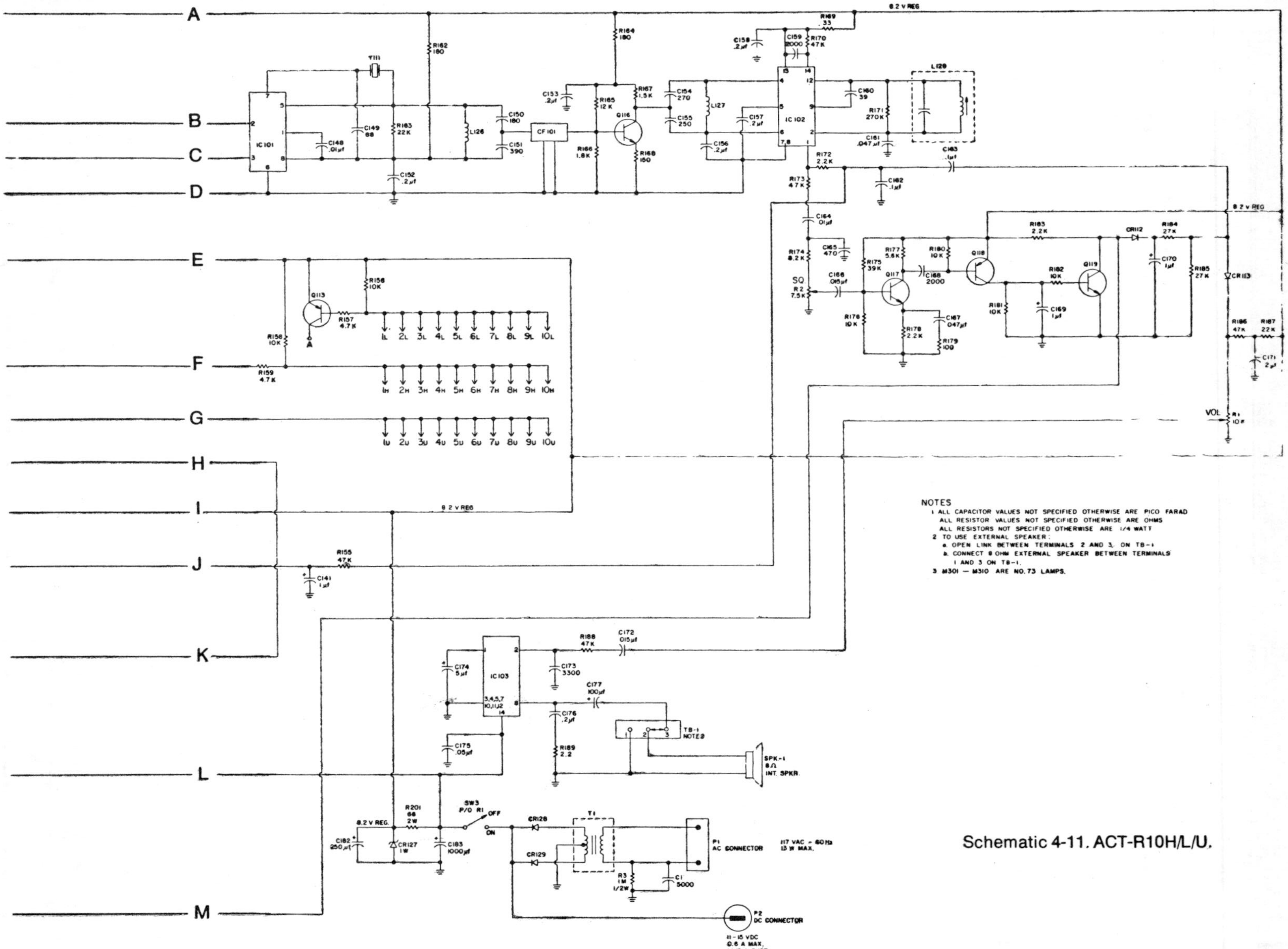

NOTES
1 ALL CAPACITOR VALUES NOT SPECIFIED OTHERWISE ARE PICO FARAD
ALL RESISTOR VALUES NOT SPECIFIED OTHERWISE ARE OHMS
ALL RESISTORS NOT SPECIFIED OTHERWISE ARE 1/4 WATT
2 TO USE EXTERNAL SPEAKER:
a. OPEN LINK BETWEEN TERMINALS 2 AND 3, ON TB-1
b. CONNECT 8 OHM EXTERNAL SPEAKER BETWEEN TERMINALS 1 AND 3 ON TB-1.
3 M301 — M310 ARE NO. 73 LAMPS.

Schematic 4-11. ACT-R10H/L/U.

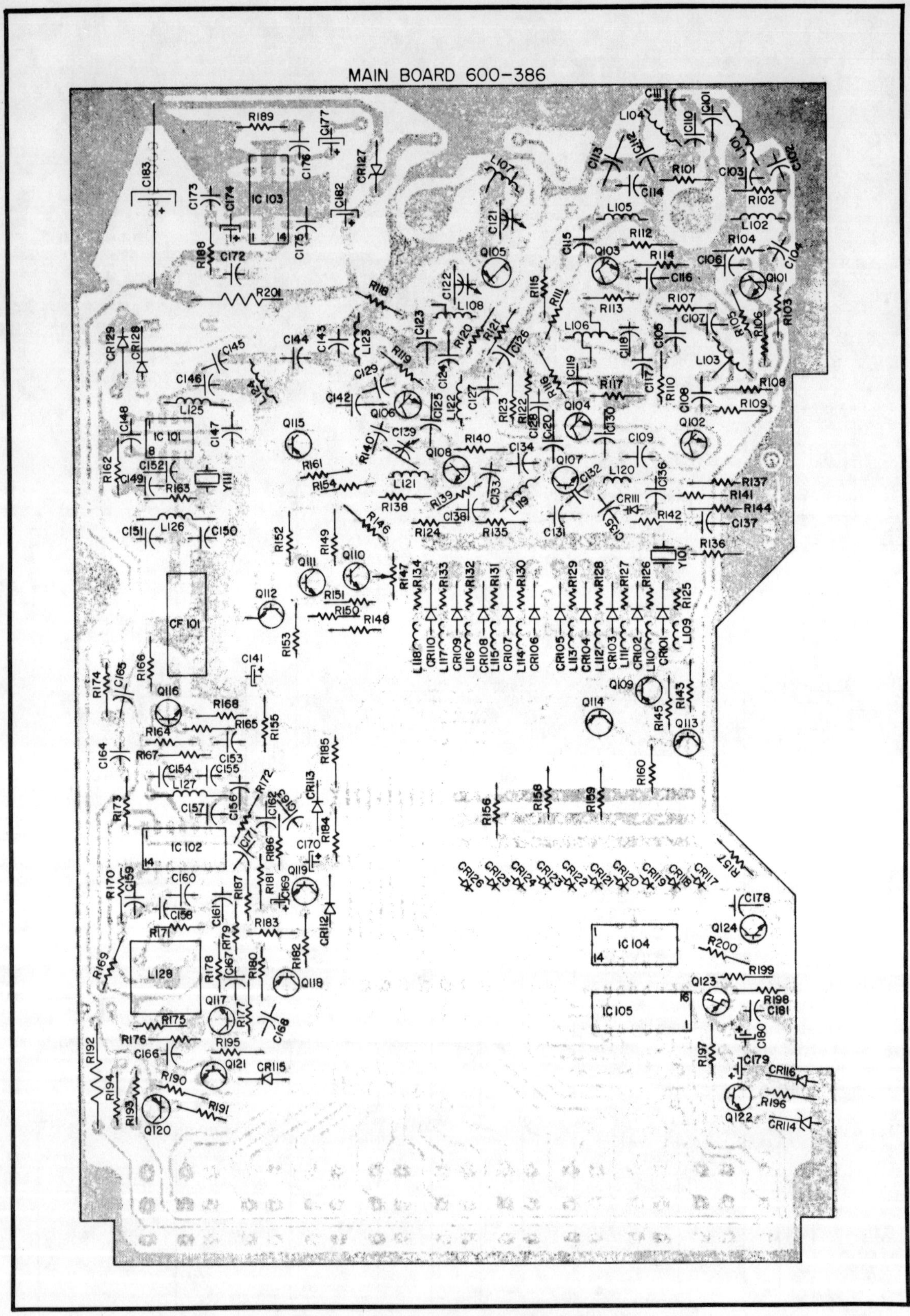

Fig. 4-24. Main board, bottom view, ACT-R10H/L/U.

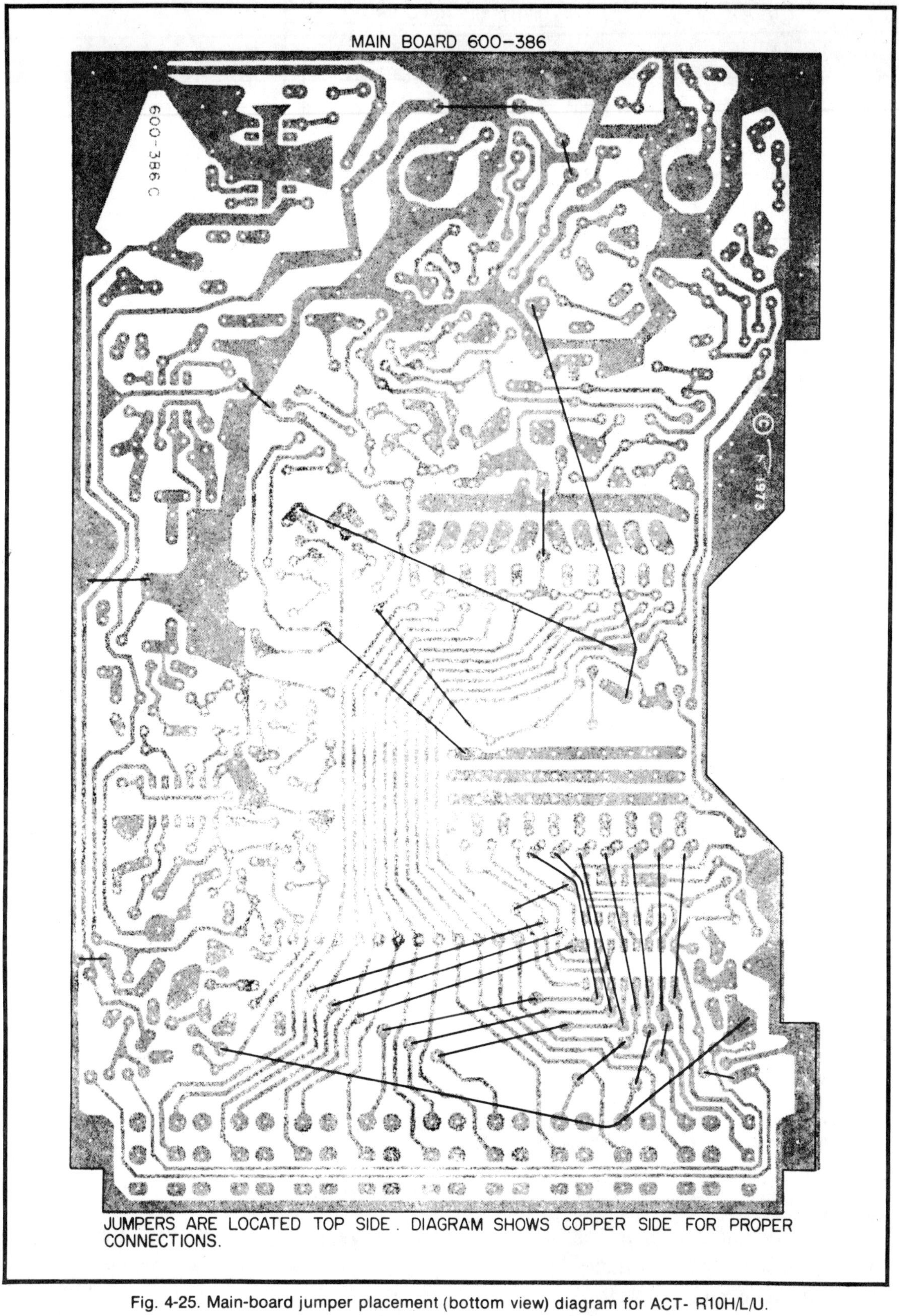

Fig. 4-25. Main-board jumper placement (bottom view) diagram for ACT- R10H/L/U.

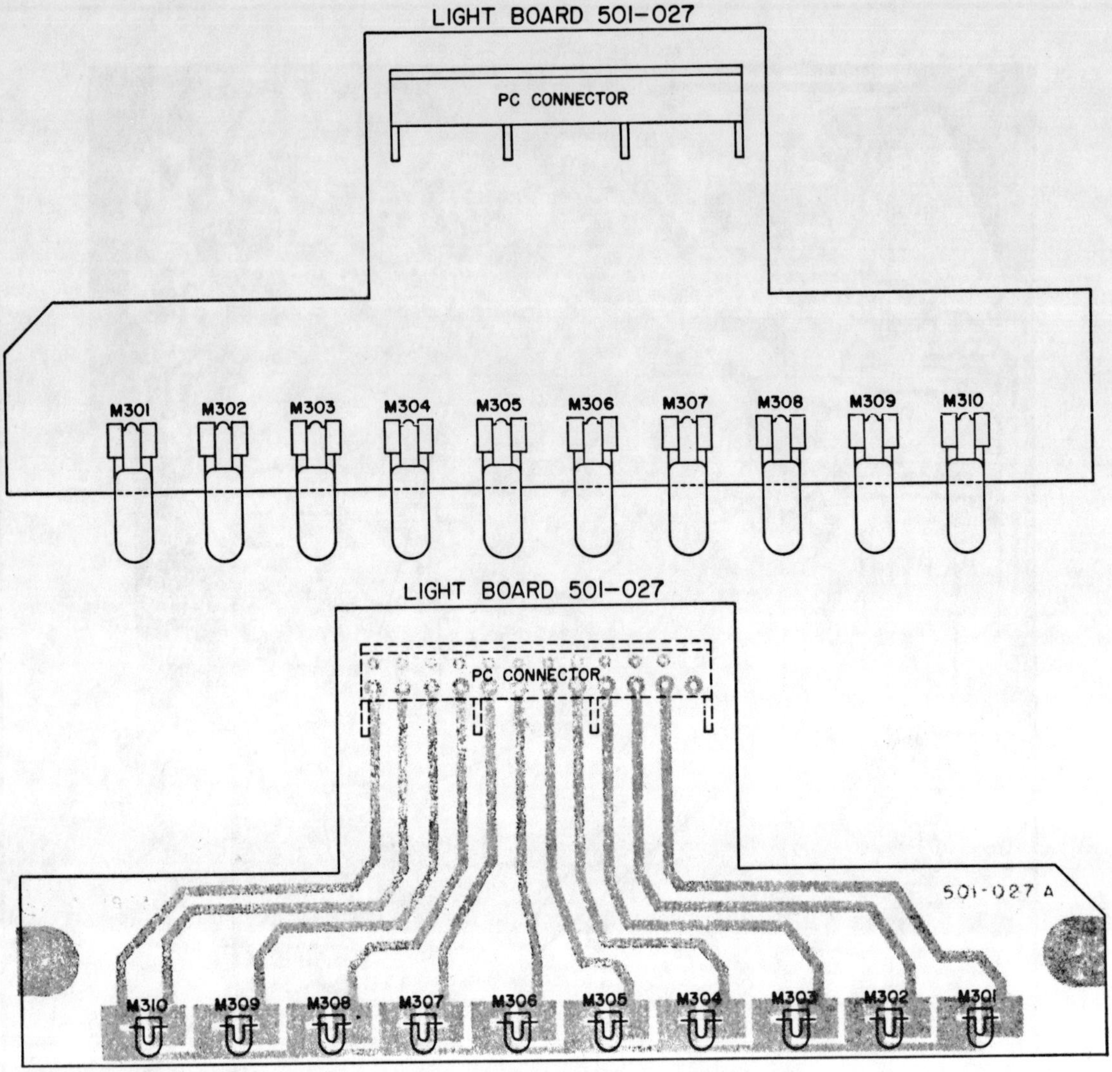

Fig. 4-26. Light board for ACT-R10H/L/U.

REGENCY TME-8H/L/U, -8H/U, -8LMH

Specifications

Receiver (Model TME-8H/L/U)

Frequency range	Low band. 30–50 MHz High band. 150–174 MHz UHF band. 450–470 MHz
Antenna impedance	50Ω
Channels	Eight. crystal-controlled
Sensitivity (at tuneup)	Low band. 0.5 μV (max.) High band. 0.6 μV (max.) UHF band. 0.7 μV (max.)
Bandwidth	
Low-band model 8H/LL/U	33–39 MHz
Low-band model 8H/LM/U	37–43 MHz
Low-band model 8H/LH/U	41–47 MHz
Model 8H/LH/U	41–47 MHz
High-band model	6 dB bandwidth. 8 MHz 10 dB bandwidth. 12 MHz
UHF-band model	6 dB bandwidth. 12 MHz
Selectivity (IF)	6 dB down. ±7 kHz (min.) 50 dB down. ±18 kHz (max.)
Spurious rejection (excluding primary image)	50 dB
AFC range (UHF board only)	Approx ±6 kHz
Modulation acceptance	±7 kHz (min.)
Intermediate frequencies	First. 10.7 MHz Second. 455 kHz
Squelch system	Noise-operated
Audio output (3.2Ω speaker)	3W at 10% or less distortion. 5W max.

Receiver (Model TME-8H/U)

Frequency range	High band. 150–174 MHz UHF band. 450–470 MHz UHF band. 450–470 MHz
Antenna impedance	50Ω
Channels	Eight. crystal-controlled
Sensitivity (at tuneup)	High band. 0.6 μV (max.) UHF band. 0.7 μV (max.)
Bandwidth	
High-band model	6 dB bandwidth. 8 MHz 10 dB bandwidth. 12 MHz
UHF-band model	6 dB bandwidth. 8 MHz 10 dB bandwidth. 12 MHz
Selectivity (IF)	6 dB down. ±7 kHz (min.) 50 dB down. ±18 kHz (max.)
Spurious rejection (excluding primary image)	50 dB
AFC range (UHF board only)	Approx ±6 kHz
Intermediate frequencies	First. 10.7 MHz Second. 455 kHz
Squelch system	Noise-operated
Audio output (3.2Ω speaker)	3W at 10% or less distortion. 5W max.

Receiver (Model TME-8LMH)

Frequency range	33–47 MHz
Antenna impedance	50Ω
Channels	Eight. crystal-controlled
Sensitivity (at tuneup)	0.5 μV
Bandwidth	
200-series RF board	40–47 MHz
300-series RF board	33–40 MHz
Selectivity (IF)	6 dB down. ±7 kHz (min.) 50 dB down. ±20 kHz (max.)
Spurious rejection (excluding primary image)	50 dB
Modulation acceptance	±7 kHz (min.)
Intermediate frequencies	First. 10.7 MHz Second. 455 kHz
Squelch system	Noise-operated
Audio output (3.2Ω speaker)	3W 10% or less distortion. 5W max.

Scanner

Scan rate	Approx 15 channels per sec
Scan delay	Approx 0.5 sec

Power

Voltage requirement	117V AC (±10%). 60 Hz. 17W max.
Current requirements at 13.8V DC	
Receiver (squelched)	180 mA
Receiver (max. audio output)	800 mA
Fuse size	1.5A. 3AG

Service Data

On Model TME-8LMH, crystals for channels above 40 MHz should be installed in the group of sockets in the 200-series board.

- For channels below 40 MHz, install the crystals in the group of sockets on the 300-series board. The important thing to remember is that a crystal should not be installed in the same channel number for both RF boards. For example, if channels 1 through 5 are programmed for the 200-series RF boards, then those crystal positions on the 300-series RF board should not be used.

- For high- and low-band crystals inserted into the 301–563 high–low RF board (Model TME-8H/L/U only), band programming will have to be performed. If the crystal inserted is for the high band (148–174 MHz), place the proper color-coded wire and socket onto the proper high-band pin; if the crystal is for the low band (30–50 MHz), place the proper wire and socket onto the proper low-band pin. Pictorial B, Fig. 4-18, illustrates how the band-selection wires are properly connected. Pictorial C shows an example of a partially programmed board.

- *Note*: If a particular channel is not used in the high–low board (in other words, there is no crystal installed for that channel), the band-selection wire must still be connected to either a high-band pin or a low-band pin. Thus, for proper scanner operation, all the band-selection wires *must* be connected, even though not all channels are used.

- The small printed circuit board with 24 wire connectors is the band programming board. By means of movable wire

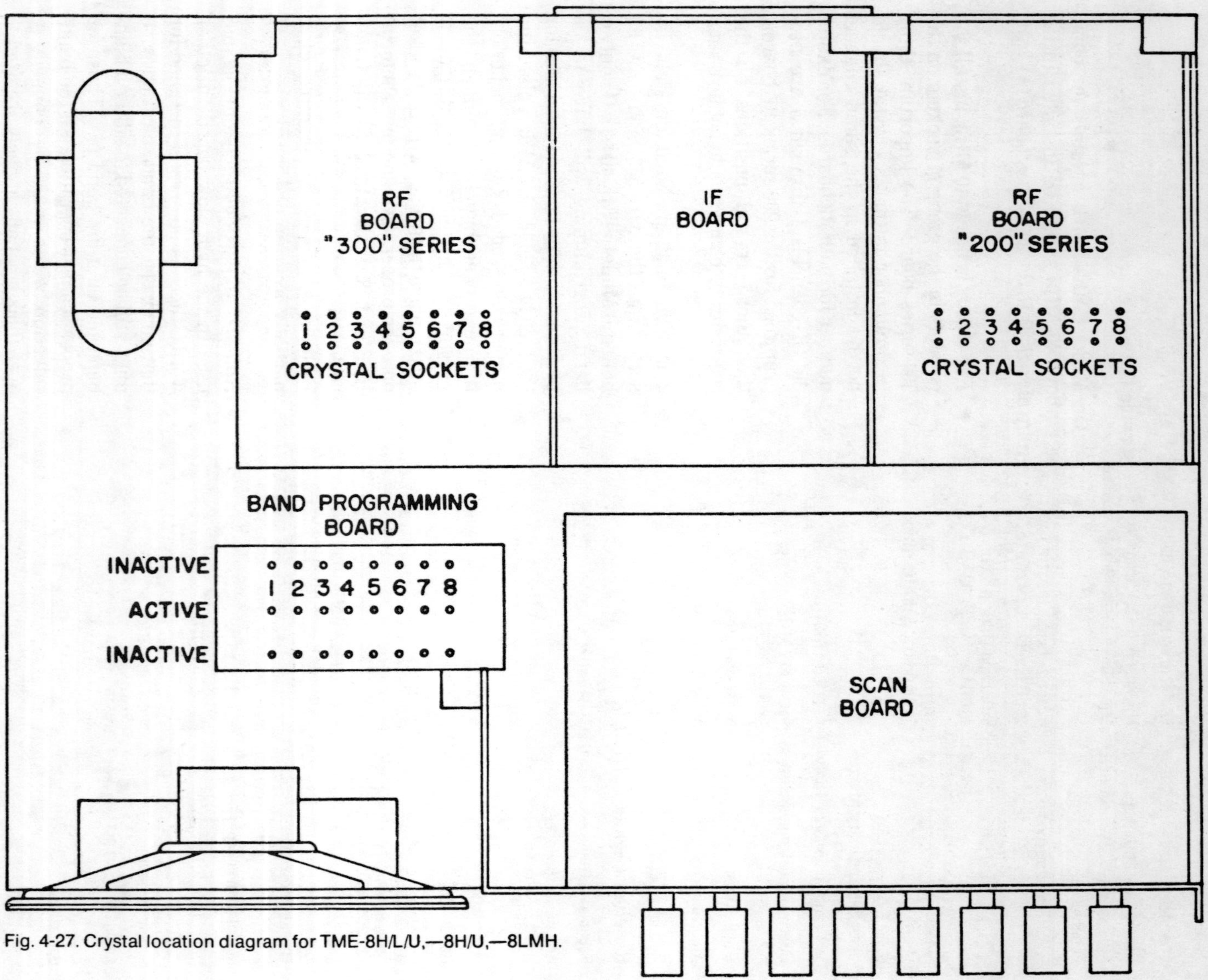

Fig. 4-27. Crystal location diagram for TME-8H/L/U,—8H/U,—8LMH.

connections, each channel can be programmed for either the 200-series RF board or the 300-series RF board. The center row of connectors is the active line. The two outer rows are the inactive lines for each band. This means that when a wire is not in one of the center connectors, it should be in its respective inactive connector. If more 200-series channels are required than are initially provided, merely move the unused 300-series channel connectors to the 300-series active row, and then move the appropriate 200-series channel connectors to the middle, or active, row. Be sure that each color-coded wire is connected to its respective channel. Reading from left to right, the color-coded wires should be in this order: brown, red, orange, yellow, green, blue, purple (or violet), and white.

- After the crystals are installed and any necessary band programming changes are completed, carefully slide the unit back into the cabinet. Screw the four feet back into place and replace the rear panel. Push the VOLUME and SQUELCH knobs back on their shafts, and the unit is again ready for operation.

Crystal layout	Figs. 4-18, 4-27
Voltage data	Tables 4-17, 4-18, 4-19
Schematics	Schematics 4-12, 4-13, 4-14
Internal layout, RF	Figs. 4-28, 4-29 (LMH only)
RF	Figs. 4-30, 4-31 (H/U only)
RF	Figs. 4-32, 4-33 (H/L/U only)
RF	Figs. 4-34, 4-35 (UHF board)
IF	Figs. 4-6, 4-7
Scanner	Figs. 4-36, 4-37
Alignment data	Steps A, N (omit for TME-8H/L/U and TME-8H/U), O, P, (TME-8H/L/U only), Q (TME-8H/U only), R (TME-8/L/U and TME-8H/U only), S (TME-8LMH only), T (TME-8H/L/U and TME-8H/U only), E (TME-8LMH only)

REGENCY TME-16H/LL/U, -16H/LM/U, -16H/LH/U, -16H/L

Specifications

Receiver (Model TME-16H/L/U)

Frequency range	Low band, 30–50 MHz High band, 150–174 MHz UHF band, 450–470 MHz
Antenna impedance	50Ω
Channels	Sixteen, crystal-controlled
Sensitivity (at tuneup)	Low band, 0.6 μV (max.) High band, 0.6 μV (max.) UHF band; 0.7 μV (max.)
Bandwidth	
Low band model H/LL/U	33–39 MHz
Low-band model H/LM/U	37–43 MHz
Low-band model H/LH/U	41–47 MHz
High-band model	6 dB bandwidth, 8 MHz 10 dB bandwidth, 12 MHz
Selectivity (IF)	6 dB down, ±7 kHz (min.) 50 dB down, ±18 kHz (max.)
Spurious rejection (excluding primary image)	50 dB
AFC range (UHF board only)	Approx ±6 kHz
Modulation acceptance	±7 kHz (min.)
Intermediate frequencies	First, 10.7 MHz Second, 455 kHz
Squelch system	Noise-operated
Audio output (3.2Ω speaker)	3W at 10% or less distortion, 5W max.

Receiver (Model TME-16H/L)

Frequency range	Low band, 30–50 MHz High band, 150–174 MHz
Antenna impedance	50Ω
Channels	Sixteen, crystal-controlled
Sensitivity (at tuneup)	Low band, 0.5 μV (max.) High band, 0.6 μV (max.)
Bandwidth	
Low-band model:	
200-series RF board	40–47 MHz
300-series RF board	33–40 MHz
High-band model	6 dB bandwidth, 8 MHz 10 dB bandwidth, 12 MHz
Selectivity (IF)	6 dB down, ±7 kHz (min.) 50 dB down, ±20 kHz (max.)
Spurious rejection (excluding primary image)	50 dB

Table 4-17. Voltage Data—TME-8H / L / U, -8H / U, -8LMH RF Boards.

NOTE: All voltages are nominal and are measured with a VTVM. SCAN indicates the unit is scanning. MAN indicates the unit is not scanning and is stopped at channel 1. A "P" beside a voltage indicates that the meter reading is pulsating (fluctuating) because the scanner section of the unit is operating.

VOLTAGE DATA – TRANSISTORS:

	TRANSISTOR	EMITTER (Source)	BASE (Gate)	COLLECTOR (Drain)	
RF Board					
No. 301-563	Q201	3.1	3.8	7.0	Low Band Activated
Note: Model		0	0	7.6	High Band Activated
TME - 8 H/L/U	Q202	0	0	7.6	Low Band Activated
		3.1	3.8	7.0	High Band Activated
	Q203	1.6	2.3	7.1	Low Band Activated
		1.6	0	7.1	High Band Activated
	Q204	1.6	0	7.1	Low Band Activated
		1.6	2.3	7.1	High Band Activated
	Q205	7.8	7.4	7.6	Low Band Activated
		7.8	11.0	0^{V}	High Band Activated
	Q206	7.8	11.0	0	Low Band Activated
		7.8	7.0	7.6	High Band Activated
	Q207	3.4	4.1	7.0	
RF Board					
No. 301-563	Q201	3.1	3.8	7.0	
Note: Model	Q202	1.6	2.3	7.1	
TME - 8 H/U	Q203	3.4	4.1	7.0	
RF Board					
No. 301-598	Q301 (FET)	0	0	5.5	
Note: Model	Q302	1.5	2.2	6.8	
TME - 8 H/L/U	Q303	.25	3.1	6.8	
and	Q304	7.8	7.2	3.0-5.0	
TME - 8 H/U	Q305	2.9	3.6	7.4	
	Q306	2.9	3.6	7.2	
	Q307	0	0.2	6.8	
RF Board					
No.301-563	Q201	3.1	3.8	7.0	Activated
Note: Model		0	0	7.6	Unactivated
TME - 8 LMH					
has "300" Series	Q202	1.6	2.3	7.1	Activated
RF Board Voltages		0	0	7.6	Unactivated
Identical To					
"200" Series	Q203	3.4	4.1	7.0	
RF Voltages	Q204	7.8	7.4	7.6	Activated
		7.8	11.0	0	Unactivated

Modulation acceptance	±7 kHz (min.)
Intermediate frequencies	First 10.7 (min.) Second 455 kHz
Squelch system	Noise-operated
Audio output (3.2Ω speaker)	3W at 10% or less distortion. 5W max.

Scanner

Scan rate	Approx 15 channels per sec
Scan delay	Approx 0.5 sec

Power

Voltage requirement	117V AC (±10%). 60 Hz. 17W max.; or 13.8V DC (±10%)

Table 4-18. Voltage Data—TME-8H/L/U, -8H/U, -8LMH IF and Scanner Boards.

	TRANSISTOR	EMITTER (Source)	BASE (Gate)	COLLECTOR (Drain)	
IF Board No. 500-858	Q101	2.3	3.0	5.8	
	Q102	1.0	1.7	4.8	
	Q103 (PNP)	8.2	8.2	0	(unsquelched)
		8.2	8.2	1.0	(squelched)
		8.2	8.2	1.5	Min. (tight squelch)
	Q104	0	0	7.2	(unsquelched)
		0	.80	.30	(squelched)
		0	.80	.10	(tight squelch)
	Q105	1.4	1.9	5.1	(unsquelched)
		1.1	.10	8.2	(tight squelch)
	Q106	0.7	1.3	12.4	
	Q107 (PNP)	13.8	13.1	7.2	
	Q108 (PNP)	6.9	6.6	.10	
	Q109	6.9	7.2	13.8	
	Q110	0	.10	6.9	
Scan Board	Q401	0	.70	.10	(SCAN)
		0	.70	0	(MAN)
No. 500-850	Q402	0	.70	.10	
	Q403	0	.10	3.6	(SCAN)
		0	.10	1.2	(MAN)
	Q404	0	.10	2.8	(SCAN)
		0	.70	0.2	(MAN)
	Q405	3.1	3.6	3.2	(SCAN)
		0.7	1.2	0.7	(MAN)
	Q407 (PNP)	5.1	4.5	.20	(SCAN)
		5.1	4.6	.10	(MAN)
	Q408	0	.10	1.5	
		Base 1	**Emitter**	**Base 2**	
	Q406	.20	3.2	4.6	(SCAN)
	(unijunction)	.20	0.7	4.6	(MAN)

Table 4-19. Voltage Data—TME-8H/L/U, -8H/U, -8LMH Integrated Circuits.

NOTE: A "P" beside a voltage indicates that the meter reading is pulsating (fluctuating) because the scanner section of the unit is operating. MAN indicates the unit is not scanning and is at channel 1 (M401 is lighted).

	IC No.	1	2	3	4	5	6	7	8	9	10	11	12	13	14
IF Board 500-858	IC 101	4.2	0.7	0.7	4.2	7.8	0	4.2	7.8	–		–	–	–	–
	IC 102	4.0	3.5	0	1.3	1.3	1.3	0	0	0.2	1.4	2.9	3.5	7.6	5.0
Scan Board 500-850	IC 401	.7P	.7P	9P	9P	.7P	.7P	0	.7P	.7P	9P	9P	.7P	.7P	5.1 (Scan)
		3.3	3.3	0.5	10.8	0.2	0.2	0	0.2	0.2	10.8	10.8	0.2	0.2	5.1 (Man)
	IC 402	.7P	.7P	9P	9P	.7P	.7P	0	.7P	.7P	9P	9P	.7P	.7P	5.1 (Scan)
		0.2	0.2	10.8	10.8	0.2	0.2	0	0.2	0.2	10.8	10.8	0.2	0.2	5.1 (Man)
	IC 403	0.1	0	0	0	0	0	0	1.5	1.5	.7P	.7P	.7P	.7P	5.1 (Scan)
		0	0	0	0	0	0	0	1.5	1.5	0.2	0.2	0.2	3.3	5.1 (Man)
	IC 404	.7P	0	0	0	0	0	0	1.5	1.5	.7P	.7P	.7P	.7P	5.1 (Scan)
		0.2	0	0	0	0	0	0	1.5	1.5	0.2	0.2	0.2	0.2	5.1 (Man)

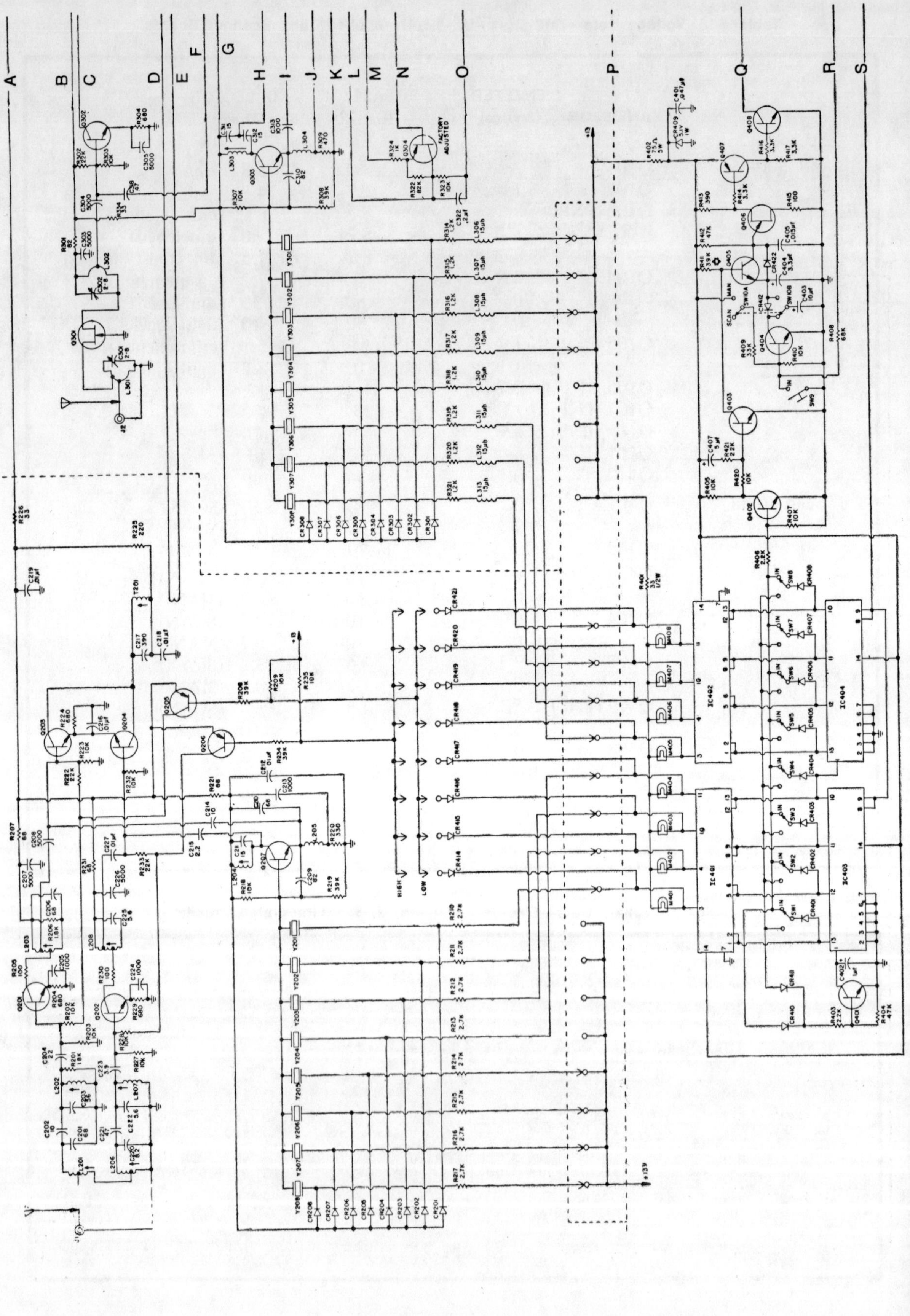
A
B
C
D
E
F
G
H
I
J
K
L
M
N
O
P
Q
R
S

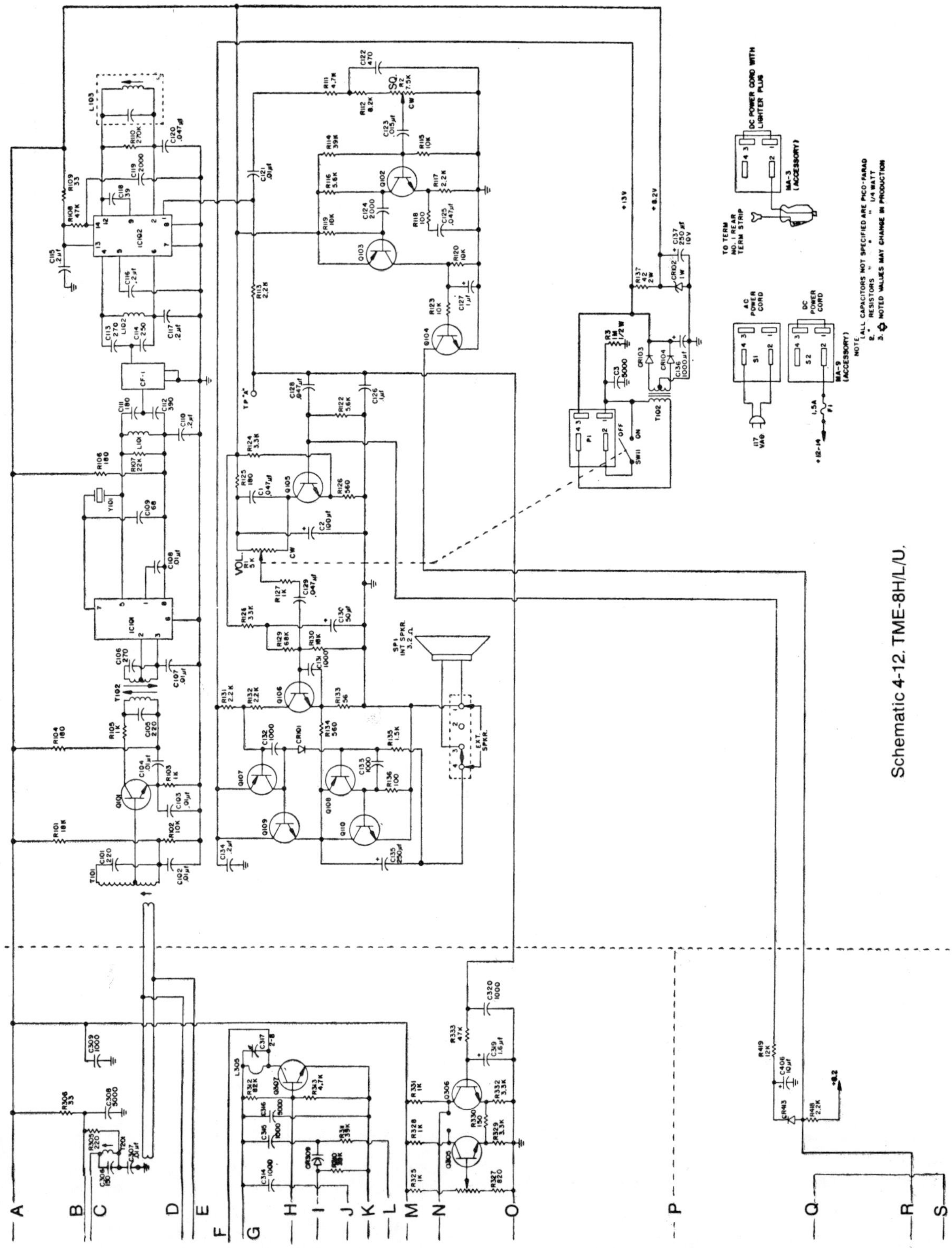

Schematic 4-12. TME-8H/L/U.

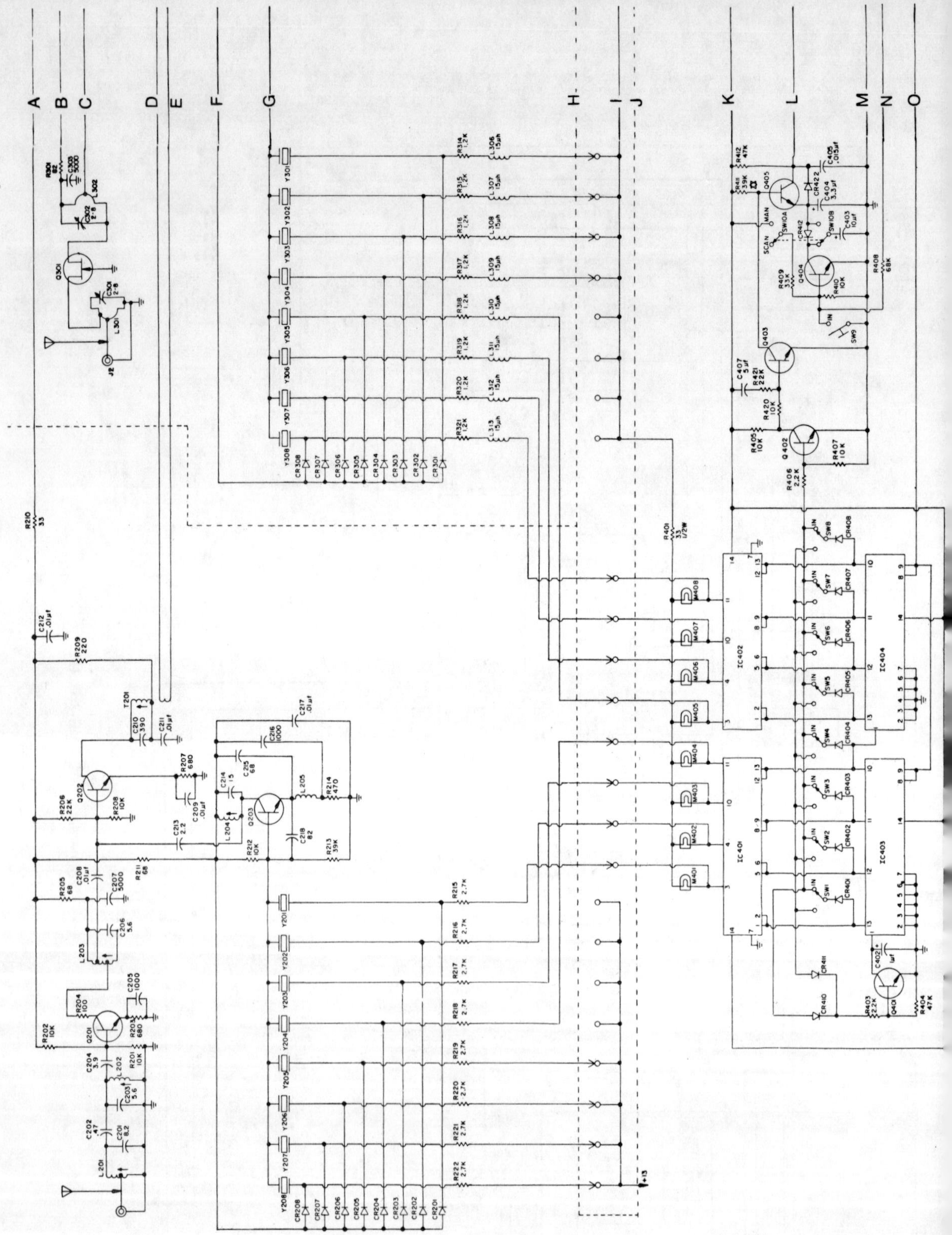
A
B
C
D
E
F
G
H
I
J
K
L
M
N
O
IC401
IC402
IC403
IC404

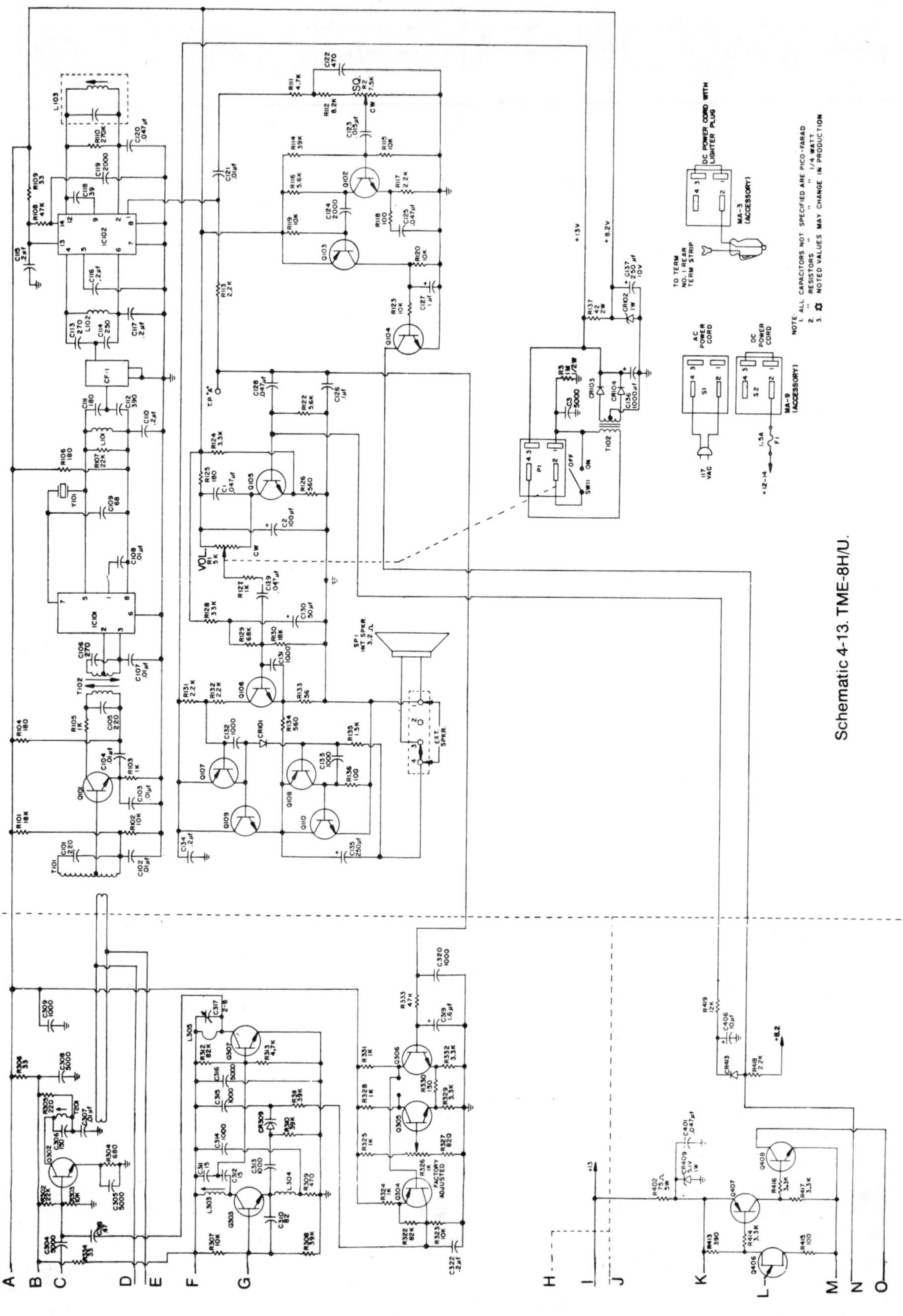

Schematic 4-13. TME-8H/U.

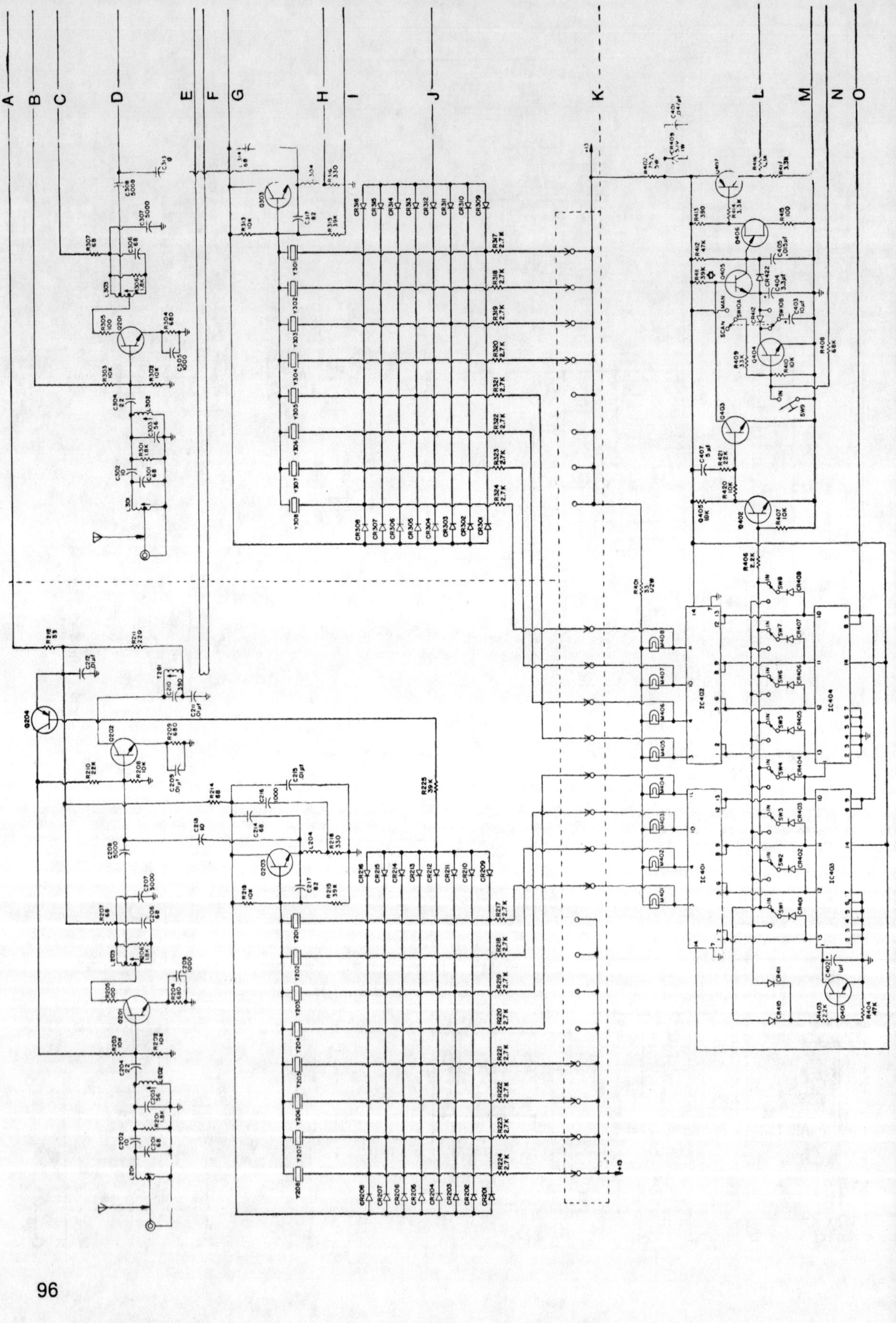
A
B
C
D
E
F
G
H
I
J
K
L
M
N
O

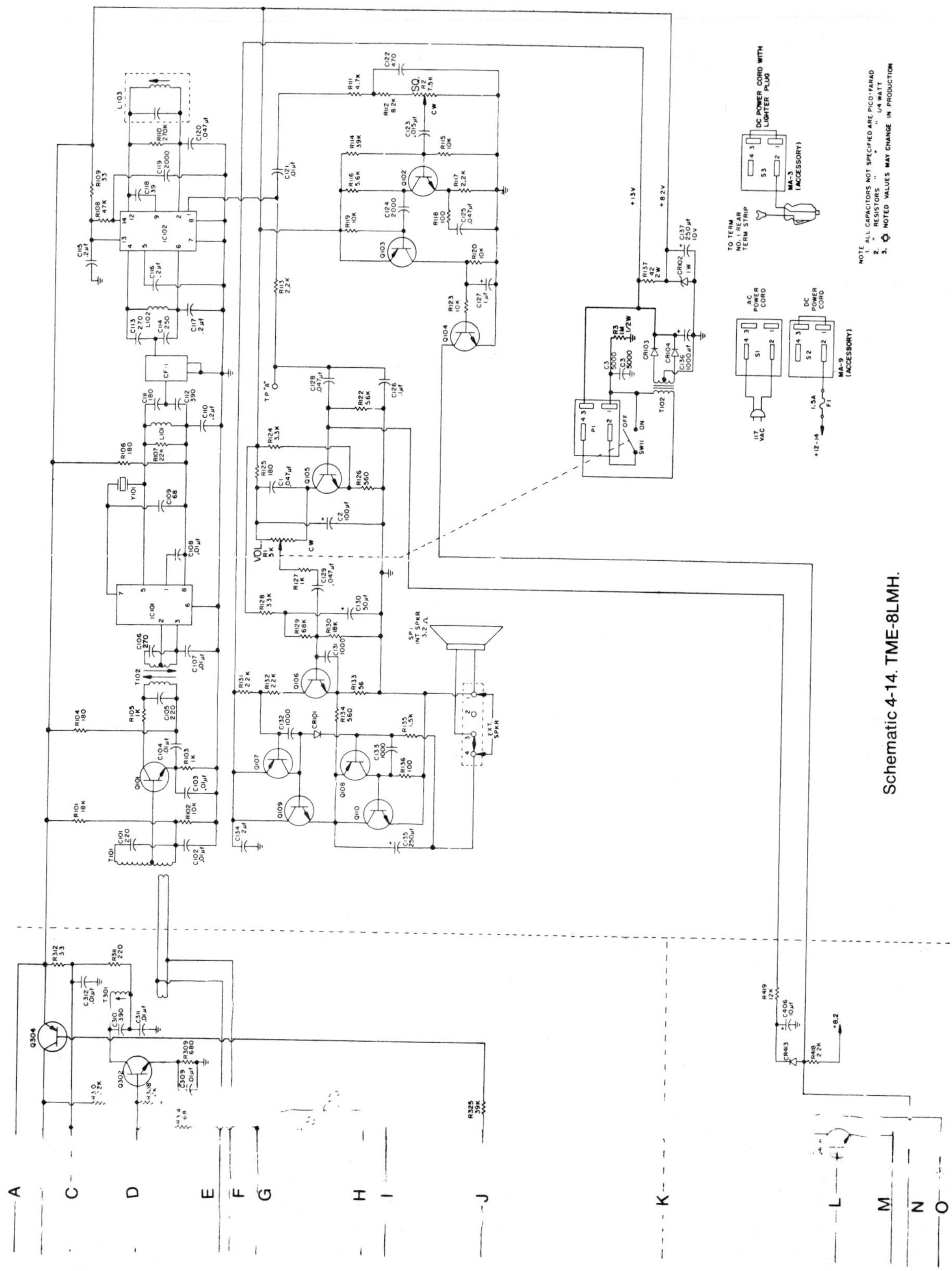

Schematic 4-14. TME-8LMH.

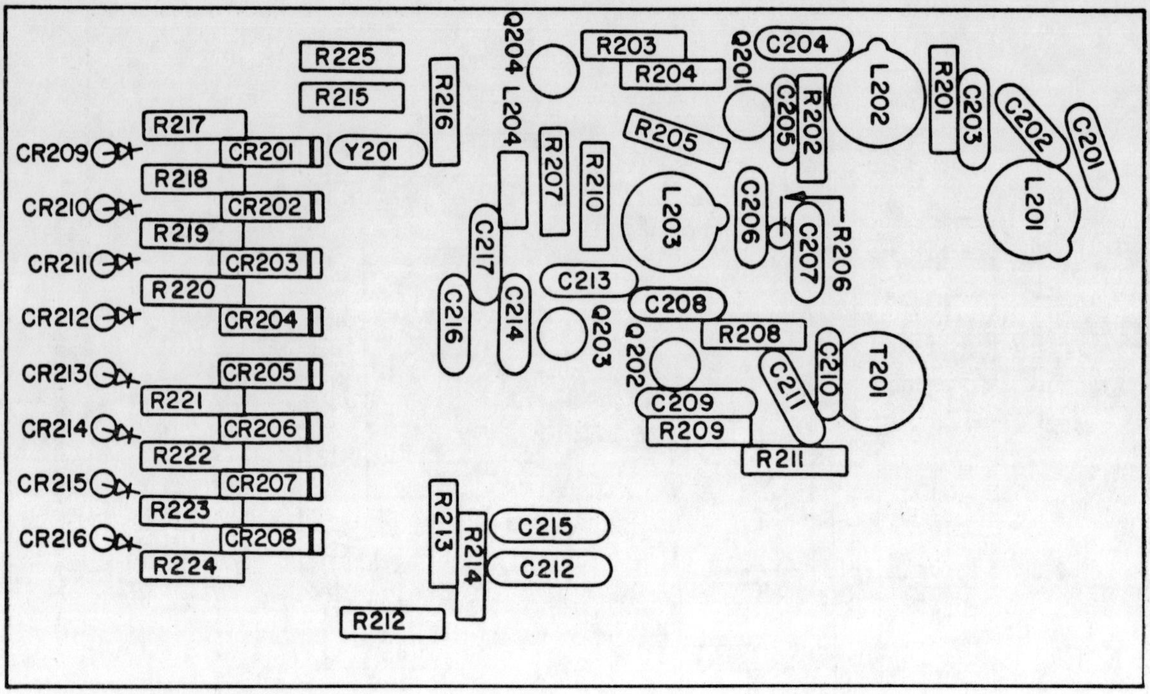

Fig. 4-28. RF board 301-563 parts placement diagram. TME-8LMH only: both RF boards are identical except for one being a 200-series board and the other a 300-series board.

Current requirements at 13.8V DC	
(squelched)	180 mA
(max. audio output)	800 mA
Fuse size	1.5A, 3AG

General

Front-panel size	$3\frac{7}{8} \times 13$ in.
Depth (including knobs and rear-panel connectors)	$9\frac{1}{2}$ in.
Antenna connector	Motorola type
Power connector	4-pin, polarized
Speaker size	4 in., square

Service Data

Band Programming

Prior to installing a crystal, the receiver will have to be partially pulled out of its cabinet.

- First, remove the telescopic antennas if they are installed. Second, remove the two knobs (VOLUME and SQUELCH). Third, remove the rear panel (cover) by removing the four mounting screws. Fourth, remove the four rubber feet by unscrewing each one. The receiver may then be slid rearward from the cabinet until the crystal socket pins are accessible.
- Insert the crystal in the proper socket pins as indicated in Fig. 4-19. The sockets are numbered, in pairs, according to the channel number on the front panel.
- For high- and low-band crystals inserted into the 301–563 high–low RF board, band programming will have to be performed. If the crystal inserted is for the high band (148–174 MHz), place the proper color-coded wire and socket onto the proper high-band pin; if the crystal is for the low band (30–50 MHz), place the proper wire and socket onto the proper low-band pin. Pictorial A, Fig. 4-18, shows how the band-selection wires are properly connected B illustrates how the band-selection wires are properly connected. Pictorial C shows an example of a partially programmed board. (See Fig. 4-18.)
- *Note*: If a particular channel is not used in the high–low board (in other words, there is no crystal installed for that channel), the band-selection wire must still be connected to either a high-band pin or a low-band pin. Thus, for proper scanner operation, all of the

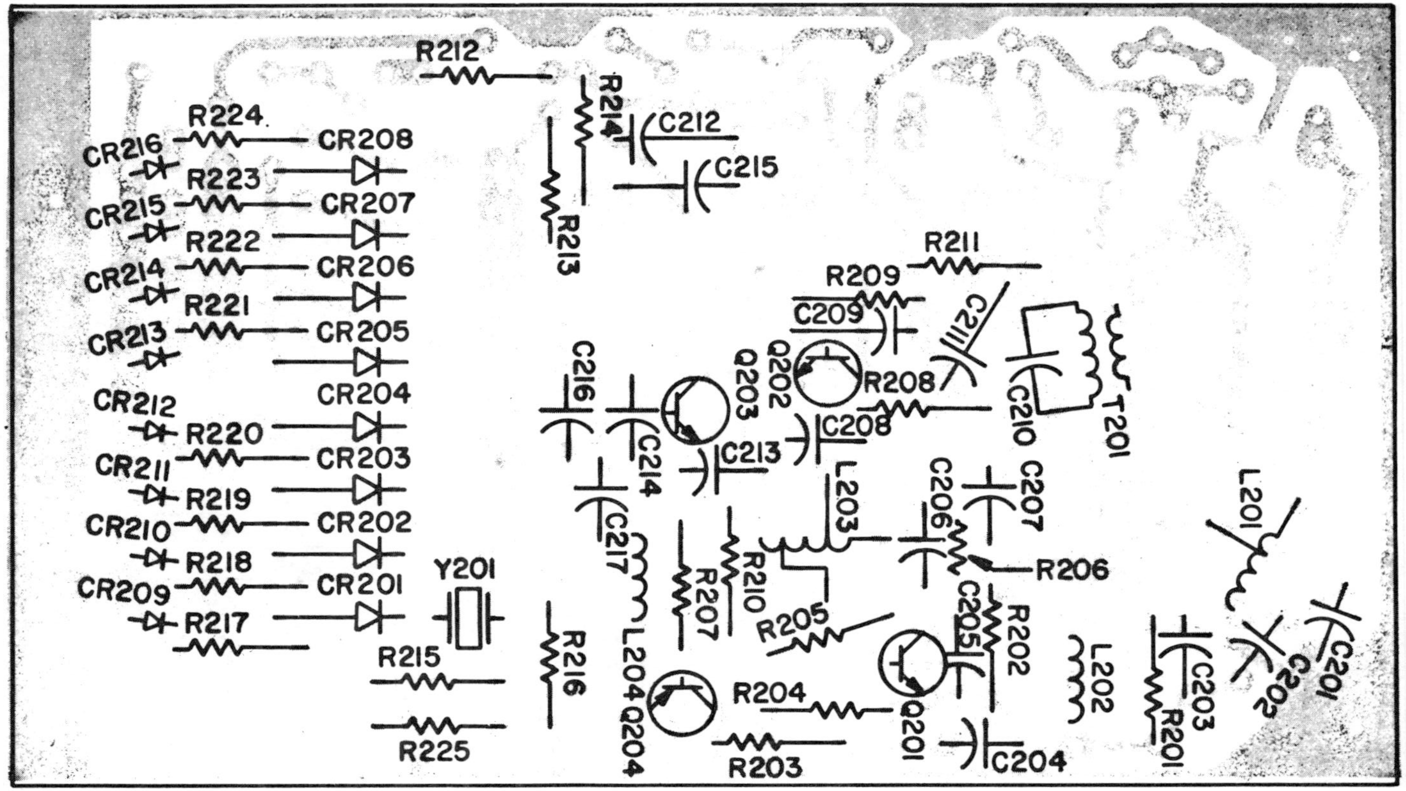

Fig. 4-29. RF board 301-563, bottom view. TME-8LMH only: Both RF boards are identical except for one being a 200-series board and the other a 300-series board.

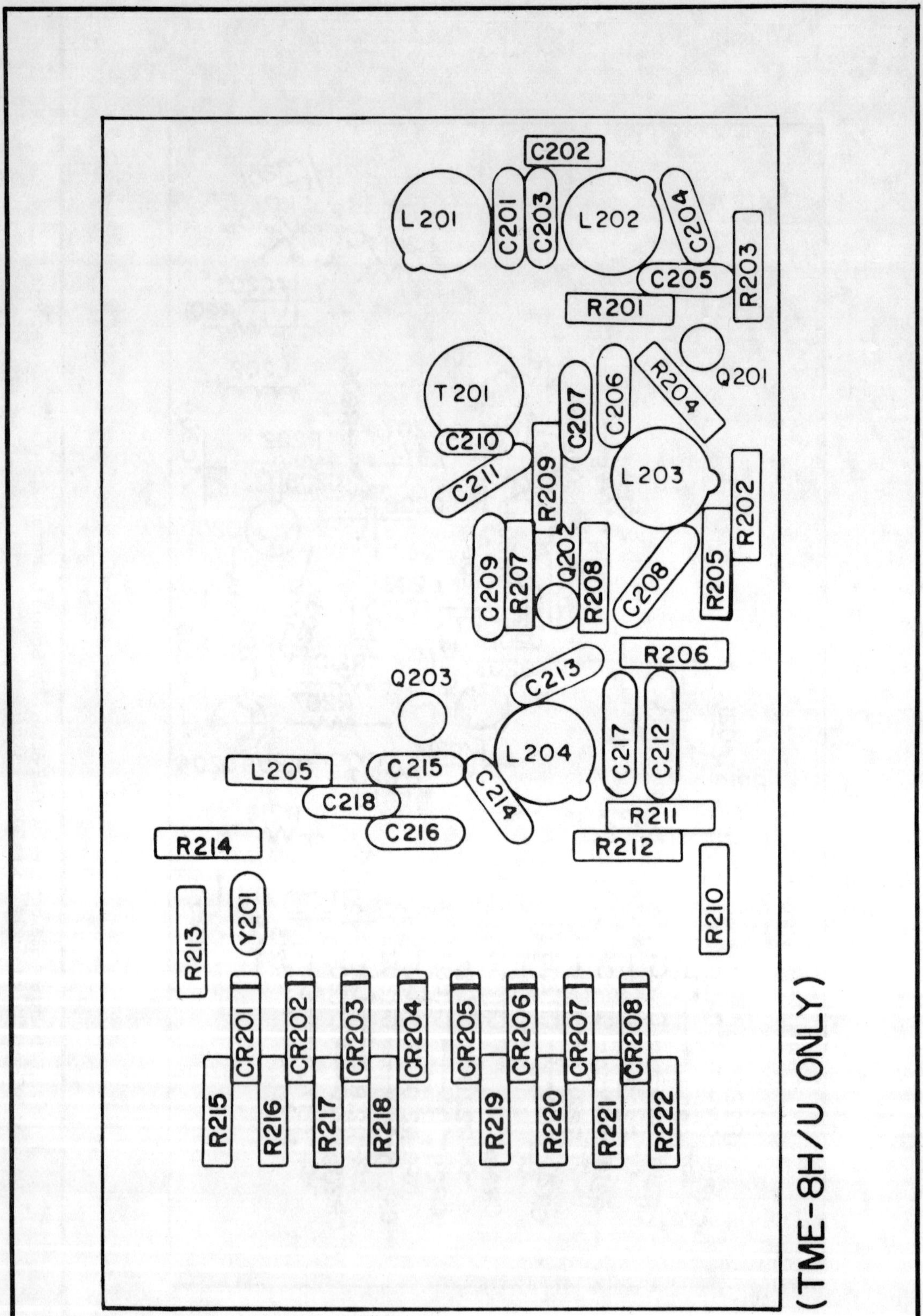

Fig. 4-30. RF board 301-563 parts placement diagram (high-band board).

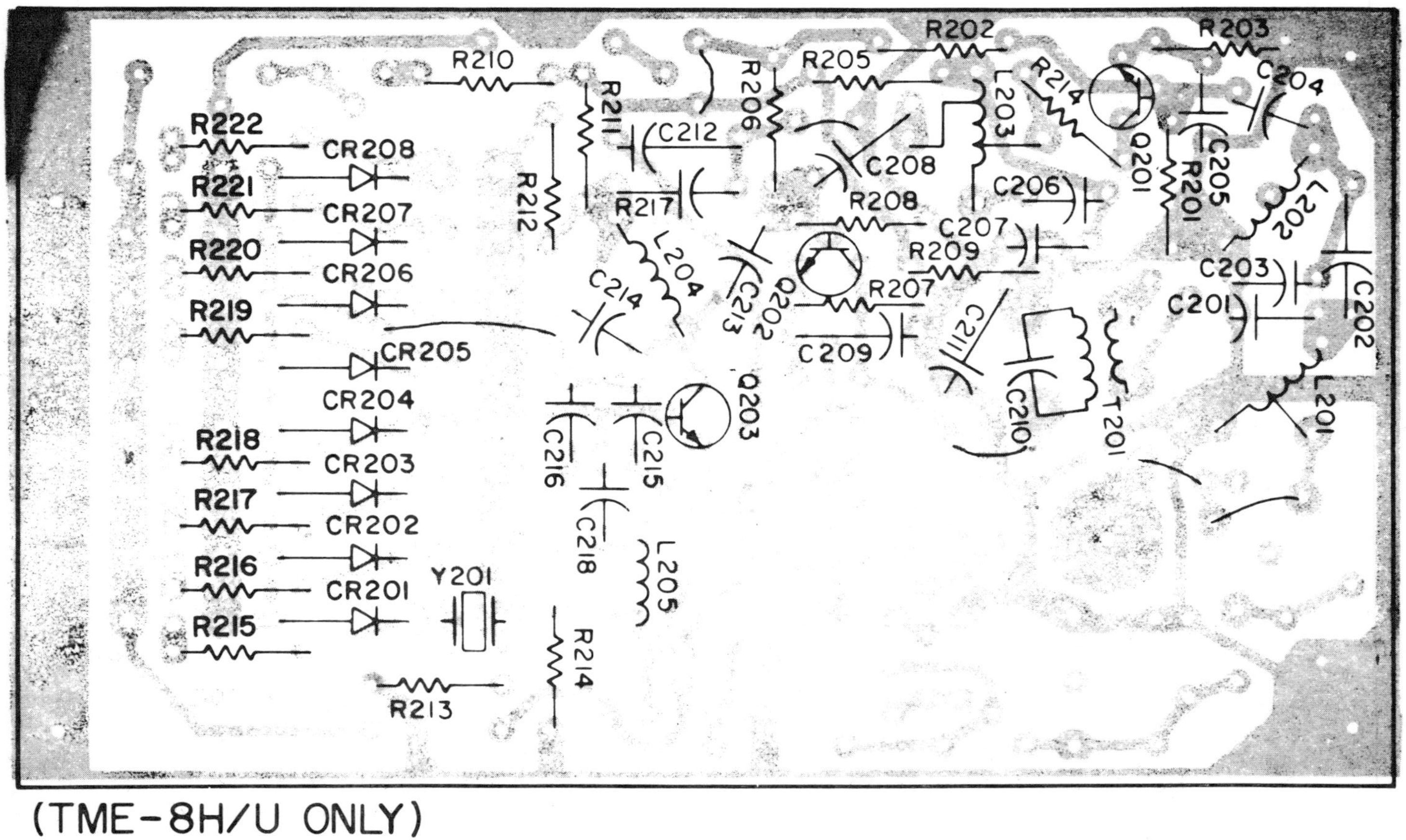

Fig. 4-31. RF board 301-563, bottom view (high-band board).

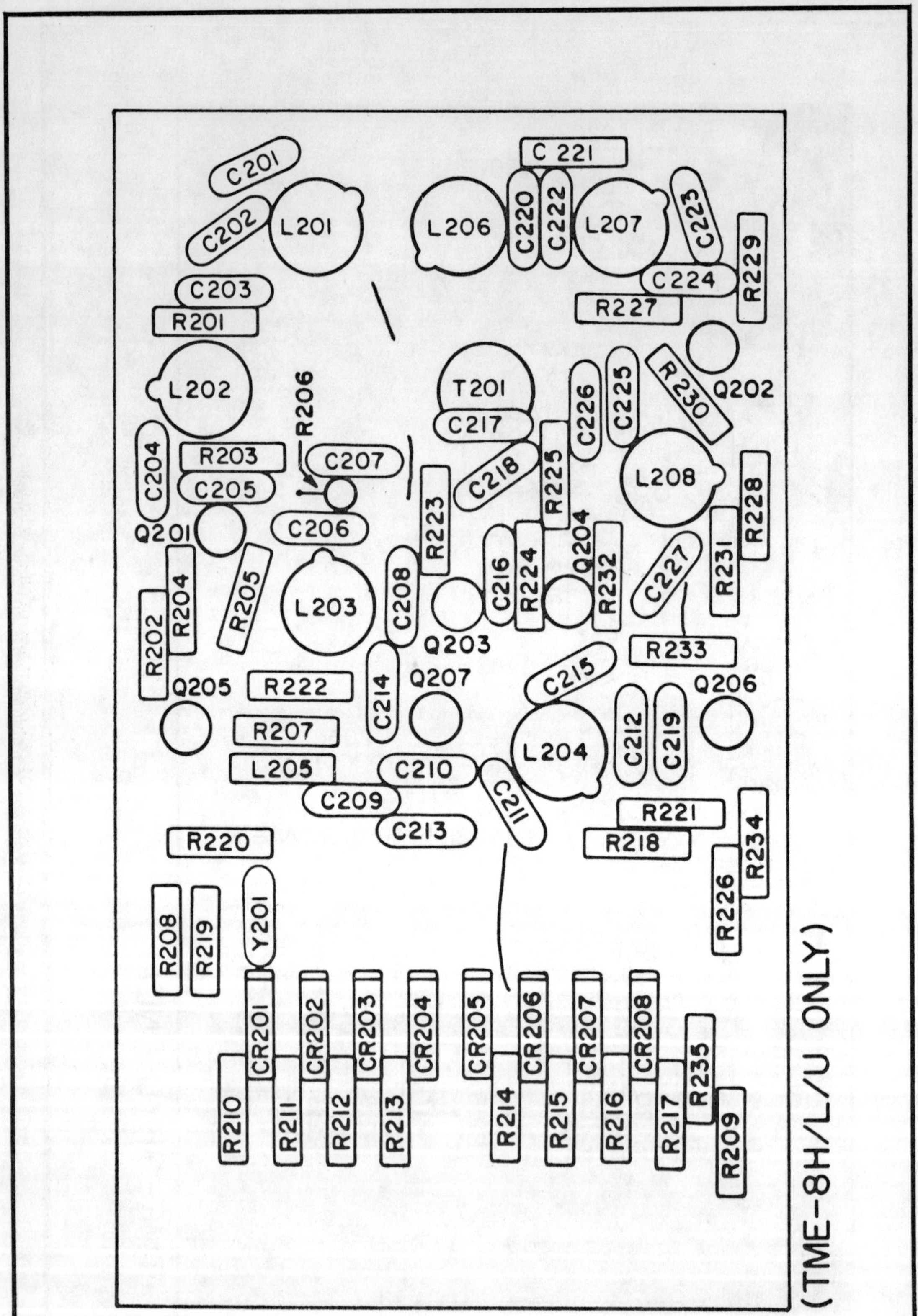

Fig. 4-32. RF board 301-563 parts placement diagram (high–low board).

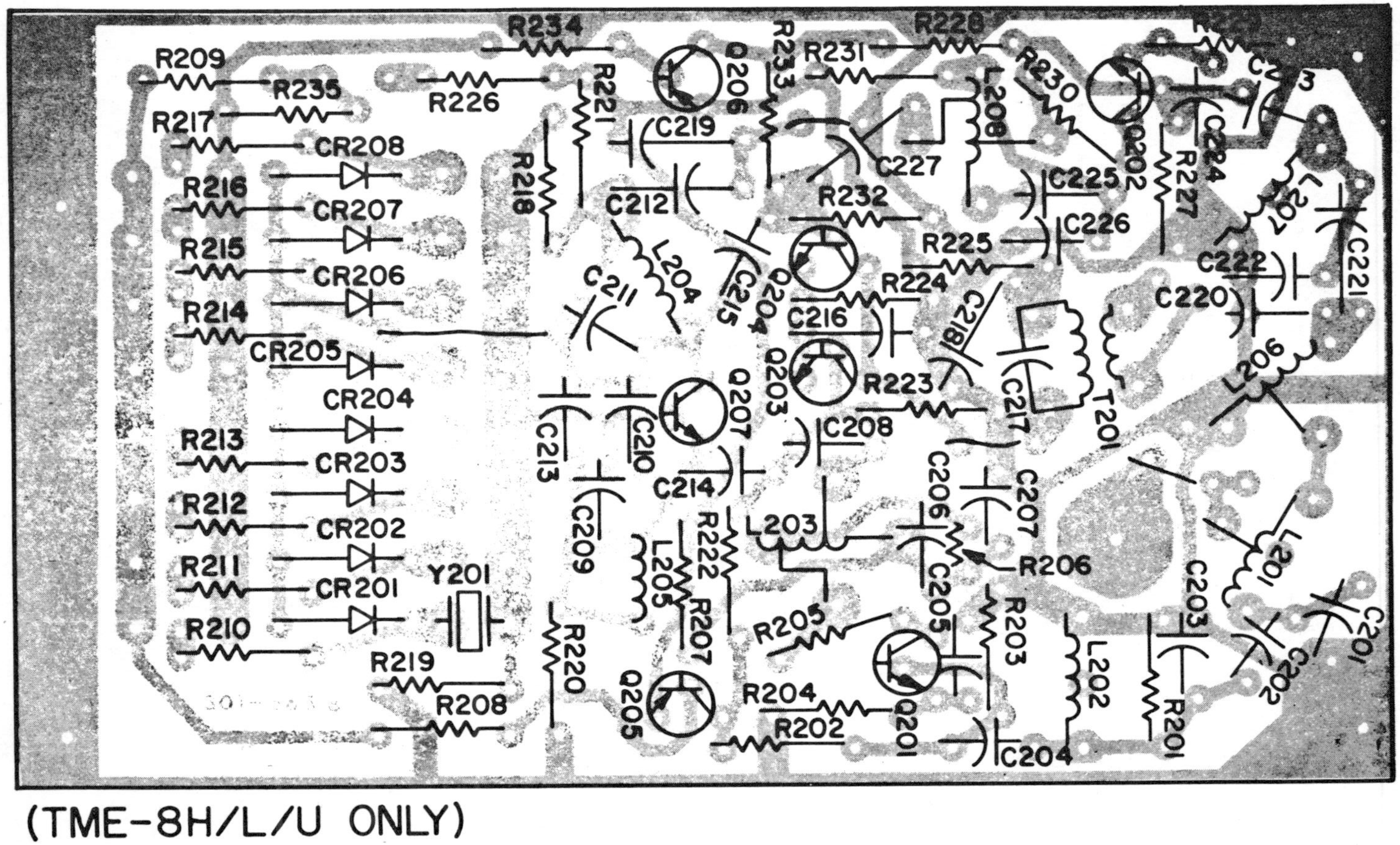

Fig. 4-33. RF board 301-563, bottom view (high – low board).

Table 4-20. Voltage Data—TME-16 Transistors.

NOTE: All voltages are nominal and are measured with a VTVM. SCAN indicates the unit is scanning. MAN indicates the unit is not scanning and is stopped at channel 1. A "P" beside a voltage indicates that the meter reading is pulsating (fluctuating) because the scanner section of the unit is operating.

VOLTAGE DATA – TRANSISTORS:

	TRANSISTOR	EMITTER (Source)	BASE (Gate)	COLLECTOR (Drain)	
RF Board					
No. 301-563	Q201	3.1	3.8	7.0	Low Band Activated
		0	0	7.6	High Band Activated
Note: Model	Q202	0	0	7.6	Low Band Activated
TME-16H/L		3.1	3.8	7.0	High Band Activated
has "300"	Q203	1.6	2.3	7.1	Low Band Activated
Series RF Board		1.6	0	7.1	High Band Activated
voltages identical	Q204	1.6	0	7.1	Low Band Activated
to "200" Series		1.6	2.3	7.1	High Band Activated
RF Board voltages	Q205	7.8	7.4	7.6	Low Band Activated
		7.8	11.0	0V	High Band Activated
	Q206	7.8	11.0	0	Low Band Activated
		7.8	7.0	7.6	High Band Activated
	Q207	3.4	4.1	7.0	
RF Board					
No. 301-598	Q301(FET)	0	0	5.5	
	Q302	1.5	2.2	6.8	
NOTE: Model	Q303	.25	3.1	6.8	
TME-16H/L/U	Q304	7.8	7.2	3.0-5.0	
only.	Q305	2.9	3.6	7.4	
	Q306	2.9	3.6	7.2	
	Q307	0	.2	6.8	
IF Board					
No. 500-858	Q101	2.3	3.0	5.8	
	Q102	1.0	1.7	4.8	
	Q103 (PNP)	8.2	8.2	0	(unsquelched)
		8.2	8.2	1.0	(squelched)
		8.2	8.2	1.5	Min. (tight squelch)
	Q104	0	0	7.2	(unsquelched)
		0	.80	.30	(squelched)
		0	.80	.10	(tight squelch)
	Q105	1.4	1.9	5.1	(unsquelched)
		1.1	.10	8.2	(tight squelch)
	Q106	0.7	1.3	12.4	
	Q107 (PNP)	13.8	13.1	7.2	
	Q108 (PNP)	6.9	6.6	.10	
	Q109	6.9	7.2	13.8	
	Q110	0	.10	6.9	
Scan Board	Q401	0	.70	.10	(SCAN)
		0	.70	0	(MAN)
No. 500-850	Q402	0	.70	.10	
	Q403	0	.10	3.6	(SCAN)
		0	.10	1.2	(MAN)
	Q404	0	.10	2.8	(SCAN)
		0	.70	0.2	(MAN)
	Q405	3.1	3.6	3.2	(SCAN)
		0.7	1.2	0.7	(MAN)
	Q407 (PNP)	5.1	4.5	.20	(SCAN)
		5.1	4.6	.10	(MAN)
	Q408	0	.10	1.5	
		Base 1	**Emitter**	**Base 2**	
	Q406	.20	3.2	4.6	(SCAN)
	(unijunction)	.20	0.7	4.6	(MAN)

band-selection wires *must* be connected, even though not all channels are used.

- After the crystals are installed and any necessary band programming changes are completed, carefully slide the unit back into the cabinet. Screw the four feet back into place and replace the rear panel. Push the VOLUME and SQUELCH knobs back onto their shafts, and the unit is again ready for operation.

Crystal layout	Figs. 4-18, 4-19
Voltage	Tables 4-20, 4-21
Schematic	Schematics 4-15, 4-16
Internal layout, RF	Figs. 4-34, 4-35 (-16-H/L/U series)
RF	Figs. 4-2, 4-3 (-16 H/L)
IF	Figs. 4-6, 4-7
Scanner	Figs. 4-36, 4-37, 4-38, 4-39
Alignment data	Steps A, N (TME-16 H/L only), O, P (200 series board), Q (300-series board H/U), R (300-series board H/L/U), T (H/L/U only), E (H/L only)

A-1—Alignment Equipment and Preliminaries

During all steps of alignment, the SQUELCH control should be in the maximum clockwise position (minimum squelch action). All receivers should be aligned to the channel nearest the center of the frequency range over which they will operate.

The following equipment is required:

1. FM signal generator
2. Oscilloscope
3. AC VTVM or FETVOM
4. Noise generator

A-2—Alternate Alignment Setup

During all steps of the alignment, the SQUELCH control should be in the maximum clockwise position (minimum squelch action).

All receivers should be aligned to the channel nearest the center of the frequency range over which they will operate.

The following equipment is required:

1. FM signal generator
2. AC VTVM or FETVOM

B—Alignment, IF

1. Connect AC voltmeter across speaker terminals.
2. Adjust VOLUME control for 0.5V reading on AC voltmeter.
3. Peak T102 (bottom core and top core, in that order) for maximum noise (maximum meter reading on AC voltmeter). If circuit is not badly misaligned, the correct point should be within two turns of the slugs' present position.
   Note: Coils will have two peaks. Adjust core to peak away from the center of form.
4. Prealign quadrature detector by tuning L103 for maximum noise (AC voltmeter).
5. Adjust VOLUME control for 1V noise reading on AC voltmeter.
6. Connect the RF signal generator to the antenna input jack. Turn modulation off. Set the generator to the operating crystal frequency.
7. Adjust the signal generator output until the voltmeter reads 0.2V.
8. Adjust T101 and T201 (in that order) for maximum quieting (lowest meter reading). Adjust signal generator to maintain reading on AC voltmeter between 0.1V and 0.2V.
   Note: Coils will have two peaks. Adjust core of T101 to peak away from center of the coil form, and adjust core of T201 to peak nearest the center of the coil form.
9. Set the generator frequency to the secondary-image frequency. This is 910 kHz below the channel frequency. *Note*: Some receivers have a second oscillator of 11.155 MHz; in this case the image frequency is 910 kHz *above* the channel frequency. Check the frequency marked on top the crystal. The AFC circuit action will be reversed when tuning to the 910 kHz image frequency. This means that the image will be sharp, or narrow, with respect to tuning the RF signal generator.
10. Adjust the signal generator output until voltmeter reads 0.2V.
11. Adjust T102 (bottom core), T102 (top core), T101, and T201 (in that order) for

Table 4-21. Voltage Data—TME-16 Integrated Circuits.

NOTE: A "P" beside a voltage indicates that the meter reading is pulsating (fluctuating) because the scanner section of the unit is operating. MAN indicates the unit is not scanning and is at channel 1 (M401 is lighted).

	IC No.	1	2	3	4	5	6	7	8	9	10	11	12	13	14
IF Board 500-858	IC 101	4.2	0.7	0.7	4.2	7.8	0	4.2	7.8	—	—	—	—	—	—
	IC 102	4.0	3.5	0	1.3	1.3	1.3	0	0	0.2	1.4	2.9	3.5	7.6	5.0
Scan Board 500-850	IC 401	.7P	.7P	9P	9P	.7P	.7P	0	.7P	.7P	9P	9P	.7P	.7P	5.1 (Scan)
		3.3	3.3	0.5	10.8	0.2	0.2	0	0.2	0.2	10.8	10.8	0.2	0.2	5.1 (Man)
	IC 402	.7P	.7P	9P	9P	.7P	.7P	0	.7P	.7P	9P	9P	.7P	.7P	5.1 (Scan)
		0.2	0.2	10.8	10.8	0.2	0.2	0	0.2	0.2	10.8	10.8	0.2	0.2	5.1 (Man)
	IC 403	0.1	0	0	0	0	0	0	1.5	1.5	.7P	.7P	.7P	.7P	5.1 (Scan)
		0	0	0	0	0	0	0	1.5	1.5	0.2	0.2	0.2	3.3	5.1 (Man)
	IC 404	.7P	0	0	0	0	0	0	1.5	1.5	.7P	.7P	.7P	.7P	5.1 (Scan)
		0.2	0	0	0	0	0	0	1.5	1.5	0.2	0.2	0.2	0.2	5.1 (Man)
Scan Board Secondary 500-850	IC 501	.7P	.7P	9P	9P	.7P	.7P	0	.7P	.7P	9P	9P	.7P	.7P	5.1 (Scan)
		0.2	0.2	10.8	10.8	0.2	0.2	0	0.2	0.2	10.8	10.8	0.2	0.2	5.1 (Man)
	IC 502	.7P	.7P	9P	9P	.7P	.7P	0	.7P	.7P	9P	9P	.7P	.7P	5.1 (Scan)
		0.2	0.2	10.8	10.8	0.2	0.2	0	0.2	0.2	10.8	10.8	0.2	0.2	5.1 (Man)
	IC 503	.7P	0	0	0	0	0	0	1.5	1.5	.7P	.7P	.7P	.7P	5.1 (Scan)
		0.2	0	0	0	0	0	0	1.5	1.5	0.2	0.2	0.2	0.2	5.1 (Man)
	IC 504	.7P	0	0	0	0	0	0	1.5	1.5	.7P	.7P	.7P	.7P	5.1 (Scan)
		0.2	0	0	0	0	0	0	1.5	1.5	0.2	0.2	0.2	0.2	5.1 (Man)

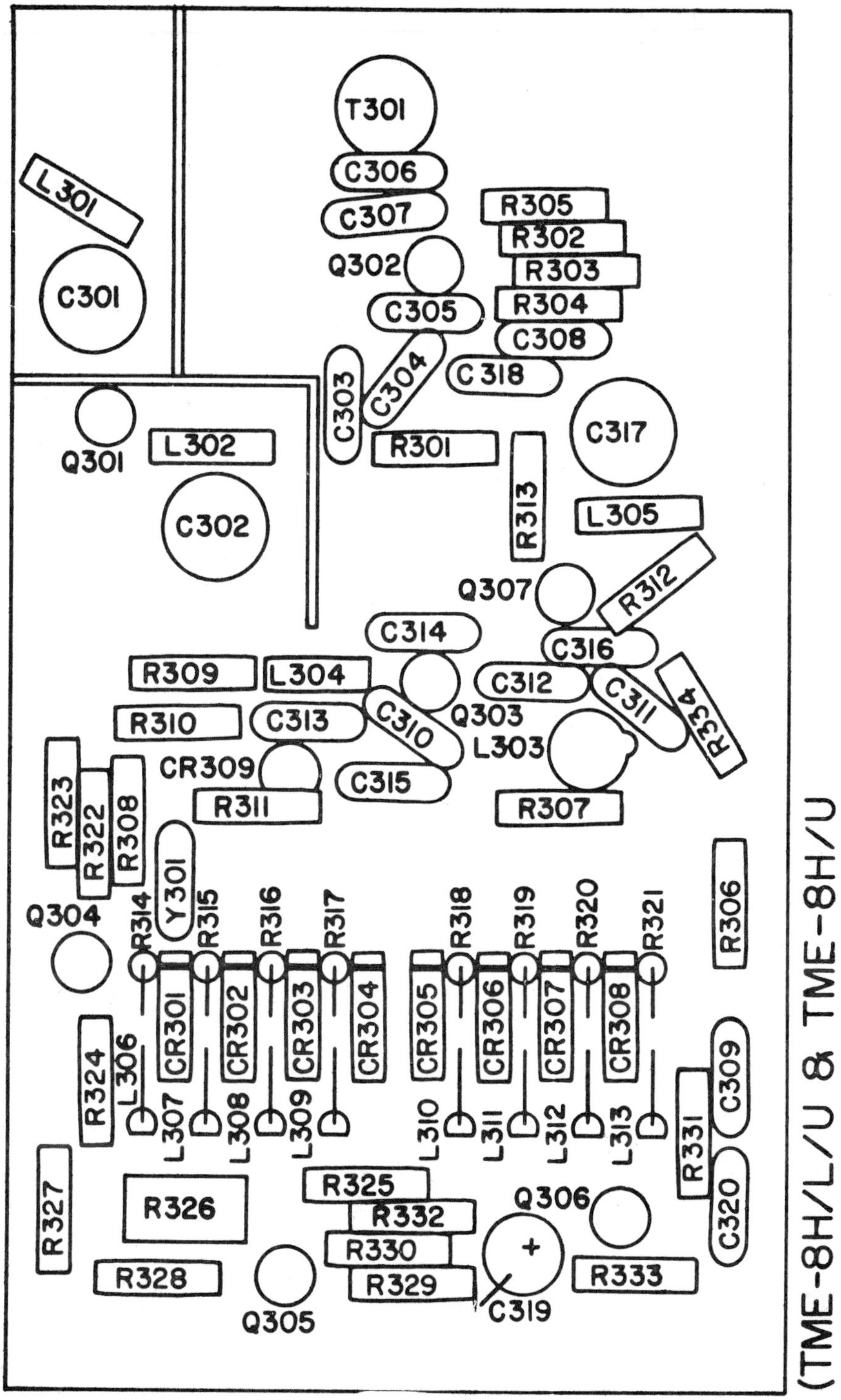

Fig. 4-34. RF board 301-598 parts placement diagram (UHF board).

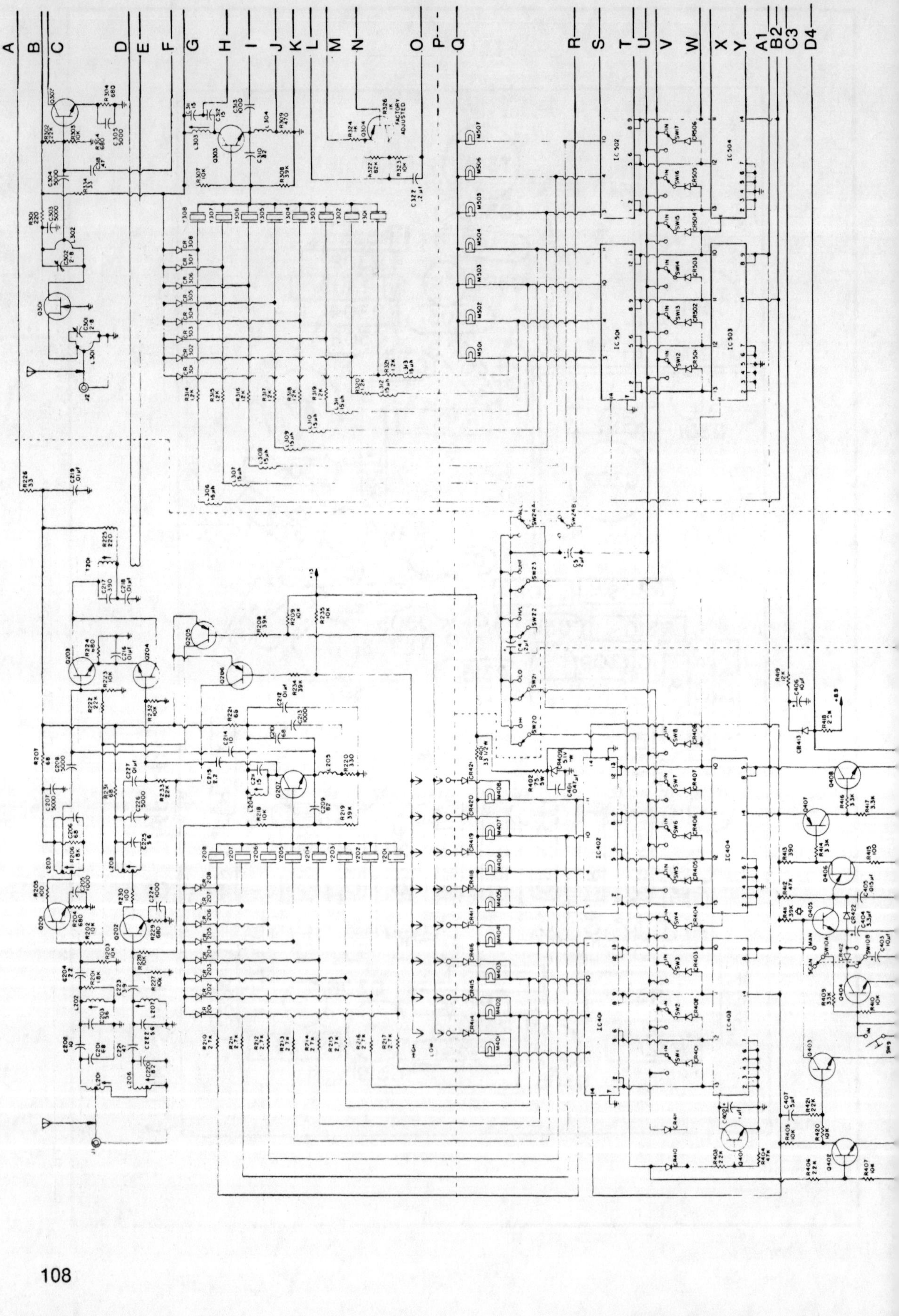
A
B
C
D
E
F
G
H
I
J
K
L
M
N
O
P
Q
R
S
T
U
V
W
X
Y
A1
B2
C3
D4

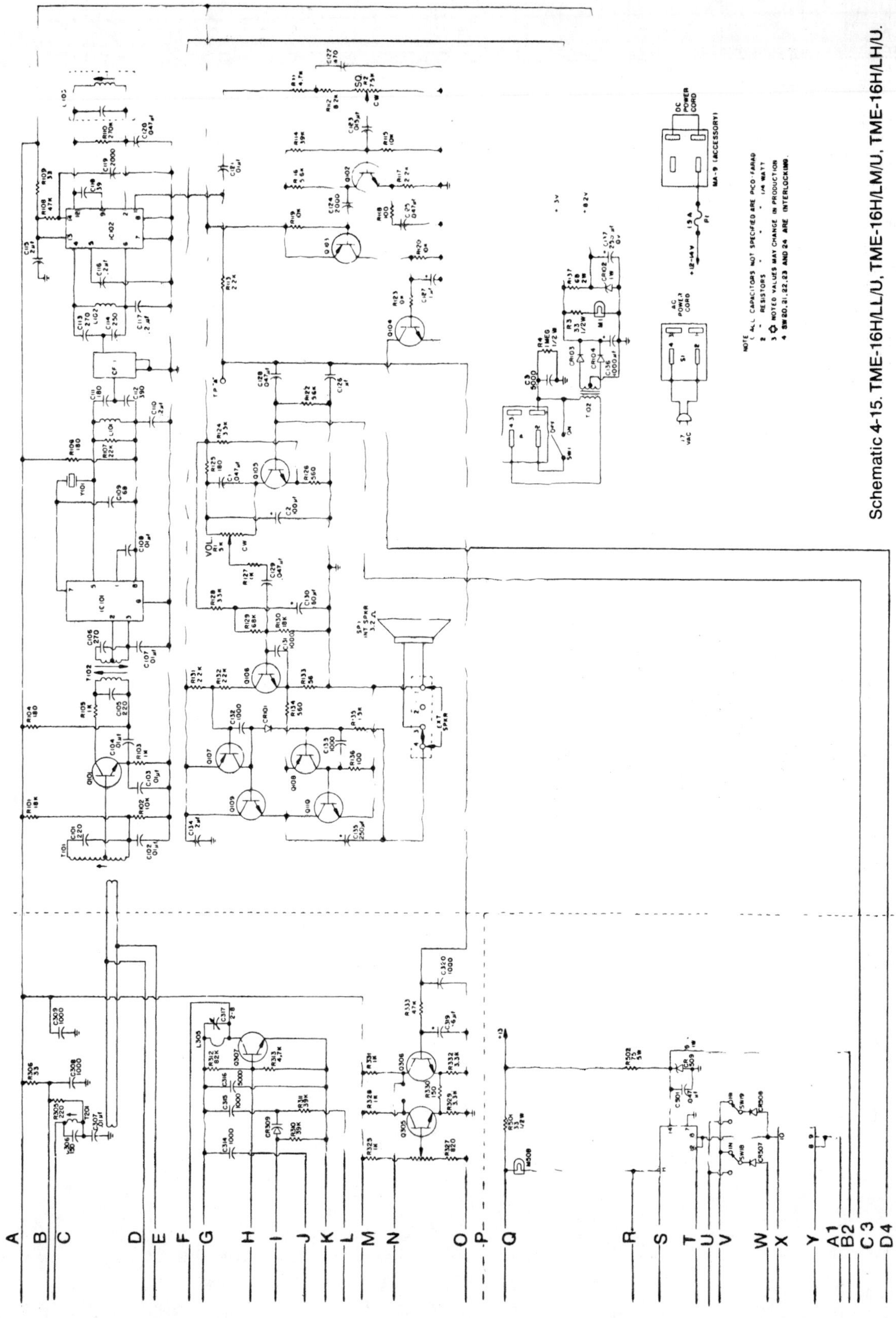

Schematic 4-15. TME-16H/LL/U, TME-16H/LM/U, TME-16H/LH/U.

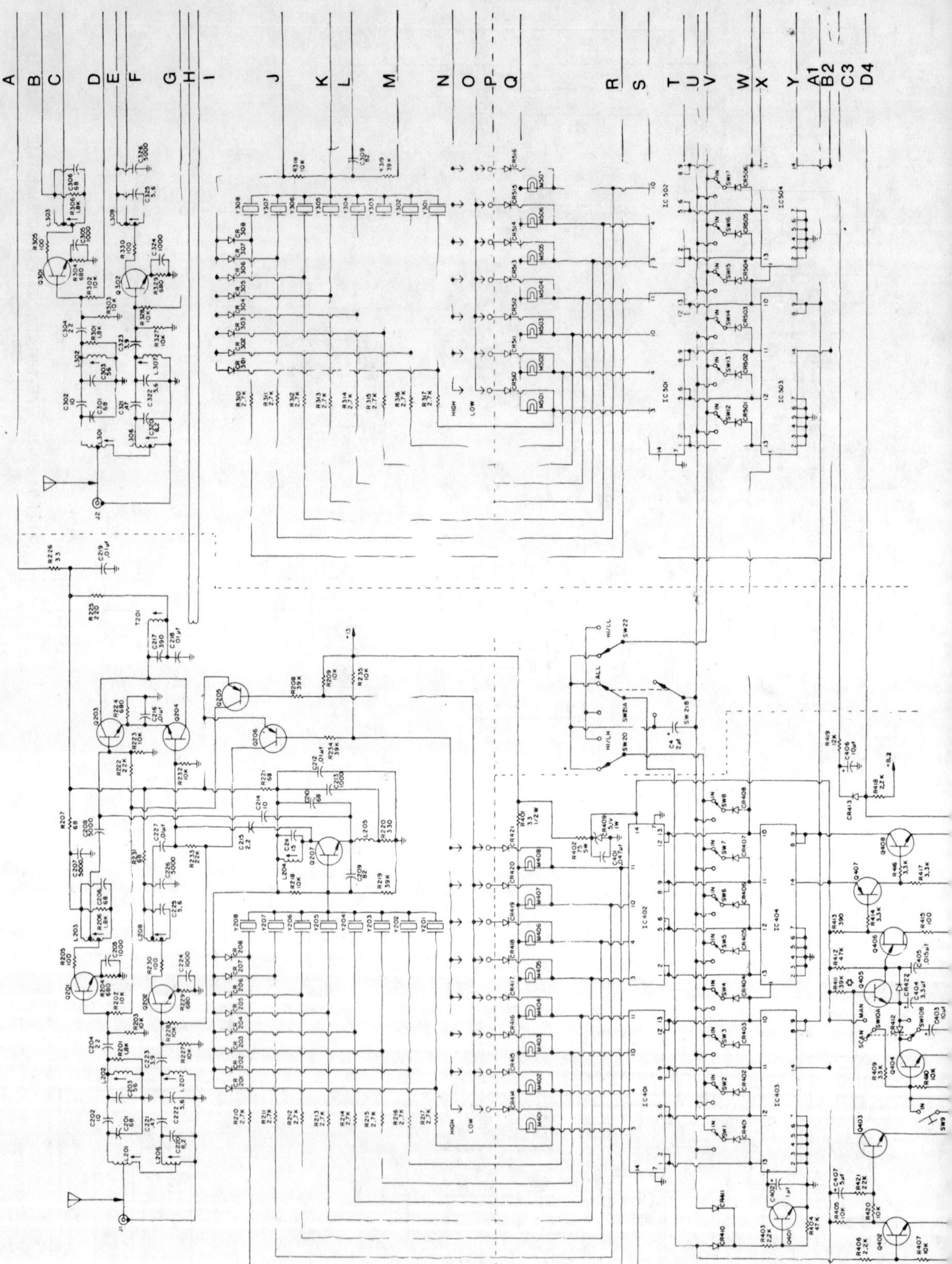

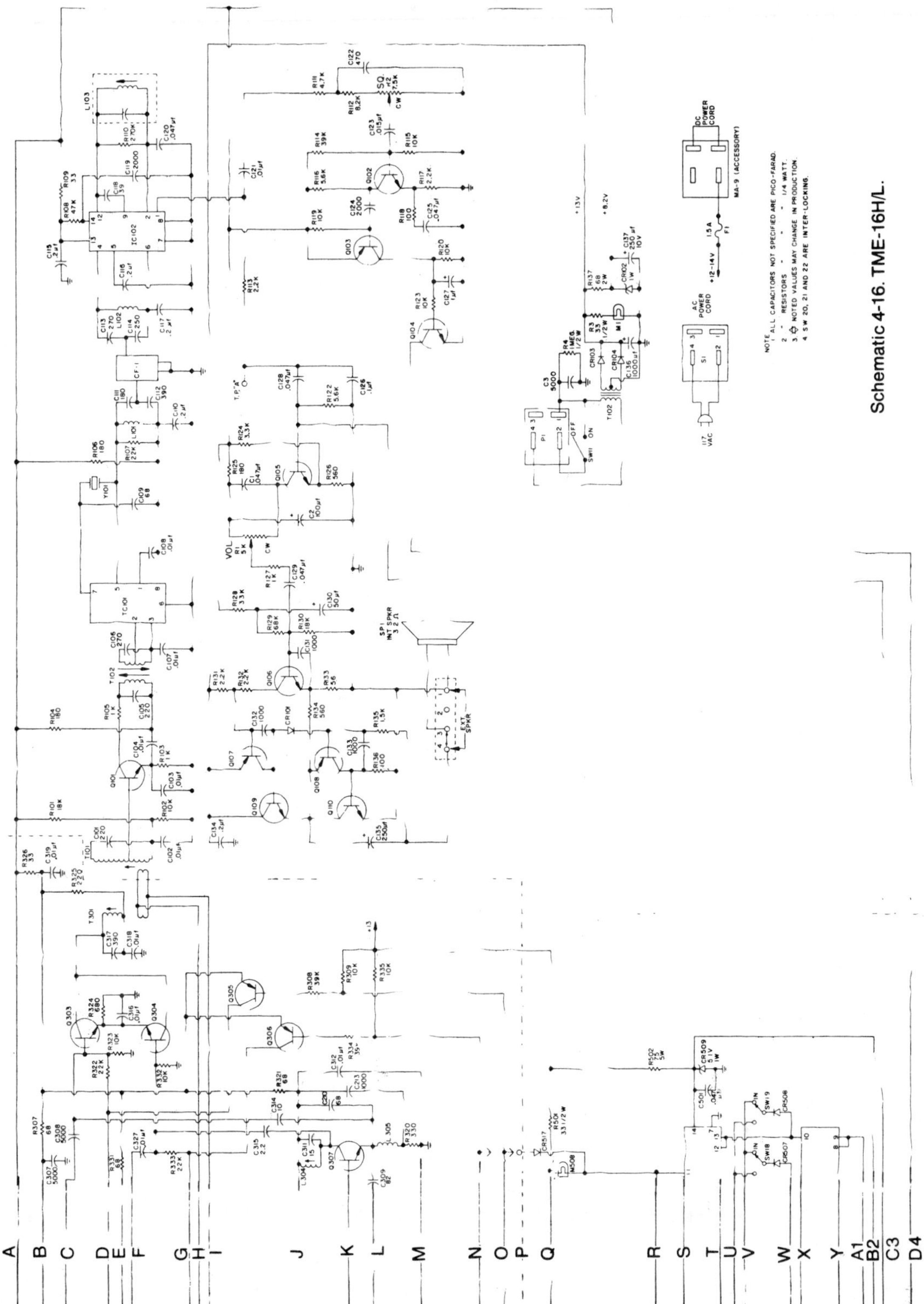

Schematic 4-16. TME-16H/L.

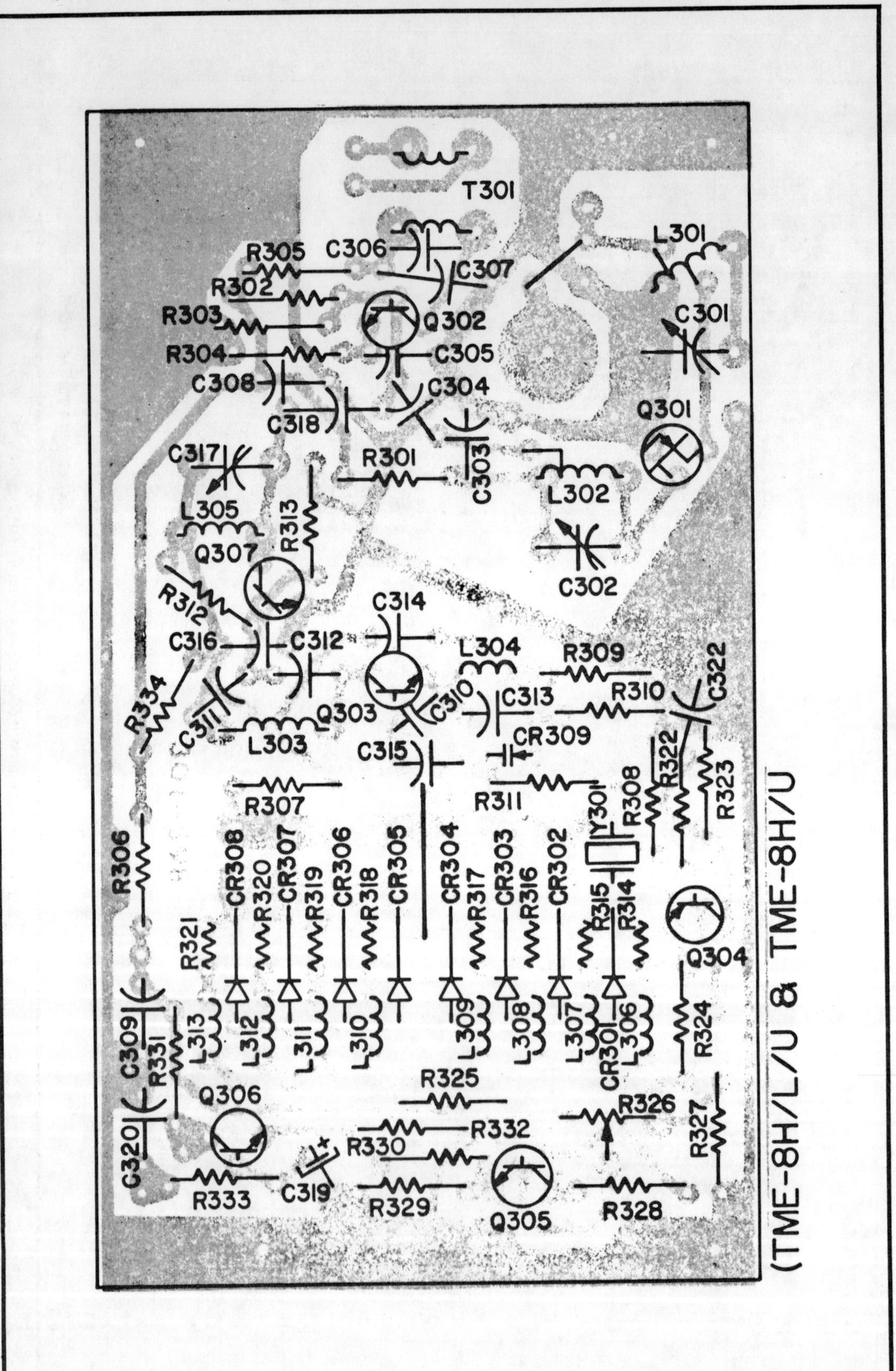

Fig. 4-35. RF board 301-598, bottom view (UHF board).

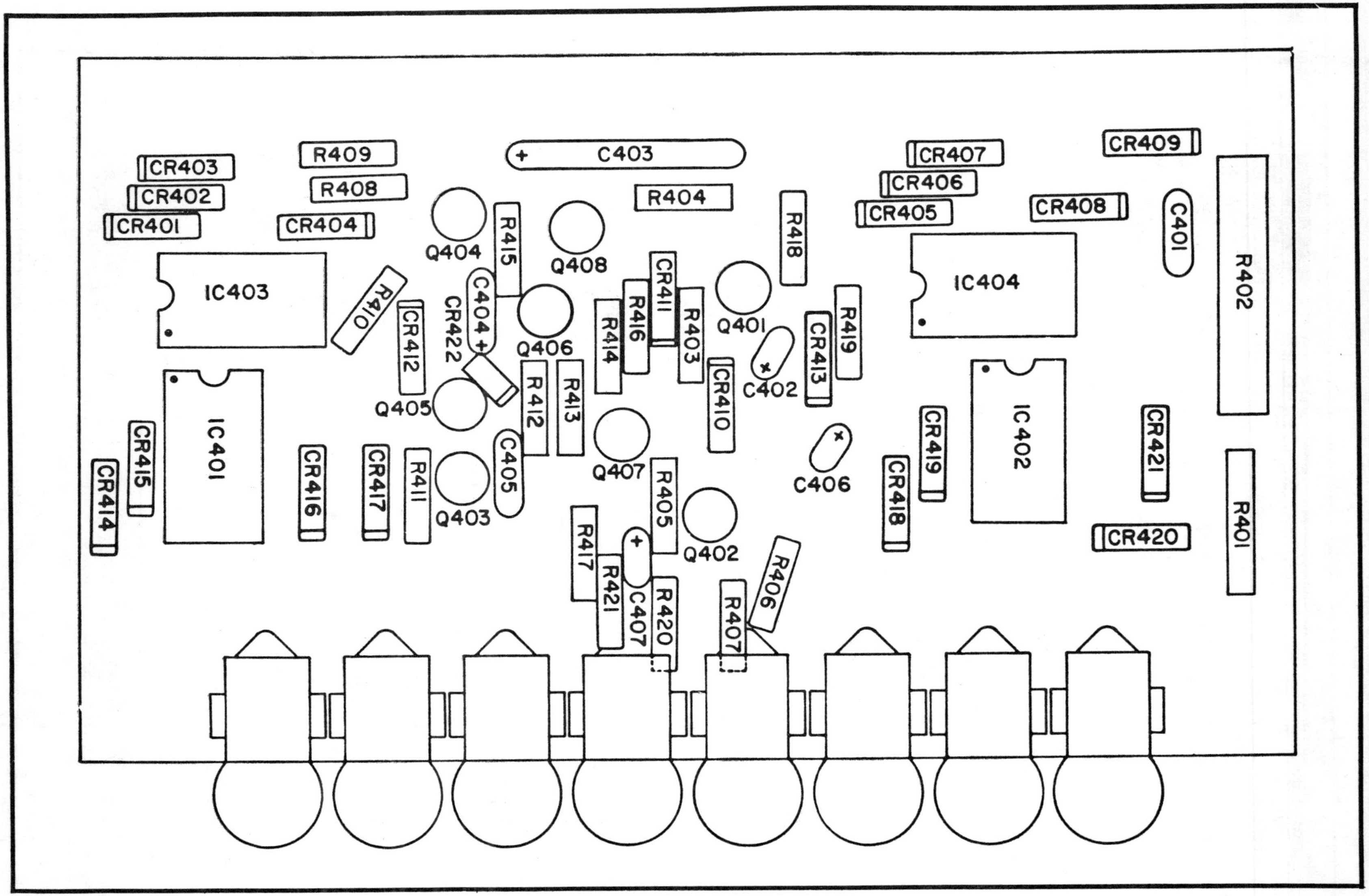

Fig. 4-36. Scanner board 500-850 parts placement diagram.

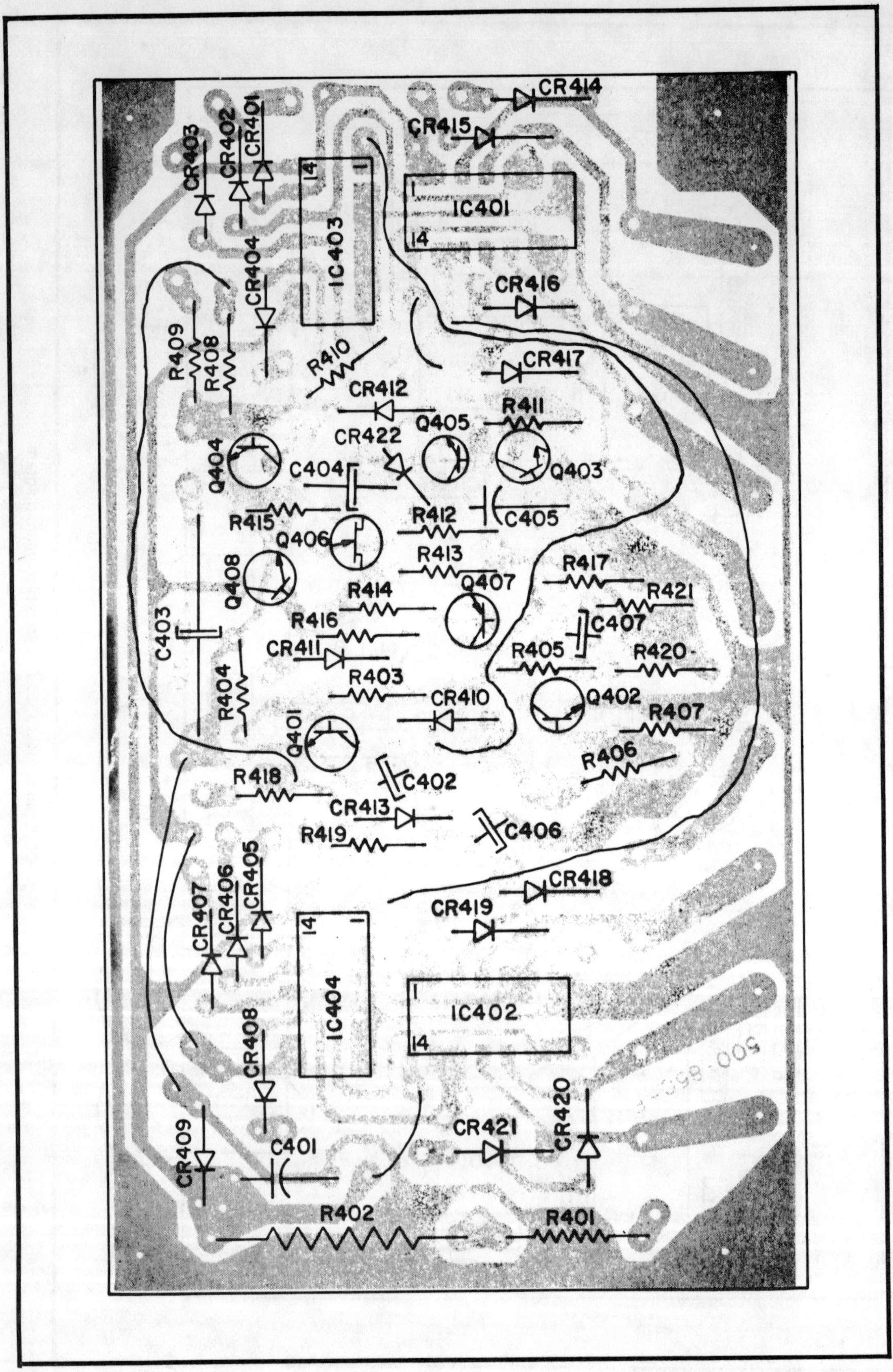

Fig. 4-37. Scanner board 500-850, bottom view.

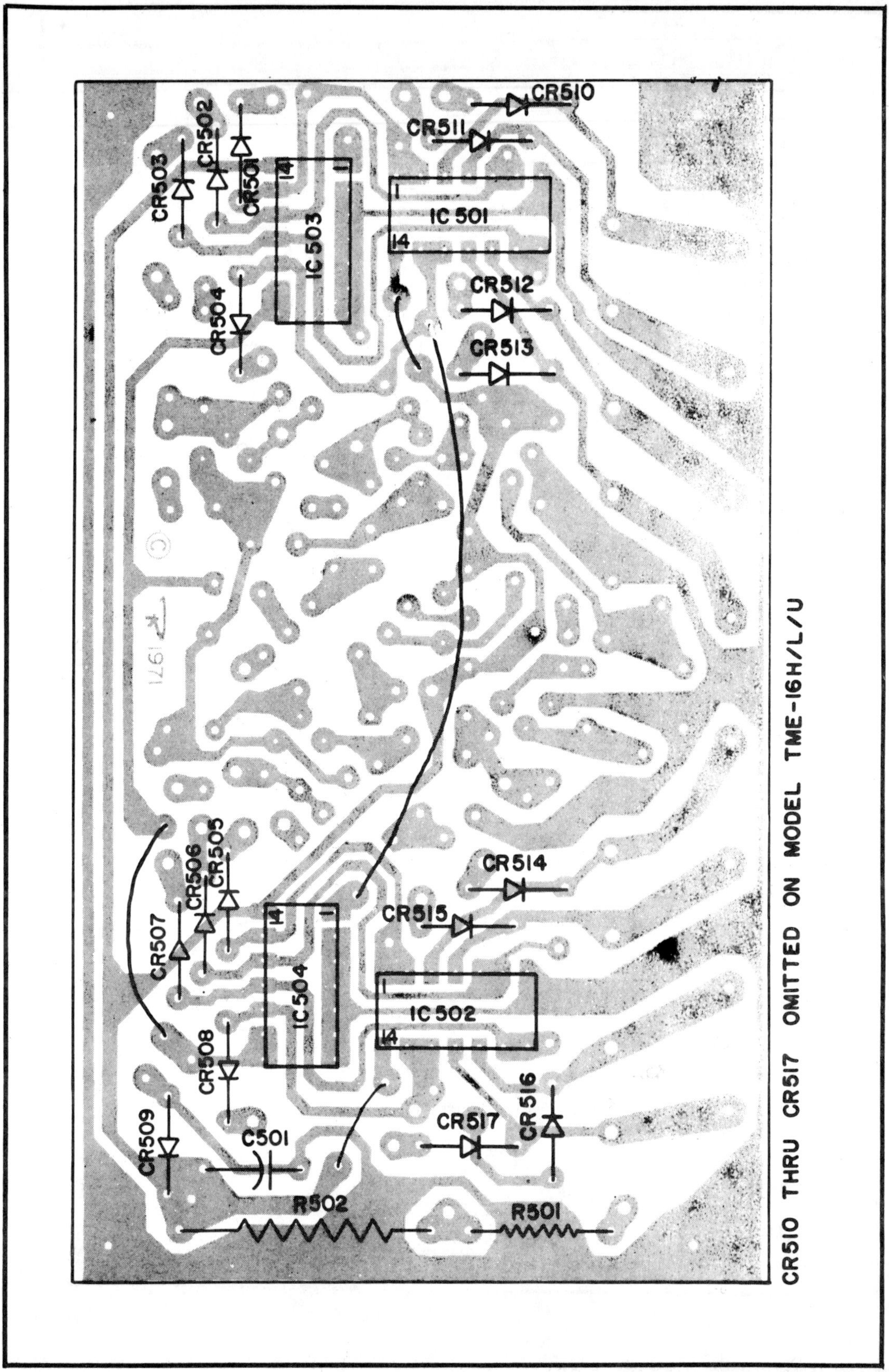

Fig. 4-38. Scanner board 500-850 (secondary) bottom view.

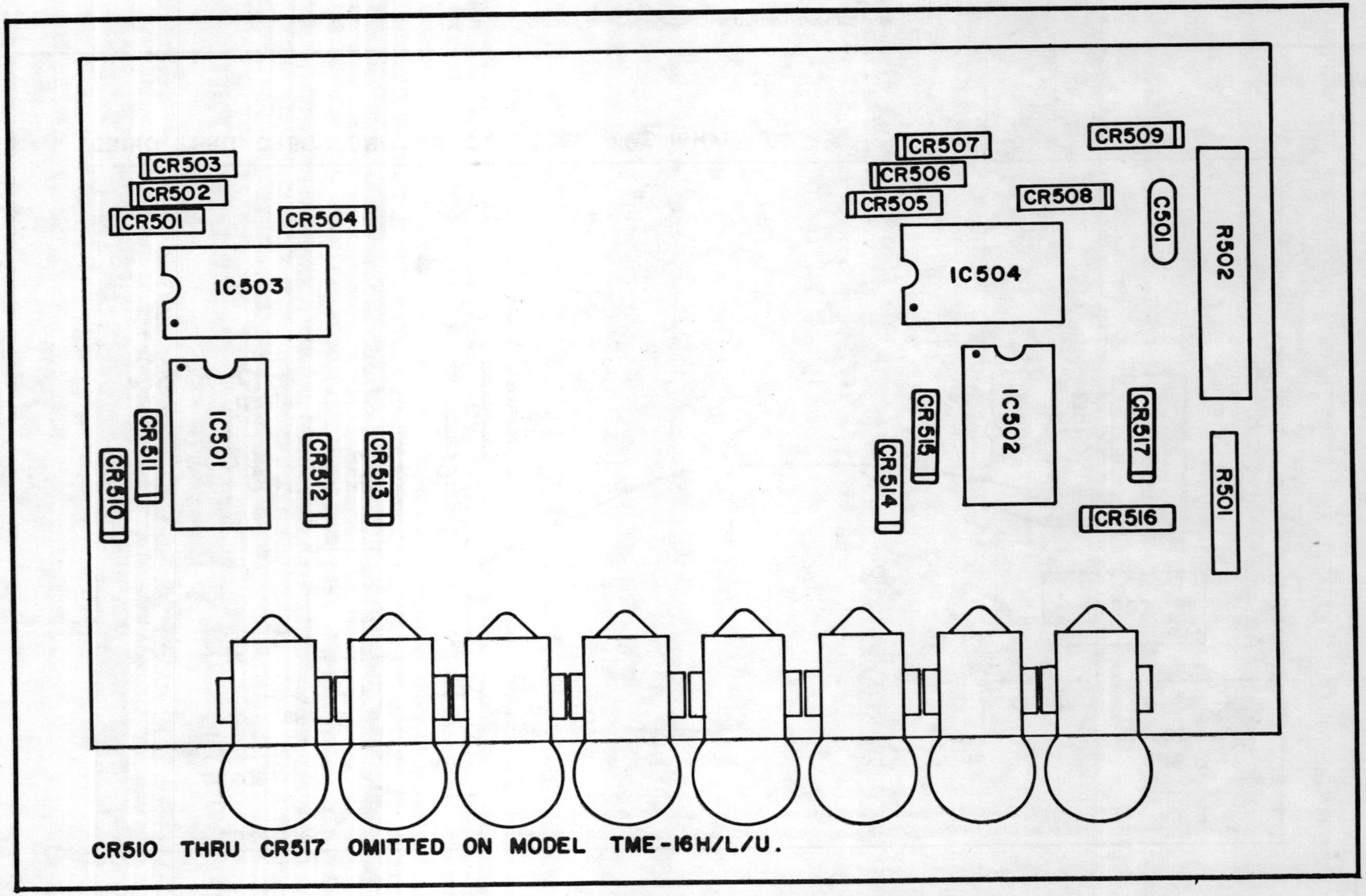

Fig. 4-39. Scanner 500-850 board (secondary) parts placement diagram.

maximum quieting degradation (highest meter reading). Adjust signal generator output to maintain voltmeter reading between 0.1V and 0.2V. The correct position for the slugs should be within two turns of the position in steps B-3 and B-8.

C—Alignment, RF

1. Preset slugs L201, L202, L203, and L204 four turns from the outer ends of the coil form.
2. Connect AC voltmeter across speaker terminals.
3. With nothing connected to the antenna input, adjust the VOLUME control until AC voltmeter reads 1.0V of noise.
4. Activate channel nearest to center of frequencies over which the unit will operate.
5. Connect signal generator to antenna input jack. Set generator accurately to the frequency of the channel being used. Turn modulation off.
6. Adjust output signal generator until AC voltmeter reads 0.2V.
7. Adjust L201, L202, L203, and L204, in that order, for maximum quieting (lowest meter reading). Adjust signal generator to maintain reading on AC voltmeter between 0.1V and 0.2V. Repeat adjustment until no further improvements can be made.

D—Alignment, RF

1. Preadjust trimmer capacitor (C217) so the silvered half-moon section is nearest to the rear of the receiver.
2. Connect AC voltmeter across the speaker terminals.
3. With nothing connected to the antenna input, adjust the VOLUME control until AC voltmeter reads 1V of noise.
4. Connect signal generator to antenna input jack. Set generator accurately to the frequency being used for alignment. This should be nearest to the center of all frequencies being used in set.
5. Adjust output of signal generator until AC voltmeter reads 0.2V.
6. Adjust C201 and C202 (trimmer capacitors) in that order, for maximum quieting (lowest meter reading). Adjust signal generator output to maintain voltmeter reading between 0.1V and 0.2V. Repeat adjustments until no further improvements can be made.

   Use nonmetallic screwdriver for trimmer adjustments. Peaks are very sharp—tune with care. Two peaks can be observed. Tune to peak with silver moon section away from IF board.
7. Adjust the core of L203 for maximum quieting (lowest meter reading). Adjust signal generator to maintain reading on AC voltmeter between 0.1 and 0.2V. To properly adjust L203, C217 must be preadjusted as in steps 1, 2, and 3.
8. Adjust C217 counterclockwise with nonmetallic screwdriver for maximum quieting (lowest meter reading). Adjust signal generator to maintain reading on AC voltmeter between 0.1V and 0.2V. *Do not* readjust L203 after C217 is adjusted.

E—Noise Balance Adjustment

This adjustment may be required when excessive ignition noise is encountered. Usually the noise problem is caused by improper or inadequate noise suppression of the vehicle's ignition system.

1. Using a *T*-connector, connect the FM signal generator and the noise generator to the antenna input jack. If a *T*-connector is not available, connect the FM generator to the antenna jack and feed in the noise signal by means of a 3- or 4-turn loop coupled to the input coil, L201.
2. Connect the oscilloscope to the junction of Q109's emitter and Q110's collector, or to the speaker terminals.
3. Apply a 3–10 μV signal as accurately as can be set to the exact channel frequency (carrier only, no modulation), and adjust the output of the noise generator until spikes are clearly seen in the audio output as viewed on the os-

cilloscope. The noise spikes will be either mostly positive or negative if an unbalanced condition exists.

4. Tune L103 (quadrature detector coil) until the noise spikes are equally positive and negative in amplitude. The overall amplitude of these spikes should be much less as a balance is achieved. Usually only one-fourth turn or less is needed to obtain the proper adjustment for best noise balance. If a proper balance cannot be achieved, repeat the IF and RF alignments and then try the noise balance adjustment again.

F—Alignment, AFC

This adjustment requires an accurate 10.7 MHz ±1 kHz oscillator or 455 kHz ±500 Hz oscillator to be used as a reference signal. If none is available, proceed to step F-5.

1. Prealign quadrature detector by tuning L103 for maximum noise (AC voltmeter reading) at the speaker terminals.
2. With a coupling loop, inject reference signal (either 10.7 MHz or 455 kHz) to produce good quieting (more than 30 dB quieting). Adjust R226 for a reading of 3.8V to 4.0V at the collector of Q204.
3. Remove the reference signal and have the unit squelched and receiving no signal. The voltage on the collector of Q204 should be between 3.2V and 4.6V. If not, note voltage and proceed to step F-4. If voltage is between 3.2V and 4.6V, AFC alignment is complete.

   Further adjustments made to L103 and R226 will require AFC to be readjusted.
4. Inject reference signal and monitor voltage on collector of Q204. Adjust L103 for same voltage as noted in step 3. Readjust R266 for a voltmeter reading of 3.8V to 4.0V. Repeat step F-3.
5. If an accurate IF signal source is not available, an approximate AFC alignment can be made by adjusting L103 for maximum noise (AC voltmeter reading) at the speaker terminals; with unit squelched and receiving no signal, adjust R226 for a voltmeter reading of 3.2V to 4.6V on the collector of Q204. Do not adjust L103 more than one-fourth turn at a time.

   Units equipped with a 10.245 MHz crystal have the jumper in the AFC circuit connected between the base of Q204 and collector of Q206. When a 11.155 MHz crystal is used, the jumper is connected between the base of Q204 and the collector of Q205. If crystal is changed from one frequency to the other, the jumper must be changed.

G—Alignment, Quadrature Detector

1. Connect the FM signal generator to the antenna input jack. Accurately set frequency to the center of the channel being used for alignment. Modulate signal generator with 1000 Hz, 3 kHz deviation.
2. Connect the oscilloscope to test point *A* (junction of C126, C128, R113).
3. Adjust output of signal generator until all noise in scope pattern barely disappears.
4. Adjust L103 for maximum peak-to-peak amplitude while maintaining symmetry of the detected signal.

H—Alignment, Receiver

1. Remove the antenna from the receiver.
2. Using the slow-scan mode, set the receiver to the channel nearest the frequency range to be covered. Set the SCAN switch to STOP, and rotate the SQUELCH control to the maximum clockwise position (minimum squelch action), opening the audio circuits.
3. Connect the AC voltmeter across the speaker terminals (note polarity).
4. Connect the FM signal generator between the antenna post and ground.
5. Adjust the receiver VOLUME control for 0.03V RMS noise reading on AC voltmeter.
6. Increase signal generator level and tune L104 for minimum noise (maximum quieting), reducing the signal generator level as necessary to maintain a noise reading near 0.1V RMS (100 mV).

This and following steps should be performed with an unmodulated signal.

7. Adjust the slugs in L103 and L106 (starting with L106) for minimum noise, reducing the signal generator level to maintain the noise level near 0.1V RMS. Repeat this procedure several times until no further improvement is noted.
8. Adjust the slugs in L101 and L102 (starting with L101) for minimum noise, reducing the signal generator level to maintain the noise reading near 0.1V RMS. Repeat several times until no further improvement is noted.

Note: If the receiver is badly misaligned, it may be necessary to preset the slugs before beginning the alignment procedure, to insure that the coils are adjusted to the correct peak (since some will have two peaks). If this should be necessary, preset all coils as follows and adjust for the indicated peak:

Coil	Set slug to:	Adjust for:
L101, L102 L103, L104	Top of coil form	1st peak as slugs are rotated into coil form
L106	Bottom of coil form	first peak as slug is backed out

After presetting the slugs, the alignment procedure beginning with H-1 is followed.

I—Alignment, RF

1. Preset slugs L201, L202, and L203 out of the outer end of the coil form three turns. Preset L204, L206, L207, and L208 flush with the outer ends of the coil form.
2. Connect AC voltmeter across speaker terminals.
3. With nothing connected to the antenna input, adjust the VOLUME control until AC voltmeter reads 1V of noise.

High-Band Section

4. Activate high-band channel to channel nearest to center of high-band frequencies being used.
5. Connect signal generator to antenna input jack. Set generator accurately to the frequency of the channel being used. Turn modulation off.
6. Adjust output of signal generator until AC voltmeter reads 0.2V.
7. Adjust L204 for maximum quieting (lowest meter reading). Adjust signal generator to maintain reading on AC voltmeter between 0.1V and 0.2V.
8. Adjust L206, L207, and L208 (in that order) for maximum quieting. Adjust the signal generator output to maintain voltmeter reading between 0.1 and 0.2V. Repeat adjustments until no further improvement can be made.

Low-Band Section

9. Activate low-band channel nearest center of low-band frequencies being used in the set.
10. Set generator accurately to the center frequency of the channel being used for alignment. Turn modulation off.
11. Adjust output of signal generator until AC voltmeter reads 0.2V.
12. Adjust coils L201, L202, and L203 (in that order) for maximum quieting (lowest meter reading). Adjust the signal generator output to maintain voltmeter reading between 0.1V and 0.2V. Repeat adjustments until no further improvement can be made.

J—Alignment, Quadrature Detector

1. Connect the FM signal generator to the high–low antenna input jack. Accurately set the frequency to the center of the channel being used for alignment. Modulate signal generator with 1000 Hz, 3 kHz deviation.
2. Connect the oscilloscope to junction of C162, C163, and R172.
3. Adjust signal generator's output until all the noise in the scope pattern barely disappears.
4. Adjust L128 for maximum peak-to-peak amplitude while maintaining symmetry of the detected signal. When L128 is properly aligned, signal at above junction should be approximately 0.2V RMS.

K—Alignment, IF

1. Preset the cores of L122, L123, and L124 nine turns in from the outer end of the coil form.

2. Connect AC voltmeter to the junction of R167 and the collector of Q116.
3. Set AC voltmeter to the 300 mV scale.
4. With generator accurately set to the frequency of the center of the channel being used for alignment, increase signal generator's output until AC voltmeter reading is in midrange.
5. Adjust L122, L123, and L124 (in that order) for maximum AC voltmeter reading. Readjust signal generator's output to maintain voltmeter reading approximately in the midrange. Repeat adjustment until no further improvements can be made.

L—Alignment, RF

Low-Band Section

1. Preset the cores of L102 a[illegible] [illegible]03 one turn from the outer ends of the coil form. Due to the broadness of the low-band section, presetting the above cores will give you optimum performance over the entire band.

High-Band Section

2. Connect AC voltmeter to the junction of R167 and the collector of Q116.
3. Set AC voltmeter to the 300 mV scale (or similar scale).
4. Activate high-band channel nearest to center of high-band frequencies being used.
5. With signal generator accurately set to the frequency of the center of the channel being used for alignment and connected to high–low antenna input jack, increase signal generator's output until AC voltmeter reading is in midrange.
6. Adjust L119, L104, L105, and L106 (in that order) for maximum AC voltmeter reading. Readjust signal generator's output to maintain voltmeter reading approximately in the midrange. Repeat adjustment until no further improvements can be made.

Adjustment to L119 of one or more turns may require UHF section to be realigned.

UHF-Band Section

7. Connect AC voltmeter across the speaker terminals.
8. With signal generator output reduced to zero, adjust the VOLUME control until AC voltmeter reads 1.0V of noise.
9. Activate UHF channel nearest to center of UHF frequencies being used.
10. Set signal generator accurately to the channel being used and adjust output of signal generator until AC voltmeter reads 0.2V.
11. Adjust trimmers C121, C122, and C139 (in that order) for maximum quieting (lowest meter reading). Adjust the signal generator output to maintain voltmeter reading between 0.1 and 0.2V. Repeat adjustments until no further improvements can be made.

Use nonmetallic screwdriver for trimmer adjustments. Coil L119 is tuned in high-band alignment. If the UHF section sensitivity is down, a readjustment to L119 may be necessary. If more than a turn one way or the other is necessary, repeat high-band alignment.

M—Alignment, AFC

Note: This adjustment requires an accurate 10.7 MHz ±1 kHz oscillator or 455 kHz ±500 Hz oscillator to be used as a reference signal. If none is available, proceed to step M-4.

1. With a coupling loop, inject reference signal (either 10.7 MHz or 455 kHz) to produce good quieting (more than 30 dB quieting). Adjust R147 for reading of 3.8V to 4.0V at the collector of Q109.
2. Remove the reference signal and have the unit squelched and receiving no signal. The voltage on the collector of Q109 should be between 3.2V and 4.6V. If not, note voltage and proceed to step M-3. If voltage is between 3.2V and 4.6V, the AFC alignment is complete. *Note*: Any further adjustments made to L128 and R147 will require AFC to be readjusted.
3. Inject reference signal and monitor voltage on collector of Q109. Adjust

L128 for same voltage as noted in step 3. Readjust R147 for a voltmeter reading of 3.8V to 4.0V. Repeat step M-2. *Note*: Do not adjust L128 more than one-fourth turn at a time.

4. If an accurate IF signal source is not available, an approximate AFC alignment can be made by adjusting L128 on a high-band or low-band crystal as specified in *Quadrature Detector Alignment*; with the unit squelched and receiving no signal, adjust R147 for voltmeter reading of 3.2V to 4.6V on the collector of Q109.

Note: Units equipped with a 10.245 MHz crystal have the jumper in the AFC circuit connected between the base of Q109 and collector of Q111. When a 11.155 MHz crystal is used, the jumper is connected between the base of Q109 and the collector of Q110. If the crystal is changed from one frequency to the other, the jumper must be changed.

N—Alignment, Quadrature Detector

1. Connect the FM signal generator to the antenna input jack of the 200-series board. Accurately set the frequency to the center of the channel being used for alignment. Modulate signal generator with 1000 Hz, 3 kHz deviation.
2. Connect the oscilloscope to test point *A* (junction of C126, C128, and R113).
3. Adjust signal generator's output until all the noise in the scope pattern barely disappears.
4. Adjust L103 for maximum peak-to-peak amplitude while maintaining symmetry of the detected signal. When L103 is properly aligned, signal at *TP-A* should be approximately 0.2V RMS with test signal input as noted in N-1.

O—Alignment, IF

1. Disconnect RF signal generator from the antenna input.
2. Connect AC voltmeter across the speaker terminals.
3. Adjust VOLUME control for 0.5V reading on AC voltmeter.
4. Peak T102 (bottom core and top core, in that order) for maximum noise (maximum meter reading on AC voltmeter). If the circuit is not badly misaligned, the correct point should be within two turns or so of the slugs' present positions. *Note*: Coil will have two peaks. Adjust the core to peak away from the center of the form.
5. Prealign quadrature detector by tuning L103 for maximum noise (AC voltmeter).
6. Adjust VOLUME control for 1V noise reading on AC voltmeter.
7. Connect the RF signal generator to the antenna input jack of 200-series board. Turn the modulation off. On Models TME-8H/L/U and TME-8H/U set the generator to the high-band crystal frequency that will be used for high-band section alignment. On Model TME-8LMH set the generator to the low-band crystal frequency that will be used for RF section alignment.
8. Adjust the signal generator output until the voltmeter reads 0.2V.
9. Adjust T101 and T201 (in that order) for maximum quieting (lowest meter reading). Adjust signal generator to maintain reading on AC voltmeter between 0.1 and 0.2V. Coils have two peaks. Adjust core to peak away from the center of the coil form.
10. Set the generator frequency to the secondary-image frequency. This is 910 kHz below the channel frequency for high-band channels and 910 kHz above for low-band channels.

    Some receivers have the second oscillator at 11.155 MHz. In this case the image frequency is 910 kHz to the opposite side of the channel frequency previously stated. Check the frequency marked on top of the crystal.
11. Adjust the signal generator output until the voltmeter reads 0.2V.
12. Adjust T102 (bottom core), T102 (top core), T101, and T201 (in that order) for maximum quieting degradation (highest meter reading). Adjust signal generator output to maintain voltmeter reading between 0.1V and 0.2V. The correct position for the cores should be

within two turns of the position in steps O-4 and O-9.

P—Alignment, RF (200-Board of TME-8H/L/U)

1. Preset slugs L201, L202, and L203 out of the outer end of the coil form three turns. Preset L204, L206, L207, and L208 four turns from the outer ends of the coil form.
2. Connect AC voltmeter across the speaker terminals.
3. With nothing connected to the antenna input, adjust the VOLUME control until AC voltmeter reads 1.0V of noise.

High-Band Section

4. Activate high-band channel nearest the center of high-band frequencies being used.
5. Connect signal generator to antenna input jack of 200-series RF board. Set generator accurately to the frequency of the channel being used. Turn modulation off.
6. Adjust output signal generator until AC voltmeter reads 0.2V.
7. Adjust L206, L207, L208, and L204, in that order, for maximum quieting (lowest meter reading). Adjust signal generator to maintain reading on AC voltmeter between 0.1 and 0.2V. Repeat adjustment until no further improvements can be made.

Low-Band Section

8. Activate low-band channel nearest to center of low-band frequencies being used.
9. Set generator accurately to the center frequency of the channel being used for alignment. Turn modulation off.
10. Adjust output of signal generator until AC voltmeter reads 0.2V.
11. Adjust coils L201, L202, and L203 (in that order) for maximum quieting (lowest meter reading). Adjust the signal generator output to maintain voltmeter reading between 0.1V and 0.2V. Repeat adjustments until no further improvements can be made.

Q—Alignment, RF (200-Board of TME-8H/U)

1. Preset L201, L202, L203, and L204 four turns from the outer end of the coil form.
2. Connect AC voltmeter across the speaker terminals.
3. With nothing connected to the antenna input, adjust the VOLUME control until AC voltmeter reads 1.0V of noise.
4. Activate high-band channel nearest to center of hand-band frequencies being used.
5. Connect signal generator to antenna input jack of 200-series RF board. Set generator accurately to the frequency of the channel being used. Turn modulation off.
6. Adjust output signal generator until AC voltmeter reads 0.2V.
7. Adjust L201, L202, L203, and L204, in that order, for maximum quieting (lowest meter reading). Adjust signal generator to maintain reading on AC voltmeter between 0.1 and 0.2V. Repeat adjustment until no further improvements can be made.

R—Alignment, RF (300-Board of TME-8H/L/U, -8H/L)

1. Connect the RF signal generator to the antenna input jack of the 300-series board. Set the generator to the operating crystal frequency.
2. Adjust output of signal generator until AC voltmeter reads 0.2V.
3. Preadjust trimmer capacitor C317 so the silvered half-moon section is nearest the rear of the receiver.
4. Adjust T301 for maximum quieting (lowest meter reading). Adjust signal generator output to maintain voltmeter reading between 0.1V and 0.2V. Adjust core to peak nearest the center of the coil form.
5. Adjust C301 and C302 (trimmer capacitors), in that order, for maximum quieting (lowest meter reading). Adjust signal generator output to maintain voltmeter reading between 0.1 and 0.2V. Repeat adjustments until no further improvements can be made.

Note: Use nonmetallic screwdriver for trimmer adjustments. Peaks are very sharp, so tune with care. Two peaks can be observed. Tune to peak with silver moon section away from IF board.

6. Adjust the core of L303 for maximum quieting (lowest meter reading). Adjust signal generator to maintain reading on AC voltmeter between 0.1V and 0.2V.
   Note: To properly adjust L303, C317 must be preadjusted as in R-3.
7. Adjust C317 counterclockwise with nonmetallic screwdriver for maximum quieting (lowest meter reading). Adjust signal generator to maintain reading on AC voltmeter between 0.1V and 0.2V. *Do not* readjust L303 after C317 is adjusted. Readjust C301, C302, and C317 (only) for maximum quieting (lowest meter reading). Adjust signal generator to maintain voltmeter reading between 0.1V and 0.2V. Repeat adjustments until no further improvements can be made.

S—Alignment, RF (TME-8LMH)

1. Preset L201, L202, L203, L301, L302, and L303 out of the outer end of the coil form three turns.
2. Connect AC voltmeter across speaker terminals.
3. With nothing connected to the antenna input, adjust the VOLUME control until AC voltmeter reads 1.0V of noise.
4. Activate channel on 200-series board that will be used for alignment.
5. Connect signal generator to antenna input jack of 200-series RF board. Set generator accurately to the frequency of the channel being used. Turn modulation off.
6. Adjust output of signal generator until AC voltmeter reads 0.2V.
7. Adjust coils L201, L202, and L203 (in that order) for maximum quieting (lowest meter reading). Adjust the signal generator output to maintain voltmeter reading between 0.1V and 0.2V. Repeat adjustments until no further improvements can be made.
8. Activate channel on 300-series board that will be used for alignment.
9. With no input signal, adjust VOLUME control for 1V of noise on AC voltmeter at speaker terminals.
10. Connect signal generator to antenna input jack of 300-series board. Set generator accurately to the frequency of the channel being used. Turn modulation off.
11. Adjust output signal generator until AC voltmeter reads 0.2V.
12. Adjust T301 for maximum quieting (lowest meter reading). Adjust signal generator output to maintain voltmeter reading between 0.1V and 0.2V. Adjust core to peak away from center of coil form.
13. Adjust coils L301, L302, and L303 (in that order) for maximum quieting (lowest meter reading). Adjust signal generator output to maintain voltmeter reading between 0.1V and 0.2V. Repeat adjustments until no further improvement can be made.

T—Alignment, AFC (TME-8H/L/U, -8H/U)

Note: This adjustment requires an accurate 10.7 MHz ±1 kHz oscillator or 455 kHz ±500 Hz oscillator to be used as a reference signal. If none is available, proceed to step T-5.

1. Prealign quadrature detector by tuning L103 for maximum noise (AC voltmeter reading) at the speaker terminals.
2. With a coupling loop inject reference signal (either 10.7 MHz or 455 kHz) to produce good quieting (more than 30 dB quieting). Adjust R326 for reading of 3.8V to 4.0V at the collector of Q304.
3. Remove the reference signal and have the unit squelched and receiving no signal. The voltage on the collector of Q304 should be between 3.2V and 4.6V. If not, note voltage and proceed to step T-4. If voltage is between 3.2V and 4.6V, AFC alignment is complete.
   Note: Any further adjustments made to L103 and R326 will require AFC to be readjusted.
4. Inject reference signal and monitor voltage on collector of Q304 and adjust

L103 for same voltage as noted in step 3. Readjust R326 for a voltmeter reading of 3.8V to 4.0V. Repeat step T-3.

Do not adjust L103 more than one-fourth turn at a time.

5. If an accurate IF signal source is not available, an approximate AFC alignment can be made by adjusting L103 for maximum noise (AC voltmeter reading) at the speaker terminals; with the unit squelched and receiving no signal, adjust R326 for voltmeter reading of 3.2V to 4.6V on the collector of Q304.

Note: Units equipped with a 10.245 MHz crystal have the jumper in the AFC circuit connected between the base of Q304 and collector of Q306. When a 11.155 MHz crystal is used, the jumper is connected between the base of Q304 and the collector of Q305. If the crystal is changed from one frequency to the other, the jumper must be changed.

Parts List

Item No.	Description	Part No.
RF BOARD (HIGH–LOW) 301-563		
	RESISTORS	
R201	1.8K, 10%, ½W	
R202	10K, 10%, ¼W	
R203	10K, 10%, ¼W	
R204	680ohm, 10%, ¼W	
R205	100ohm, 10%, ¼W	
R206	1.8K, 10%, ¼W	
R207	68ohm, 10%, ¼W	
R208	39K, 10%, ¼W	
R209	10K, 10%, ¼W	
R210	2.7K, 10%, ¼W	
R211	2.7K, 10%, ¼W	
R212	2.7K, 10%, ¼W	
R213	2.7K, 10%, ¼W	
R214	2.7K, 10%, ¼W	
R215	2.7K, 10%, ¼W	
R216	2.7K, 10%, ¼W	
R217	2.7K, 10%, ¼W	
R218	10K, 10%, ¼W	
R219	39K 10%, ¼W	
R220	330ohm, 10%, ¼W	
R221	68ohm, 10%, ¼W	
R222	22K, 10%, ¼W	
R223	10K, 10%, ¼W	
R224	680ohm, 10%, ¼W	
R225	220ohm, 10%, ¼W	
R226	33ohm, 10%, ¼W	
R227	10K, 10%, ¼W	
R228	10K, 10%, ¼W	
R229	680ohm, 10%, ¼W	
R230	100ohm, 10%, ¼W	
R231	68ohm, 10%, ¼W	
R232	10K, 10%, ¼W	
R233	22K, 10%, ¼W	
R234	39K, 10%, ¼W	
R235	10K, 10%, ¼W	
	CAPACITORS	
C201	68pf, 5%, 50V (MICA)	DM-10 or Equiv.
C202	10pf, 10% NPO (DISC)	RMC-Type CG
C203	56pf, 5% 50V (MICA)	DM-10 or Equiv.
C204	22pf, 10% NPO (DISC)	RMC-Type CG
C205	.001mf, +80%-20%, 500VZ5U (DISC)	RMC-Type BG
C206	68pf, 5% 50V (MICA)	DM-10 or Equiv.
C207	.005mf, +80%-20%, 500VZ5U (DISC)	RMC-Type SM
C208	.005mf, +80%-20%, 500VZ5U (DISC)	RMC-Type SM
C209	82pf, 5%, 50V (MICA)	DM-10 or Equiv.
C210	68pf, 5% 50V (MICA)	DM-10 or Equiv.
C211	15pf, 10% NPO (DISC)	RMC-Type CG
C212	.01mf, +80%-20%, 500VZ5U (DISC)	RMC-Type BG
C213	.001mf, +80%-20%, 500V (DISC)	RMC-Type BG
C214	10pf, 10% NPO (DISC)	RMC-Type CG
C215	2.2pf, 10% NPO (DISC)	RMC-Type CG
C216	.01mf, +80%-20%, 500VZ5U (DISC)	RMC-Type BG
C217	390pf, 5% 50V (MICA)	DM-10 or Equiv
C218	.01mf, +80%-20%, 500VZ5U (DISC)	RMC-Type BG
C219	.01mf, +80%-20%, 500VZ5U (DISC)	RMC-Type BG
C220	8.2pf, 10% NPO (DISC)	RMC-Type CG
C221	.47pf, 10% (composition)	
C222	5.6pf, 10% NPO (DISC)	RMC-Type CG
C223	3.9pf, 10% NPO (DISC)	RMC-Type CG
C224	.001, +80%-20%, 500VZ5U (DISC)	RMC-Type BG
C225	5.6pf, 10% NPO (DISC)	RMC-Type CG
C226	.005mf, +80%-20%, 500VZ5U (DISC)	RMC-Type SM
C227	.01MF, +80%-20%, 500V Z5U (DISC)	RMC-Type BG
	COILS	
L201	Coil, Ant. (YEL)	301-520-4
L202	Coil, RF Input (GRN)	301-520-5
L203	Coil, RF Output (BLU)	301-520-6
L204	Coil, RF Inj. (WHT)	301-520-9
L205	Coil, Osc.	102-369
L206	Coil, Ant (BRN)	301-520-1
L207	Coil R.F. Input (RED)	301-520-2
L208	Coil RF Output (ORG)	301-520-3
T201	Coil Mixer Output	102-405
	TRANSISTORS	
Q201	Silicon NPN, 2N5222	SPS-1473 RT
Q202	Silicon NPN, 2N5222	SPS-1473 RT
Q203	Silicon NPN, 2N5222	SPS-1473 RT

Item	Description	Part No.

RF BOARD 301-563

Item	Description	Part No.
Q204	Silicon NPN, 2N5222	SPS-1473 RT
Q205	Silicon PNP, 2N5227	SPS-1539 WT
Q206	Silicon PNP, 2N5227	SPS-1539 WT
Q207	Silicon NPN, 2N5230 (Low Beta)	SM-4304-S
	Note: RT=red top, WT=white top	

RESISTORS

Item	Description	Part No.
R201	1.8K, 10%, ¼ W	
R202	10K, 10%, ¼ W	
R203	10K, 10%, ¼ W	
R204	680 ohm, 10%, ¼ W	
R205	100 ohm, 10%, ¼ W	
R206	1.8K, 10%, ¼ W	
R207	68 ohm, 10%, ¼ W	
R208	10K, 10%, ¼ W	
R209	680 ohm, 10%, ¼ W	
R210	22K, 10%, ¼ W	
R211	220 ohm, 10%, ¼ W	
R212	33 ohm, 10%, ¼ W	
R213	10K, 10% ¼ W	
R214	68 ohm, 10%, ¼ W	
R215	39K, 10%, ¼ W	
R216	330 ohm, 10%, ¼ W	
R217	2.7K, 10%, ¼ W	
R218	2.7K, 10%, ¼ W	
R219	2.7K, 10%, ¼ W	
R220	2.7K, 10%, ¼ W	
R221	2.7K, 10%, ¼ W	
R222	2.7K, 10%, ¼ W	
R223	27K, 10%, ¼ W	
R224	2.7K, 10%, ¼ W	
R225	39K, 10%, ¼ W	

CAPACITORS

Item	Description	Part No.
C201	68Pf, 5% 50V (MICA)	DM-10 or Equiv.
C202	10Pf, 10%, NPO (DISC)	RMC-Type CG
C203	56Pf, 5%, 50V (MICA)	DM-10 or Equiv.
C204	22Pf, 10%, NPO (DISC)	RMC-Type CG
C205	.001 mf, +80%-20%, 500V Z5U (DISC)	RMC-Type BG
C206	68pf, 5% 50V (MICA)	DM-10 or Equiv.
C207	.005mf, +80%-20%, 500V Z5U (DISC)	RMC-Type SM
C208	.005mf, +80%-20%, 500V Z5U (DISC)	RMC-Type SM
C209	.01mf, +80%-20%, 500V Z5U (DISC)	RMC-Type BG
C210	390Pf, 5% 50V (MICA)	DM-10 or Equiv.
C211	.01mf, +80%-20%, 500V Z5U (DISC)	RMC-Type BG
C212	.01mf, +80%-20%, 500V Z5U (DISC)	RMC-Type BG
C213	15pf, 10%, NPO (DISC)	RMC-Type CG
C214	68pf, 5% 50V (MICA)	DM-10 or Equiv.
C215	.01mf, +80%-20%, 500V Z5U (DISC)	RMC-Type BG
C216	.001mf, +80%-20%, 500V Z5U (DISC)	RMC-Type BG
C217	82Pf, 5% 50V (MICA)	DM-10 or Equiv.

COILS

Item	Description	Part No.
L201	Coil, Ant (YEL)	301-5204
L202	Coil, RF Input (GRN)	301-5205
L203	Coil, RF Output (BLU)	301-5206
L204	Coil, OSC	102-369
T201	Coil, Mixer Output	102-405

TRANSISTORS

Item	Description	Part No.
Q201	Silicon NPN, 2N5222	SPS-1473 (RT)
Q202	Silicon NPN, 2N5222	SPS-1473 (RT)
Q203	Silicon NPN, 2N5130 (low Beta)	SM-4304-S
Q204	Silicon PNP, 2N5227	SPS-1539 (WT)
	NOTE: RT = Red Top WT = White Top	

DIODES

Item	Description	Part No.
CR201	Germanium-Junction, Signal	102-339
CR202	Germanium-Junction, Signal	102-339
CR203	Germanium-Junction, Signal	102-339
CR204	Germanium-Junction, Signal	102-339
CR205	Germanium-Junction, Signal	102-339
CR206	Germanium-Junction, Signal	102-339
CR207	Germanium-Junction, Signal	102-339
CR208	Germanium-Junction, Signal	102-339
CR209	Silicon, General Purpose	102-405
CR211	Silicon, General Purpose	102-339
CR212	Silicon, General Purpose	102-339
CR213	Silicon, General Purpose	102-339
CR214	Silicon, General Purpose	102-339
CR215	Silicon, General Purpose	102-339
CR216	Silicon, General Purpose	102-339

RF BOARD (UHF) 301-598

RESISTORS

Item	Description	Part No.
R301	220 ohm, 10%, ¼W	
R302	22K, 10%, ¼W	
R303	10K, 10%, ¼W	
R304	680 ohm, 10%, ¼W	
R305	220 ohm, 10%, ¼W	
R306	33 ohm, 10%, ¼W	
R307	10K, 10%, ¼W	
R308	38K, 10%, ¼W	
R309	470 ohm, 10%, ¼W	
R310	39K, 10%, ¼W	
R311	39K, 10%, ¼W	
R312	82K, 10%, ¼W	
R313	4.7K, 10%, ¼W	
R314	1.2K, 10%, ¼W	
R315	1.2K, 10%, ¼W	
R316	1.2K, 10%, ¼W	
R317	1.2K, 10%, ¼W	
R318	1.2K, 10%, ¼W	
R319	1.2K, 10%, ¼W	
R320	1.2K, 10%, ¼W	
R321	1.2K, 10%, ¼W	
R322	82K, 10%, ¼W	

Item No.	Description	Part No.
R323	10K, 10%, ¼W	
R324	1K, 10%, ¼W	
R325	1K, 10%, ¼W	
R326	Trimmer 1K, 30% (Stand Up)	X201R102B
R327	820 ohm, 10%, ¼W	
R328	1K, 10%, ¼W	
R329	3.3K, 10%, ¼W	
R330	150 ohm, 10%, ¼W	
R332	1K, 10%, ¼W	
R332	3.3K, 10%, ¼W	
R333	47K, 10%, ¼W	
R324	33 ohm, 10%, ¼W	

CAPACITORS

Item No.	Description	Part No.
C301	2-8pf, NPO, Trimmer	10S-Triko 22
C302	2-8pf, NPO, Trimmer	10S-Triko 22
C303	.005MF, +80%-20% 500V, Z5U (Disc.)	RMC-Type SM
C304	.005MF, +80%-20% 500V, Z5U (Disc.)	RMC-Type SM
C305	.005MF, +80%-20% 500V, Z5U (Disc.)	RMC-Type SM
C306	150pf, 5% 50V (Mica)	DM-10 or Equiv
C307	.01MF, +80%-20% 500V, Z5U (Disc)	RMC-Type BG
C308	.005MF, +80%-20% 500V, Z5U (Disc.)	RMC-Type SM
C309	.001MF, +80%-20% 500V, Z5U (Disc.)	RMC-Type BG
C310	82pf, 5% 50V (Mica)	DM-10 or equiv.
C311	15pf, 10%, NPO (Disc.)	RMC-Type CG
C312	15pf, 10% NPO (Disc.)	RMC-Type CG
C313	.001MF, +80%-20% 500V, Z5U (Disc.)	RMC-Type BG
C314	.001MF, +80%-20% 500V, Z5U (Disc.)	RMC-Type BG
C315	.001MF, +80%-20% 500V, Z5U (Disc.)	RMC-Type BG
C316	.005MF, +80-20% 500V, Z5U (Disc)	RMC-Type SM
C317	2-8pf, NPO, Trimmer	10S Triko 22
C318	.47pf, 10% (Composition)	
C319	1.0mf 85°C 50V (Electrolytic)	
C320	.001MF, +80%-20% 500V, Z5U (Disc.)	RMC-Type BG
C322	.2 MF, +80%-20% 12V (Disc.)	Murata Type BC-12

COILS

Item No.	Description	Part No.
L301	Coil, RF Input	301-600-1
L302	Coil, RF Output	301-600-2
L303	Coil, RF (Red)	301-520-2
L304	Coil, RF Osc.	102-369
L305	Coil, RF Osc.	301-600-3
L306	Choke, 15uh	
L307	Choke, 15uh	
L308	Choke, 15uh	
L309	Choke, 15uh	
L310	Choke, 15uh	
L311	Choke, 15uh	
L312	Choke 15uh	
L313	Choke, 15uh	
T301	Coil, Mixer	301-619

TRANSISTOR

Item No.	Description	Part No.
Q301	Silicon, N-Channel FET	2N5245
Q302	Silicon NPN, 2N5222	SPS-1473 (RT)
Q303	Silicon NPN (Low Beta) 2N5130	SM-4304-S
Q304	Silicon PNP, 2N5227	SPS-1539 (WT)
Q305	Silicon NPN, MPS-5172	SPS-952
Q306	Silicon NPN, MPS-5172	SPS-952
Q307	Silicon NPN, 2N5222	SPS-1473 (RT)

NOTE: RT = Red Top, WT = White Top

RF BOARD (HIGH) 301-563

RESISTORS

Item No.	Description	Part No.
R201	10K, 10%, ¼W	
R202	10K, 10%, ¼W	
R203	680 ohm, 10%, ¼W	
R204	100 ohm, 10%, ¼W	
R205	68 ohm, 10%, ¼W	
R206	22K, 10%, ¼W	
R207	680 ohm, 10%, ¼W	
R208	10K, 10%, ¼W	
R209	220 ohm, 10%, ¼W	
R210	33 ohm, 10%, ¼W	
R211	68 ohm, 10%, ¼W	
R212	10K, 10%, ¼W	
R213	39K, 10%, ¼W	
R214	470 ohm, 10%, ¼W	
R215	2.7K, 10%, ¼W	
R216	2.7K, 10%, ¼W	
R217	2.7K, 10%, ¼W	
R218	2.7K, 10%, ¼W	
R219	2.7K, 10%, ¼W	
R220	2.7K, 10%, ¼W	
R221	2.7K, 10%, ¼W	
R222	2.7K, 10%, ¼W	

CAPACITORS

Item No.	Description	Part No.
C201	8.2pf, 10%, NPO (DISC)	RMC-Type CG
C202	.47pf, 10% (COMPOSITION)	
C203	5.6pf, 10% NPO (DISC)	RMC-Type CG
C204	3.9pf, 10% NPO (DISC)	RMC-Type CG
C205	.001mf, +80%-20%, 500V Z5U (DISC)	RMC-Type BG
C206	5.6pf, 10% NPO (DISC)	RMC-Type CG
C207	.005mf, +80%-20%, 500V Z5U (DISC)	RMC-Type SM
C208	.01mf, +80%-20%, 500V Z5U (DISC)	RMC-Type BG
C209	.01mf, +80%-20%, 500V Z5U (DISC)	RMC-Type BG
C210	390pf, 5%, 50V (MICA)	DM-10 or Equiv.
C211	.01mf, +80%-20%, 500V Z5U (DISC)	RMC-Type BG
C212	.01mf, +80%-20%, 500V Z5U (DISC)	RMC-Type BG
C213	2.2pf, 10% NPO (DISC)	RMC-Type CG
C214	15pf, 10% NPO (DISC)	RMC-Type CG
C215	68pf, 5% NPO (MICA)	DM-10 or Equiv.
C216	.001mf, +80%-20%, 500V Z5U (DISC)	RMC-Type BG
C217	.01mf, +80%-20%, 500V Z5U (DISC)	RMC-Type BG
C218	82pf, 5% 50V (MICA)	DM-10 or Equiv.

Item No.	Description	Part No.
	COILS	
L201	Coil, Ant (BRN)	301-520-1
L202	Coil, RF Input (RED)	301-520-2
L203	Coil, RF Output (ORG)	301-520-3
L204	Coil, Inj. (WHT)	301-520-9
L205	Coil, OSC.	102-369
T201	Coil, Mixer Output	102-405
	TRANSISTORS	
Q201	Silicon NPN, 2N5222	SPS 1473 (RT)
Q202	Silicon NPN, 2N5222	SPS-1473 (RT)
Q203	Silicon NPN, 2N5130 (low Beta)	SM-4304-S
	NOTE: RT = Red Top	
	DIODES	
CR201	Germanium-Junction, Signal	102-339
CR202	Germanium-Junction, Signal	102-339
CR203	Germanium-Junction, Signal	102-339
CR204	Germanium-Junction, Signal	102-339
CR205	Germanium-Junction, Signal	102-339
CR206	Germanium-Junction, Signal	102-339
CR207	Germanium-Junction, Signal	102-339
CR208	Germanium-Junction, Signal	102-339

RF BOARD 300-598

Item No.	Description	Part No.
	RESISTORS	
R201	220Ω, 10%, ¼W	
R202	22K, 10%, ¼W	
R203	10K, 10%, ¼W	
R204	680Ω, 10%, ¼W	
R205	220Ω, 10%, ¼W	
R206	33Ω, 10%, ¼W	
R207	10K, 10%, ¼W	
R208	39K, 10%, ¼W	
R209	470Ω, 10%, ¼W	
R210	39K, 10%, ¼W	
R211	39K, 10%, ¼W	
R212	82K, 10%, ¼W	
R213	4.7K, 10%, ¼W	
R214	1.2K, 10%, ¼W	
R215	1.2K, 10%, ¼W (TMR-8U, Only)	
R216	1.2K, 10%, ¼W (TMR-8U, Only)	
R217	1.2K, 10%, ¼W (TMR-8U, Only)	
R218	1.2K, 10%, ¼W (TMR-8U, Only)	
R219	1.2K, 10% ¼W (TMR-8U, Only)	
R220	1.2K, 10%, ¼W (TMR-8U, Only)	
R221	1.2K, 10%, ¼W (TMR-8U, Only)	
R222	82K, 10%, ¼W	
R223	10K, 10%, ¼W	
R224	1K, 10%, ¼W	
R225	1K, 10%, ¼W	
R226	Trimmer 1K, 30% (Stand Up)	X201R102B

Item No.	Description	Part No.
R227	820Ω, 10%, ¼W	
R228	1K, 10%, ¼W	
R229	3.3K, 10%, ¼W	
R230	150Ω, 10%, ¼W	
R231	1K, 10%, ¼W	
R232	3.3K, 10%, ¼W	
R233	47K, 10%, ¼W	
R234	33Ω, 10%, ¼W	
	CAPACITORS	
C201	2-8pf, NPO, Trimmer	10S-Triko 22
C202	2-8pf, NPO, Trimmer	10S-Triko 22
C203	.005MF, +80% -20% 500V, Z5U (Disc.)	RMC-Type SM
C204	.005MF, +80% -20% 500V, Z5U (Disc.)	RMC Type SM
C205	.005MF, +80% -20% 500V, Z5U (Disc.)	RMC-Type SM
C206	150pf, 5% 50V (Mica)	DM-10 or Equiv.
C207	.01MF, +80% -20% 500V, Z5U (Disc)	RMC-Type BG
C208	.005MF, +80% -20% 500V, Z5U (Disc.)	RMC-Type SM
C209	.001MF, +80% -20% 500V, Z5U (Disc.)	RMC-Type BG
C210	82pf, 5% 50V (Mica)	DM-10 or equiv.
C211	15pf, 10%, NPO (Disc.)	RMC-Type CG
C212	15pf, 10%, NPO (Disc.)	RMC-Type CG
C213	.001MF, +80% -20% 500V, Z5U (Disc.)	RMC-Type BG
C214	.001MF, +80% -20% 500V, Z5U (Disc.)	RMC-Type BG
C215	.001MF, +80% -20% 500V, Z5U (Disc.)	RMC-Type BG
C216	.005MF, +80% -20% 500V, Z5U (Disc.)	RMC-Type SM
C217	2-8pf, NPO, Trimmer	10S Triko 22
C218	.47pf, 10% (Composition)	
C219	1.0mf 85°C 50V (Electrolytic)	
C220	.001MF, +80% -20% 500V, Z5U (Disc.)	RMC-Type BG
C221	5.6pf, 10% NPO (Disc.) (TMR-1U Only)	RMC-Type CG
C222	.2MF, +80% -20% 12V (Disc.)	Murata Type BC-12
	COILS	
L201	Coil, RF Input	301-600-1
L202	Coil, RF Output	301-600-2
L203	Coil, RF (Red)	301-520-2
L204	Coil, RF Osc.	102-369
L205	Coil, RF Osc.	301-600-3
L206	Choke, 15uh (TMR-8U Only)	
L207	Choke, 15uh (TMR-8U Only)	
L208	Choke, 15uh (TMR-8U Only)	
L209	Choke, 15uh (TMR-8U Only)	
L210	Choke, 15uh (TMR-8U Only)	
L211	Choke, 15uh (TMR-8U Only)	
L212	Choke, 15uh (TMR-8U Only)	
L213	Choke, 15uh (TMR-8U Only)	
T201	Coil, Mixer	301-619
	TRANSISTOR	
Q201	Silicon, N-Channel FET	2N5245
Q202	Silicon NPN, 2N5222	SPS-1473 (RT)

Item No.	Description	Part No.
Q203	Silicon NPN (Low Beta) 2N5130	SM-4304-S
Q204	Silicon PNP, 2N5227	SPS-1539 (WT)
Q205	Silicon NPN, MPS 5172	SPS-952
Q206	Silicon NPN, MPS 5172	SPS-952
Q207	Silicon NPN, 2N5222	SPS-1473 (RT)

NOTE: RT = Red Top, WT = White Top

DIODES

Item No.	Description	Part No.
CR201	Germanium-Junction, Signal	102-339
CR202	Germanium-Junction, Signal (TMR-8U Only)	102-339
CR203	Germanium-Junction, Signal	102-339
CR204	Germanium-Junction, Signal	102-339
CR205	Germanium-Junction, Signal	102-339
CR206	Germanium-Junction, Signal	102-339
CR207	Germanium-Junction, Signal	102-339
CR208	Germanium-Junction, Signal	102-339
CR209	Silicon, Varactor	Motorola MV2209

MISCELLANEOUS

Item No.	Description	Part No.
	Shield Assembly	301-602

IF – AUDIO BOARD 500-858 (EARLY)

RESISTORS

Item No.	Description	Part No.
R101	18K, 10%, ¼W	
R102	10K, 10%, ¼W	
R103	1K, 10%, ¼W	
R104	180 ohm, 10%, ¼W	
R105	1K, 10%, ¼W W	
R106	180 ohm, 10%, ¼W	
R107	22K, 10%, ¼W	
R108	47K, 10%, ¼W	
R109	33 ohm, 10%, ¼W	
R110	270K, 10%, ¼W	
R111	4.7K, 10%, ¼W	
R112	8.2K, 10%, ¼W	
R113	2.2K, 10%, ¼W	
R114	39K, 10%, ¼W	
R115	10K, 10%, ¼W	
R116	5.6, 10%, ¼W	
R117	2.2K, 10%, ¼W	
R118	100ohm, 10%, ¼W	
R119	10K, 10%, ¼W	
R120	10K, 10%, ¼W	
R122	5.6K, 10%, ¼W	
R123	10K, 10%, ¼W	
R124	3.3K, 10%, ¼W	
R125	180 ohm, 10%, ¼W	
R126	560 ohm, 10%, ¼W	
R127	1K, 10%, ¼W	
R128	33K, 10%, ¼W	
R129	68K, 10%, ¼W	
R130	18K, 10%, ¼W	
R131	2.2K, 10%, ¼W	
R132	2.2K, 10%, ¼W	
R133	56 ohm, 10%, ¼W	
R134	560 ohm, 10%, ¼W	
R135	1.5K, 10%, ¼W	
R136	100 ohm, 10%, ¼W	
R137	43 ohm, 5%, 2W (Wire Wound) IRC-Type PW-2	

CAPACITORS

Item No.	Description	Part No.
C101	220pf, 5% 50V (Mica)	DM-10 or equiv.
C102	.01mf, 10% 100V (Mylar Film)	
C103	.01mf, +80%-20% 500V Z5U (Disc.)	RMC-Type BG
C104	.01mf, 10%, 100V (Mylar Film)	
C105	220pf, 5% 50V (Mica)	DM-10 or equiv.
C106	270pf, 5% 50V (Mica)	DM-10 or equiv.
C107	.01mf+80%-20% 500V Z5U (Disc.)	RMC-Type BG
C108	.01mf+80%-20% 500V Z5U (Disc.)	RMC-Type BG
C109	68pf, 5% 50V (Mica)	DM-10 or equiv.
C110	.2mf, +80%-20% 12V (Disc.)	Murata Type BC-12
C111	180pf, 5% 50V (Mica)	DM-10 or equiv.
C112	390pf, 5% 50V (Mica)	DM-10 or equiv.
C113	270pf, 5% 50V (Mica)	DM-10 or equiv.
C114	250pf, 5% 50V (Mica)	DM-10 or equiv.
C115	.2mf, +80%-20% 12V (Disc)	Murata Type BC-1
C116	.2mf, +80%-20% 12V (Disc.)	Murata Type BC-1
C117	.2mf, +80%-20% 12V (Disc.)	Murata Type BC-1
C118	39pf, 10% NPO (Disc.)	RMC-Type CG
C119	.002mf, 20% 500V Z5U (Disc.)	RMC-Type JG
C120	.047mf, 10% 100V (Mylar Film)	
C121	.01mf, 10% 100V (Mylar Film)	
C122	470pf, 20% 500V Z5F (Disc.)	RMC-Type JG
C123	.015mf, 10% 100V (Mylar Film)	
C124	.002mf, 20% 500V Z5U (Disc.)	RMC-Type JG
C125	.047mf, 10% 100V (Mylar Film)	
C126	.1mf, 20% 12V (Disc.)	Murata Type BC-1
C127	1mf, 85°C 50V (Electrolytic)	
C128	.047mf, 10% 100V (Mylar Film)	
C129	.047mf, 10% 100V (Mylar Film)	
C130	50mf, 85°C 10V (Electrolytic)	
C131	.001+80%-20% 500V Z5U (Disc.)	RMC-Type BG
C132	.001+80%-20% 500V Z5U (Dsic.)	RMC-Type BG
C133	.001+80%-20% 500V Z5U (Disc.)	RMC-Type BG
C134	.2mf, +80%-20% 12V (Disc.)	Murata Type BC-1
C135	250mf, 85°C 16V (Electrolytic)	
C136	1000mf, 85°C 16V (Electrolytic)	
C137	250mf, 85°C 10V (Electrolytic)	

COILS

Item No.	Description	Part No.
T101	Coil, 10.7 MHz Input	102-507
T102	Coil, 10.7 MHz Output	301-730

Item No.	Description	Part No.
L101	Choke, 820uh	ES-2228
L102	Choke, 820uh	ES-2228
L103	Coil, Quadrature	301-517

INTEGRATED CIRCUITS

Item No.	Description	Part No.
IC101	Integrated Circuit	301-679-1
IC102	Integrated Circuit, MC-1357P	301-576-3
IC101	**(Some models)**	**MC1550G**

DIODES

Item No.	Description	Part No.
C R101	Diode, Silicon, 1N4148	102-412
CR102	Diode, Zener, 8.2V 5% 1W.	IN4738A
CF103	Diode, Rectifier	1N4002
CR104	Diode, Rectifier	1N4002

FILTER

Item No.	Description	Part No.
CF-1	455KHz Ceramic Filter (TME-16H/L/U only)	301-724
CF-1	455KHz Ceramic Filter (TME-16H/L only)	301-723

CRYSTAL

Item No.	Description	Part No.
Y101	10.245 MHz or	301-516-1
	11.155 MHz	301-516-2

TRANSISTORS

Item No.	Description	Part No.
Q101	Silicon NPN, MPS 5172	SPS-952
Q102	Silicon NPN, MPS 5172	SPS-952
Q103	Silicon PNP, 2N5227	SPS-1539 (WT)
Q104	Silicon NPN, MPS 5172	SPS-952
Q105	Silicon NPN, MPS 5172	SPS-952
Q106	Silicon NPN, MPS 5172	SPS-952
Q107	Silicon PNP	MPS-A55
Q108	Silicon PNP	MPS-A55
Q109	Silicon NPN, AF Power	MJE-521
Q110	Silicon NPN, AF Power	MJE-521

NOTE: WT=White Top

IF – AUDIO BOARD 500-858 (LATE)

RESISTORS

Item No.	Description	Part No.
R101	18K, 10%, ¼W	
R102	10K, 10%, ¼W	
R103	1K, 10%, ¼W	
R104	180 ohm, 10%, ¼W	
R105	1K, 10%, ¼W W	
R106	180 ohm, 10%, ¼W	
R107	22K, 10%, ¼W	
R108	47K, 10%, ¼W	
R109	33 ohm, 10%, ¼W	
R110	270K, 10%, ¼W	
R111	4.7K, 10%, ¼W	
R112	8.2K, 10%, ¼W	
R113	2.2K, 10%, ¼W	
R114	39K, 10%, ¼W	
R115	10K, 10%, ¼W	
R116	5.6, 10%, ¼W	
R117	2.2K, 10%, ¼W	
R118	100ohm, 10%, ¼W	
R119	10K, 10%, ¼W	
R120	10K, 10%, ¼W	
R122	5.6K, 10%, ¼W	
R123	10K, 10%, ¼W	
R124	3.3K, 10%, ¼W	
R125	180 ohm, 10%, ¼W	
R126	560 ohm, 10%, ¼W	
R127	1K, 10%, ¼W	
R128	33K, 10%, ¼W	
R129	68K, 10%, ¼W	
R130	18K, 10%, ¼W	
R131	2.2K, 10%, ¼W	
R132	2.2K, 10%, ¼W	
R133	56 ohm, 10%, ¼W	
R134	560 ohm, 10%, ¼W	
R135	1.5K, 10%, ¼W	
R136	100 ohm, 10%, ¼W	
R137	68 ohm, 10% 2W (Wire Wound) IRC-Type BWH	

CAPACITORS

Item No.	Description	Part No.
C101	220pf, 5% 50V (Mica)	DM-10 or equiv.
C102	.01mf, 10% 100V (Mylar Film)	
C103	.01mf, +80%-20% 500V Z5U (Disc.)	RMC-Type BG
C104	.01mf, 10%, 100V (Mylar Film)	
C105	220pf, 5% 50V (Mica)	DM-10 or equiv.
C106	27pf, 5% 50V (Mica)	DM-10 or equiv.
C107	.01mf+80%-20% 500V Z5U (Disc.)	RMC-Type BG
C108	.01mf+80%-20% 500V Z5U (Disc.)	RMC-Type BG
C109	68pf, 5% 50V (Mica)	DM-10 or equiv.
C110	.2mf, +80%-20% 12V (Disc.)	Murata Type BC-12
C111	180pf, 5% 50V (Mica)	DM-10 or equiv.
C112	390pf, 5% 50V (Mica)	DM-10 or equiv.
C113	270pf, 5% 50V (Mica)	DM-10 or equiv.
C114	250pf, 5% 50V (Mica)	DM-10 or equiv.
C115	.2mf, +80%-20% 12V (Disc)	Murata Type BC-12
C116	.2mf, +80%-20% 12V (Disc.)	Murata Type BC-12
C117	.2mf, +80%-20% 12V (Disc.)	Murata Type BC-12
C118	39pf, 10% NPO (Disc.)	RMC-Type CG
C119	.002mf, 20% 500V Z5U (Disc.)	RMC-Type JG
C120	.047mf, 10% 100V (Mylar Film)	
C121	.01mf, 10% 100V (Mylar Film)	
C122	470pf, 20% 500V Z5F (Disc.)	RMC-Type JG
C123	.015mf, 10% 100V (Mylar Film)	
C124	.002mf, 20% 500V Z5U (Disc.)	RMC-Type JG
C125	.047mf, 10% 100V (Mylar Film)	
C126	.1mf, 20% 12V (Disc.)	Murata Type BC-12
C127	1mf, 85°C 50V (Electrolytic)	

Item No.	Description	Part No.
C128	.047mf, 10% 100V (Mylar Film)	
C129	.047mf, 10% 100V (Mylar Film)	
C130	50mf, 85°C 10V (Electrolytic)	
C131	.001+80%-20% 500V Z5U (Disc.)	RMC-Type BG
C132	.001+80%-20% 500V Z5U (Dsic.)	RMC-Type BG
C133	.001+80%-20% 500V Z5U (Disc.)	RMC-Type BG
C134	.2mf, +80%-20% 12V (Disc.)	Murata Type BC-12
C135	250mf, 85°C 16V (Electrolytic)	
C136	1000mf, 85°C 16V (Electrolytic)	
C137	250mf, 85°C 10V (Electrolytic)	

COILS

Item No.	Description	Part No.
T101	Coil, 10.7 MHz Input	102-507
T102	Coil, 10.7 MHz Output	301-730
L101	Choke, 820uh	ES-2228
L102	Choke820uh	ES-2228
L103	Coil, Quadrature	301-517

INTEGRATED CIRCUITS

Item No.	Description	Part No.
IC101	Integrated Circuit	301-679-1
IC102	Integrated Circuit, MC-1357P	301-576-3

DIODES

Item No.	Description	Part No.
C R101	Diode, Silicon, 1N4148	102-412
CR102	Diode, Zener, 8.2V 5% 1W.	IN4738A
CF103	Diode, Rectifier	1N4002
CR104	Diode, Rectifier	1N4002

FILTER

Item No.	Description	Part No.
CF-1	455KHz Ceramic Filter	301-724
CF-1	455KHz Ceramic Filter	301-723

CRYSTAL

Item No.	Description	Part No.
Y101	10.245 MHz or	301-516-1
	11.155 MHz	301-516-2

TRANSISTORS

Item No.	Description	Part No.
Q101	Silicon NPN, MPS 5172	SPS-952
Q102	Silicon NPN, MPS 5172	SPS-952
Q103	Silicon PNP, 2N5227	SPS-1539 (WT)
Q104	Silicon NPN, MPS 5172	SPS-952
Q105	Silicon NPN, MPS 5172	SPS-952
Q106	Silicon NPN, MPS 5172	SPS-952
Q107	Silicon PNP	MPS-A55
Q108	Silicon PNP	MPS-A55
Q109	Silicon NPN, AF Power	MJE-521
Q110	Silicon NPN, AF Power	MJE-521

NOTE: WT=White Top

Item No.	Description	Part No.

SCANNER BOARD 500-841

RESISTORS

Item No.	Description	Part No.
R301	33 ohm, 10%, ½W	
R302	150 ohm, 10%, 2W	
R303	1.2K, 10%, ¼W (TMR-8H only)	
R304	1.2K, 10%, ¼W (TMR-8H only)	
R305	10K, 10%, ¼W (TMR-8H only)	
R306	10K, 10%, ¼W (TMR-8H only)	
R307	10K, 10%, ¼W	
R308	10K, 10%, ¼W	
R309	1.2K, 10%, ¼W	
R310	1.2K, 10%, ¼W	
R311	10K, 10%, ¼W	
R312	10K, 10%, ¼W	
R313	1.2K, 10%, ¼W	
R314	1 Meg., 10%, ¼W	
R315	10K, 10%, ¼W	
R316	33K, 10%, ¼W	
R317	2.2K, 10%, ¼W	
R318	12K, 10%, ¼W	
R319	39 ohm, 10%, ¼W	

CAPACITORS

Item No.	Description	Part No.
C301	.001mf +80%-20% 500V Z5U (DISC) (TMR-8H only)	RMC-Type BG
C302	.001mf +80%-20% 500V Z5U (DISC) (TMR-8H only)	RMC-Type BG
C303	.001mf, +80%-20% 500V Z5U (DISC)	RMC-Type BG
C304	.001mf, +80%-20% 500V Z5U (DISC)	RMC-Type BG
C305	20mf, 10% 85°C 16V (Electrolytic)	
C306	20mf, 10% 85°C 16V (Electrolytic)	
C308	.001mf, +80%-20% 500V Z5U (DISC)	RMC-Type BG
C309	.001mf, +80%-20% 500V Z5U (DISC)	RMC-Type BG
C310	200mf, 85°C 10V (Electrolytic)	
C312	470pf ± 20% 500V Z5F (DISC)	RMC-Type JG
C313	10mf, 85°C 10V (Electrolytic)	
C314	.047mf, 10% 100V (Mylar Film)	

INTEGRATED CIRCUITS

Item No.	Description	Part No.
IC301	Power Nand Gate (TMR-8H only)	301-576-2
IC302	Power Nand Gate	301-576-2
IC303	Quad-2 Input Nand Gate	301-576-4
IC304	Quad-2 Input Nand Gate	301-576-4

TRANSISTORS

Item No.	Description	Part No.
Q301	Silicon NPN, MPS5172 (TMR-8H only)	SPS-952
Q302	Silicon NPN, MPS5172 (TMR-8H only)	SPS-952
Q303	Silicon NPN, MPS5172	SPS-952

DIODES

Item No.	Description	Part No.
CR309	Silicon, Zener, 6.0V, 5%, ½W	MZ92-6.OB
CR311	Silicon, Signal, 1N4148	102-412
CR312	Silicon, Signal, 1N4148	102-412
CR313	Silicon, Signal, 1N4148	102-412

Item No.	Description	Part No.
	LAMPS	
M301	Incandescent, 14.4V, 120 MA.	No. 53
M302	Incandescent, 14.4V, 120 MA.	No. 53
M303	Incandescent, 14.4V, 120 MA.	No. 53
M304	Incandescent, 14.4V, 120 MA.	No. 53
M305	Incandescent, 14.4V, 120 MA. (TMR-8H only)	No. 53
M306	Incandescent, 14.4V, 120 MA. (TMR-8H only)	No. 53
M307	Incandescent, 14.4V, 120 MA. (TMR-8H only)	No. 53
M308	Incandescent, 14.4V, 120 MA. (TMR-8H only)	No. 53
	DIODES	
CR101	Diode, Silicon, 1N4148	102-412
CR102	Diode, Zener, 8.2V 5% 1W	IN4738A
CR103	Diode, Rectifier	1N4002
CR104	Diode, Rectifier	1N4002
	FILTER	
CF-1	455 KHz Ceramic Filter	301-723
	CRYSTAL	
Y101	10.245 MHz or	301-516-1
	11.155 MHz	301-516-2
	TRANSISTORS	
Q101	Silicon NPN, MPS 5172	SPS-952
Q102	Silicon NPN, MPS 5172	SPS-952
Q103	Silicon PNP, 2N5227	SPS-1539 (WT)
Q104	Silicon NPN, MPS 5172	SPS-952
Q105	Silicon NPN, MPS 5172	SPS-952
Q106	Silicon NPN, MPS 5172	SPS-952
Q107	Silicon PNP	MPS-A55
Q108	Silicon PNP	MPS-A55
Q109	Silicon NPN, AF Power	MJE-521
Q110	Silicon NPN, AF Power	MJE-521

NOTE: WT = White Top

SCANNER BOARD 500-850 (PRIMARY BOARD)

RESISTORS

Item No.	Description	Part No.
R401	33ohms, 10%, ¼W	
R402	75 ohms, 10%, 5 W Wirewound	
R403	2.2K, 10%, ¼W	
R404	47K, 10%, ¼W	
R405	10K, 10%, ¼W	
R406	2.2K, 10%, ¼W	
R407	10K, 10%, ¼W	
R408	68K, 10%, ¼W	
R409	33K, 10%, ¼W	
R410	22K, 10%, ¼W	
R411	39K, 10%, ¼W	
R412	47K, 10%, ¼W	
R413	390 ohms, 10%, ¼W	
R414	3.3K, 10%, ¼W	
R415	100 ohms, 10%, ¼W	
R416	3.3K, 10%, ¼W	
R417	3.3K, 10%, ¼W	
R418	2.2K, 10%, ¼W	
R419	12K, 10%, ¼W	
R420	10K, 10%, ¼W	
R421	22K, 10%, ¼W	
	CAPACITORS	
C401	.047mf, 20%, 100V (Mylar Film)	
C402	1mf, 80°C. 25V (Electrolytic)	
C403	10mf, 85°C. 25V (Electrolytic)	
C404	3.3mf, 10%, 16V Tant. (Electrolytic)	TAG43212-14133
C405	.015mf, 20%, 100V (Mylar Film)	
C406	10 mf, 85°C. 25V (Electrolytic)	
C407	5mf, 85°C. 10V (Electrolytic)	
	INTEGRATED CIRCUITS	
IC401	Power NAND Gate	301-576-2
IC402	Power NAND Gate	301-576-2
IC403	4-Bit Shift Register	301-576-6
IC404	4-Bit Shift Register	301-576-6
	TRANSISTORS	
Q401	NPN Silicon, General Purpose, MPS5172	SPS-952
Q402	NPN Silicon, General Purpose, MPS5172	SPS-952
Q403	NPN Silicon, General Purpose, MPS5172	SPS-952
Q404	NPN Silicon, General Purpose, MPS5172	SPS-952
Q405	NPN Silicon, General Purpose, MPS5172	SPS-952
Q406	PN Silicon, Unijunction	2N4871
Q407	PNP Silicon, 2N227	SPS-1539 (WT)
Q408	NPN Silicon, General Purpose, MPS5172	SPS-952

NOTE: WT - White Top

Item No.	Description	Part No.
	DIODES	
CR401	Silicon, Signal, IN4148	102-412
CR402	Silicon, Signal, IN4148	102-412
CR503	Silicon, Signal IN4148	102-412
CR404	Silicon, Signal, IN4148	102-412
CR406	Silicon, Signal, IN4148	102-412
CR407	Silicon, Signal, IN4148	102-412
CR409	Silicon, Zener, 5.1V, 5%, 1W	IN4733A
CR410	Silicon, Signal, IN4148	102-412
CR411	Silicon, Signal, IN4148	102-412
CR413	Silicon, Signal, IN4148	102-412
CR414	Silicon, Signal, IN4148	102-412

Item No.	Description	Part No.
CR415	Silicon, Signal, IN4148	102-412
CR416	Silicon, Signal, IN4148	102-412
CR417	Silicon, Signal, IN4148	102-412
CR418	Silicon, Signal, IN4148	102-412
CR419	Silicon, Signal, IN4148	102-412
CR420	Silicon, Signal, IN4148	102-412
CR421	Silicon, Signal, IN4148	102-412
CR422	Silicon, Signal, IN4148	102-412

LAMPS

Item No.	Description	Part No.
M401	Incandescent, 14.4V, 120 MA.	No. 53
M402	Incandescent, 14.4V, 120 MA.	No. 53
M403	Incandescent, 14.4V, 120 MA.	No. 53
M404	Incandescent, 14.4V, 120 MA.	No. 53
M405	Incandescent, 14.4V, 120 MA.	No. 53
M406	Incandescent, 14.4V, 120 MA.	No. 53
M407	Incandescent, 14.4V, 120 MA.	No. 53
M408	Incandescent, 14.4V, 120 MA.	No. 53

SCANNER BOARD 500-850 (SECONDARY BOARD)

RESISTORS

Item No.	Description	Part No.
R501	33 ohms, 10%, ¼W	
R502	75 ohms, 5%, 5W Wirewound	

CAPACITORS

Item No.	Description	Part No.
C501	.047 mf, 20%, 100V (Mylar Film)	

INTEGRATED CIRCUITS

Item No.	Description	Part No.
IC501	Power NAND Gate	301-576-2
IC502	Power NAND Gate	301-576-2
IC503	4-Bit Shift Register	301-576-6
IC504	4-Bit Shift Register	301-576-6

DIODES

Item No.	Description	Part No.
CR501	Silicon, Signal, IN4148	102-412
CR502	Silicon, Signal, IN4148	102-412
CR503	Silicon, Signal, IN4148	102-412
CR504	Silicon, Signal, IN4148	102-412
CR505	Silicon, Signal, IN4148	102-412
CR506	Silicon, Signal, IN4148	102-412
CR507	Silicon, Signal, IN4148	102-412
CR508	Silicon, Signal, IN4148	102-412
CR509	Silicon, Zener, 5.1V, 5%, 1W	IN4733A
CR510	Silicon, Signal, IN4148 (TME-16H/L only)	102-412
CR511	Silicon, Signal, IN4148 (TME-16H/L only)	102-412
CR512	Silicon, Signal, IN4148 (TME-16H/L only)	102-412
CR513	Silicon, Signal, IN4148 (TME-16H/L only)	102-412
CR514	Silicon, Signal, IN4148 (TME-16H/L only)	102-412
CR515	Silicon, Signal, IN4148 (TME-16H/L only)	102-412
CR516	Silicon, Signal, IN4148 (TME-16H/L only)	102-412
CR517	Silicon, Signal, IN4148 (TME-16H/L only)	102-412

LAMPS

Item No.	Description	Part No.
M501	Incandescent, 14.4V, 120 MA.	No. 53
M502	Incandescent, 14.4V, 120 MA.	No. 53
M503	Incandescent, 14.4V, 120 MA.	No. 53
M504	Incandescent, 14.4V, 120 MA.	No. 53
M505	Incandescent, 14.4V, 120 MA.	No. 53
M506	Incandescent, 14.4V, 120 MA.	No. 53
M507	Incandescent, 14.4V, 120 MA.	No. 53
M508	Incandescent, 14.4V, 120 MA.	No. 53

CHASSIS ASSEMBLY—TME-16H/L/U, —16H/L

ELECTRICAL COMPONENTS

Item No.	Description	Part No.
R1	5K, Volume Control	Stackpole R72-00691
R2	7.5K, Squelch Control	Stackpole R72-00692
R3	33 ohms, 10%, ¼W	
R4	1 Meg, 10%, 112W	
C1	.047mf, 10%, 100V (Mylar Film)	
C2	250mf, 85°C 16V (Electrolytic)	
C3	.005mf, +80%-20% 1400 Z5U (Disc.)	RMC-Type U
C4	2mf, 85°C, 50V (Electrolytic)	
T1	Transformer, Power	301-699
M1	Lamp, No. 53	
Y200	See Section 1–2	
Y300	See Section 1–2	
SW1-8, SW12-19	1P2T, 8 Stations on Single Frame, P-P	UID500-874-3 Prom. 301-551-3
SW-11	1P2T, Push-push (UL)	UID 500-874-6 Prom. 301-551-7
SW9-10	2P2T	UID 500-874-19 Prom. 301-551-20
SW20-22	2P2T (TME-16H/Lonly)	UID 500-874-13 Prom. 301-551-14
SW20-24	(TME-16H/L/U only)	UID 500-874-12 Prom. 301-551-6
Ant.-1	Telescoping Antenna (UHF)	1201-0000-003
Ant.-1	Telescoping Antenna (H/L)	1201-0000-002
Spk.-1	Speaker, 3.2 ohm, 3½" Square Assembly	301-793
J1	Antenna, Connector	Cinch 201-24-01-002
P1	Connector, Chassis, Power	Beauchaine P-3304-AB

Item No.	Description	Part No.
S1	Connector, Cable, Power	Beauchaine S-3304-FHT
	AC Power Cord Assembly	MA-1
	Cable, Coaxial 50ohm, Teflon	RG-188/U
	MECHANICAL COMPONENTS	
	Panel, Front (TME-16H/L)	600-330
	Panel, Front (TME-16H/L/U)	600-307
	Panel, Back	301-674
	Knob, Volume and Squelch	Plasticware 301-670
	Lens, Green, Power Indicator	1759
	Lens, Red, Channel Lamps	102-353
	Socket, Lamp (Power)	300-269-4
	Socket, Pins, Crystal	T35-362
	Terminal Board, 4-Lug (Rear Apron)	301-079-14
	Foot, Rubber	Lavelle 705R
	Cabinet/Wrap Assembly	600-311
	Manual, Owner's Instruction (TME-16H/L/U)	IS-10-324-1
	Manual, Owner's Instruction (TME-16H/L)	IS-10-324-2
	Manual, Service ($5.00 Prepaid)	SM-10-324-1

CHASSIS ASSEMBLY TMR-1U, -8U, 8H/L

ELECTRICAL COMPONENTS

Item No.	Description	Part No.
R1	5K, Volume Control/Switch (SW 1) (TMR-8U Only)	102-479-3
R1	5K, Volume Control/Switch (SW 1) (TMR-1U Only)	102-303-5
R2	7.5K, Squelch Control (TMR-8UOnly)	102-479-2
R2	7.5K, Squelch Control (TMR-1U Only)	102-303-6
C1	.047mf, 10%, 100V (Mylar Film)	
C2	100mf, 85°C 10V (Electrolytic)	
C3	.005mf, +80% -20% 1400V Z5U (Disc.)	RMC-Type U
TI	Transformer, Power	301-515
M1	Lamp, No. 53 (TMR-1U Only)	
Y200	Crystals, Receive (Specify frequency)	see text
SW1-8	1P2T, 8 Stations on Single Frame	UID 500-874-3 Prom. 301-551-3
SW9	1P2T, Push-push	UID 500-874-1 Prom. 301-551-1
SW10	1P2T, Momentary On Same Frame with SW9	UID 500-874-1 Prom. 301-551-1
Ant.-1	Telescoping Antenna	Reg.-1
Spk.-1	Speaker, 3.2 Ω, 4 inch Square Assembly	301-537-1
J1	Antenna, Connector	Cinch 201-24-01-002
P1	Connector, Chassis, Power	Beauchaine P-3304-AB
S1-3	Connector, Cable, Power	Beauchaine S-3304-FHT
F1	Fuse, 1½ Ampere, 3AG	
	DC Power Cord Assembly	MA-9
	AC Power Cord Assembly	MA-1
	Cable, Coaxial 50 Ω, Teflon	RG-188/U

Item No.	Description	Part No.
	MECHANICAL COMPONENTS	
	Front Panel (Casting)	500-813
	Face Plate (TMR-8U Only)	301-513
	Face Plate (TMR-1U Only)	301-535
	Knob, Volume and Squelch (TMR-8U Only)	Plasticware 27500
	Knob, Volume and Squelch (TMR-1U Only)	Phillips Series No. 3-9
	Lens, Red, Channel Lamps	102-353-1
	Socket Pins, Crystals	T35-362
	Terminal Board, 4 Lug (Rear Panel)	300-079-14
	Cabinet/Wrap Assembly	600-259
	Feet, Rubber	Atlantic India 734
	Bracket, Mobile Mounting	301-431
	Manual, Owner's Instruction (TMR-8U)	IS-10-270-3
	Manual, Owner's Instruction (TMR-1U)	IS-10-271-3
	Manual, Service ($3.00 Prepaid)	SM-10-270-3

CHASSIS ASSEMBLY—TME-8H/L, -8H

ELECTRICAL COMPONENTS

Item No.	Description	Part No.
R1	5K, Volume Control	102-479-3
R2	7.5K, Squelch Control	102-479-2
R3	1M.E.G, 10%, ½W	
C1	.047mf, 10%, 100V (Mylar Film)	
C2	100mf, 85°C 10V (Electrolytic)	
C3	.005mf, +80%-20% 1400V Z5U (Disc.)	RMC-Type U
T1	Transformer, Power	301-515
Y200	See Section 1-2	
SW1-8, SW9-10	IP2T, 8 Stations on Single Frame, P-P 2P2T	UID 500-874-19 Prom. 301-551-20
Ant.-1	Telescoping Antenna (TME-8H/L)	TAD-392450-AD
Ant.-1	Telescoping Antenna (TME-8H)	P-6-125/102
Spk.-1	Speaker, 3.2 ohms, 3½" Square Assembly	301-793
J1	Antenna, Connector	Cinch 201-24-01-002
P1	Connector, Chassis, Power	Beauchaine P-3304-AB
S1	Connector, Cable, Power	Beauchaine S-3304-FHT
	AC Power Cord Assembly	MA-1
	Cable, Coaxial 50 ohms, Teflon	RG-188/U
	MECHANICAL COMPONENTS	
	Panel, Front	600-339
	Panel, Back	301-675
	Knob, Volume and Squalch	Plasticware 40600CC
	Lens, Red, Channel Lamps	102-353-1
	Socket, Pins, Crystal	T35-362
	Terminal Board, 4-Lug (Rear Apron)	301-079-14
	Foot, Rubber	Lavelle 705R
	Cabinet/Wrap Assembly	600-312-2
	Manual, Owner's Instruction (TME-8H	IS-10-326-1
	Manual, Owner's Instruction (TME-8H/L)	IS-10-326-2
	Manual, Service ($5.00 Prepaid)	SM-10-326-1

Item No.	Description	Part No.

CHASSIS ASSEMBLY—TMR-8H, -4H, -1H

ELECTRICAL COMPONENTS

Item No.	Description	Part No.
R1	5K, Volume Control/Switch (SW 1) (TMR-8H & 4H only)	102-479-3
R1	5K, Volume Control/Switch (SW 1) (TMR-1H only)	102-303-5
R2	7.5K, Squelch Control (TMR-8H & 4H only)	102-479-2
R2	7.5K, Squelch Control (TMR-1H only)	102-303-6
R3	1 Meg., 10%, ½W	
C1	.047mf, 10%, 100V (Mylar Film)	
C2	100mf, 85°C 10V (Electrolytic)	
C3	.005mf, +80%-20% 1400V Z5U (DISC)	RMC-Type U
T1	Transformer, Power	301-515
M1	Lamp, No. 53 (TMR-1H only)	
Y200	Crystals, Receive (Specify frequency)	301-532
SW1-8	1P2T, 8 Stations on Single Frame (TMR-8H only)	UID 500-874-3 Prom. 301-551-3
SW1-4	1P2T, 4 Stations on Single Frame (TMR-4H only)	UID 500-874-2 Prom. 301-551-2
SW9	1P2T, Push-push (TMR-8H & 4H only)	UID 500-874-1 Prom. 301-551-1
SW10	1P2T, Momentary (TMR-8H & 4H only) On Same Frame with SW9	UID 500-874-1 Prom. 301-551-1
Ant.-1	Telescoping Antenna	P-6-125/102
Spk.-1	Speaker, 3.2 ohm, 4-in. Sq. Assembly	301-537-1
J1	Antenna, Connector	Cinch 201-24-01-002
P1	Connector, Chassis, Power	Beauchaine P-3304-AB
S1-3	Connector, Cable, Power	Beauchaine S-3304-FHT
F1	Fuse, 1½ Ampere, 3AG	
	DC Power Cord Assembly	MA-9
	AC Power Cord Assembly	MA-1
	Cable, Coaxial 50 ohm, Teflon	RG-188/U

MECHANICAL COMPONENTS

Description	Part No.
Front Panel (Casting)	500-813
Face Plate (TMR-8H only)	301-513
Face Plate (TMR-4H only)	301-538
Face Plate (TMR-1H only)	301-535
Knob, Volume and Squelch (TMR-8H & 4H only)	Plasticware 27500
Knob, Volume and Squelch (TMR-1H only)	Phillips Series No. 3-9
Lens, Red, Channel Lamps	102-353-1
Socket Pins, Crystals	T35-362
Terminal Board, 4 Lug (Rear Panel)	300-079-14
Cabinet/Wrap Assembly	600-259
Feet, Rubber	Atlantic India 734
Bracket, Mobile Mounting	301-431
Manual, Owner's Instruction (TMR-8H)	IS-10-270-1
Manual, Owner's Instruction (TMR-4H)	IS-10-275-1
Manual, Owner's Instruction (TMR-1H)	IS-10-271-1
Manual, Service ($5.00 Prepaid)	SM-10-270-3

RECEIVER BOARD 501-055

RESISTORS

All Resistors are ¼ watt, 10% unless otherwise specified.

Item No.	Description	Part No.
R101	33K	
R102	68K	
R103	1.8K	
R104	27 ohm	
R105	1.5K	
R106	33K	
R107	33K	
R108	10K	
R109	4.7K	
R110	33 ohm	
R111	2.2K	
R112	4.7K	
R113	2.2K	
R115	33 ohm	
R116	22K	
R117	8.2K	
R118	22K	
R119	330 ohm	
R120	330 ohm	
R121	330 ohm	
R122	330 ohm	
R123	220 ohm	
R124	4.7K	
R125	2.2K	
R126	2.2K	
R127	270 ohm	
R128	1.8K	
R129	22 ohm	
R130	39 ohm	
R131	680 ohm	
R132	1K	
R133	4.7K	
R134	22K	
R135	2.2K	
R136	180 ohm	
R137	10K	
R138	4.7K	
R139	Res. Var. 50K 302-201-2	4750-3220-102
R140	39K	
R141	82K	
R142	82K	
R143	Res. Var 10K w/sw 302-201-1	4750-3220-101
R144	27K	
R145	180 ohm	
R146	5.6K	
R147	1.2K	
R148	3.3 Megohm	
R149	3.3K	
R150	3.3K	
R151	100 ohm ½ watt, 10%	

Item No.	Description	Part No.
R152	3.3 ohm	
R153	1K	
R154	100K	
R155	470K	
R156	10 Meg. ohm	
R157	2.2K	
R158	22K	
R159	100 ohm	
R160	15K	
R161	150K	
R162	22 ohm	
R163	2.2K	

CAPACITORS

Item No.	Description	Part No.
C101	5.6pf 10% 500V NPO (Disc.)	1500-0569-905
C102	15pf 10% 50V NPO (Disc.)	1500-0150-650
C103	0.47pf 10% (Composition)	1510-0478-900
C104	5.6 pf 10% 500V NPO (Disc.)	1500-0529-905
C105	8.2 pf 10% 500V NPO (Disc.)	1503-0502-005
C106	.005mf +80%-20% 50v Z5U (Disc)	1503-0502-005
C107	.001 +80%020% 50V, Z5U (Disc.)	1503-0102-003
C108	30pf 10% 50V NPO (Disc.)	1500-0300-650
C109	0.2 mf + 80%-20% 12V (Disc.)	1502-0204-006
C110	.001 mf, +80%-20% 50V, Z5U (Disc.)	1503-0102-003
C111	.005 mf, +80%-20% 50V, Z5U (Disc.)	1503-0502-005
C112	2.2 pf 10% 500V NPO (Disc.)	1500-0229-905
C113	22pf 10% 50V NPO (Disc.)	1500-0220-650
C114	270 pf 5%, 50V (Mica)	1506-0271-550
C115	.01mf +80%-20%, 25V Z5U (Disc.)	1502-0103-004
C116	10mf, 85^{0}C, 10V (Electrolytic)	1511-0100-001
C117	680 pf 5% 50V (Mica)	1506-0681-550
C118	.01mf +80%-20%, 25V Z5U (Disc.)	1502-0103-004
C119	100 pf 5% 50V (Mica)	1506-0101-550
C120	100 pf 5% 50V (Mica)	1506-0101-550
C121	68 pf 5% 50V NPO (Disc.)	1524-0680-002
C122	0.2 mf +80%-20% 12V (Disc.)	1502-0204-006
C123	220 pf 5% 50V (Mica)	1506-0221-550
C124	0.1 mf 20% 12V (Disc.)	1502-0104-005
C125	0.1 mf 20% 12V (Disc.)	1502-0104-005
C126	82 pf 5% 50V NPO (Disç.)	1524-0820-002
C127	68 pf 5% 50V NPO (Disc.)	1524-0680-002
C128	.001 mf +80%-20% 50V Z5U (Disc.)	1503-0102-003
C129	390 pf 5% 50V (Mica)	1506-0391-550
C130	180 pf 5% 50V (Mica)	1506-0181-550
C131	100 mf 85°C 10V (Electrolytic)	1513-0101-001
C132	0.2 mf +80%-20% 12V (Disc.)	1502-0204-006
C133	0.2 mf +80%-20% 12V (Disc.)	1502-0204-006
C134	0.001 +80%-20% 50V Z5U (Disc.)	1503-0102-003
C135	470 pf 20% 50V Z5U (Disc.)	1523-0471-002
C136	100 mf 85°C 10V (Electrolytic)	1513-0101-001
C137	0.1 mf 20% 12V Z5U (Disc.)	1502-0104-005
C138	0.1 mf 20% 12V Z5U (Disc.)	1502-0104-005
C139	0.1 mf 20% 12V Z5U (Disc.)	1502-0104-005
C140	0.1 mf 20% 12V Z5U (Disc.)	1502-0104-005
C141	0.1 mf 20% 12V Z5U (Disc.)	1502-0104-005
C142	100 mf 85°C 10V (Electrolytic)	1513-0101-001
C143	220 pf 5% 50V (Mica)	1506-0221-004
C144	100 pf 5% 50V (Mica)	1506-0101-004
C145	.01 mf +80%-20% 25V Z5U (Disc.)	1502-0103-004
C146	25mf 85°C 10V (Electrolytic)	1513-0250-001
C147	.001 20% 50V Z5U (Disc.)	1523-0102-002
C148	0.01 mf +80%-20% 25V Z5U (Disc.)	1502-0103-004
C149	0.001 20% 50V Z5U (Disc.)	1523-0102-002
C150	0.05 mf + 80%-20% 12V (Disc.)	1502-0503-006
C151	0.01 mf +80%-20% 25V Z5U (Disc.)	1502-0103-004
C152	8.2 pf 10% 500V NPO (Disc.)	1500-0829-905
C153	220 mf 85°C 10V (Electrolytic)	1513-0101-001
C154	0.01 mf + 80%-20% 25V (Disc.)	1502-0103-004
C155	100 mf 85°C 10V (Electrolytic)	1513-0101-001
C156	0.047 mf 10% 100V (Mylar)	1508-0473-610
C157	0.01 mf +80%-20% 25V Z5U (Disc.)	1502-0103-004
C158	100 mf 85°C 10V (Electrolytic)	1513-0101-001
C159	0.01 mf +80%-20% 25V Z5U (Disc.)	1502-0103-004
C160	0.01 mf +80%-20% 25V Z5U (Disc.)	1502-0103-004
C161	250 mf 85°C 16V (Electrolytic)	1513-0251-002
C162	0.2 mf +80%-20% 12V Z5U (Disc.)	1502-0204-006
C163	25 mf 85°C 10V (Electrolytic)	1513-0250-001
C164	0.1 mf 20% 12V Z5U (Disc.)	1502-0104-005
C165	0.01 mf +80%-20% 25V Z5U (Disc.)	1502-0103-004
C166	3.9 pf 10% 500V NPO (Disc.)	1500-0399-905
C167	0.002 mf 20% 50V Z5F (Disc.)	1523-0202-002
C168	0.2 mf +80%-20% 12V Z5U (Disc.)	1502-0204-006
C169	0.2 mf +80%-20% 12V Z5U (Disc.)	1502-0204-006
C170	330 pf 20% 50V Z5U (Disc.)	1523-0331-002
C171	100 pf 5% 50V (Mica)	1506-0101-004

COILS

Item No.	Description	Part No.
L101	Coil RF Input (GRN)	1800-5100-508
L102	Coil RF Input (GRN)	1800-5100-508
L103	Coil RF AMP Output (Yel.)	1800-5100-504
L104	Coil 1st Mixer Output (Red)	1800-5100-402
L105	Choke 3.3 uhy	1802-0339-007
L106	Coil Oscillator Output (Orange)	1800-5100-503
L107	Coil Oscillator	1801-1236-900
L108	Choke 820 uhy	1800-0000-002

INTEGRATED CIRCUITS

Item No.	Description	Part No.
IC101	IC 10.7 1F	3130-3167-901
IC102	IC 455 KHZ 1F	3130-3157-618
IC103	IC Multifunction	3130-3157-617

TRANSISTORS

Item No.	Description	Part No.
Q101	Silicon PNP 2N4258	4801-1284-100
Q102	Silicon NPN 2N3563/2N5130	4801-0000-018
Q103	Silicon NPN 2N3563/2N5130	9801-0000-018
Q104	Silicon NPN (RT)	4801-0000-035
Q105	Silicon NPN MPS 5172	4801-0000-100
Q106	Silicon PNP (WT)	4801-0000-060
Q107	Silicon PNP (WT)	4801-0000-060
Q108	Silicon PNP (WT)	4801-0000-060
Q109	Silicon NPN MPS 5172	4801-0000-100
Q110	Silicon NPN MPS 5172	4801-0000-100
Q111	Silicon PNP (WT)	4801-0000-060
Q112	Silicon PNP (WT)	4801-0000-060
Q113	Silicon PNP (WT)	4801-0000-060

(WT) = White Top (RT) = Red Top

DIODES

Item No.	Description	Part No.
CR101	Silicon, IN4148	4805-1241-200
CR102	Silicon, IN4148	4805-1241-200
CR103	Silicon, IN4148	4805-1241-200
CR104	Silicon, IN4148	4805-1241-200

Item	Description	Part No.
CR105	Silicon, IN4148	4805-1241-200
CR106	Silicon, IN4148	4805-1241-200
CR107	Silicon, IN4148	4805-1241-200
CR108	Silicon, IN4148	4805-1241-200
CR109	Silicon, IN4148	4805-1241-200
CR110	Germanium, 102-339	4807-1233-900

FILTERS

Item	Description	Part No.
CF101	Ceramic, 10.7 MHZ	2700-3209-600
CF102	Ceramic 455 KHZ	2700-0000-007

CRYSTALS

Item	Description	Part No.
Y101	Crystal, Receive (High Band)	2302-0000-000
Y102	Crystal, Receive (High Band)	2302-0000-000
Y103	Crystal, Receive (High Band)	2302-0000-000
Y104	Crystal, Receive (High Band)	2302-0000-000
Y105	10.245 MHZ	2301-3151-601

MECHANICAL

Item	Description	Part No.
	Antenna Bracket	1400-3216-700
	Clip, XTAL Mtg.	2830-3216-400

RECEIVER ASSEMBLY—ACT-P4H

Item	Description	Part No.
	Spacer, Threaded, HEX	2800-1280-501
	Screw, 6-32 x ¼ Binding	2804-0250-001
	Plascrew #5 x 3/8 Phil Pan Hd.	2816-3229-601
	Case	1411-7010-000
	Grill Cover	1411-7010-800
	Battery Door	1411-6038-700
	Plascrew #5	2816-3221-800
	Pushnut Palnut PB094070	2853-0000-006
J2	Jack, Earphone 2.5MM	2101-0000-007
J1	Plug, Charger, Coax	2104-3221-900
	Inlay, Top	2509-3216-801
	Inlay, Front	2509-3216-901
	Thumbwheel	2402-3222-400
	Screw, Metric 1.7MM	2820-1285-800
	Battery Connectors	2105-1277-200
	Antenna, Telescoping	1201-0000-001
	Speaker, 2¼" 16 ohm	1301-3213-402
	Pushnut, Palnut (Spkr. Mtg.)	2853-0000-005
	Insulator 302-305	3106-3230-500

SCANNER BOARD 501-056

RESISTORS

All Resistors are ¼ watt, 10% unless otherwise specified.

Item	Description	Part No.
R201	22K	
R202	22K	
R203	22K	
R204	22K	
R205	6.8K	
R206	33K	
R207	33K	
R208	6.8K	
R209	33K	
R210	33K	
R211	6.8K	
R212	33K	
R213	33K	
R214	6.8K	
R215	33K	
R216	33K	
R217	1 Meg.	
R218	33K	
R219	330K	
R220	3.3 Megohm	
R221	680 ohm	

CAPACITORS

Item	Description	Part No.
C201	100 mf 85°C 10V (Electrolytic)	1513-0101-001
C202	1 mf 85°C 16V (Electrolytic)	1513-0010-002
C203	0.1 mf 20% 12V Z5U (Disc.)	1502-0104-005

INTEGRATED CIRCUITS

Item	Description	Part No.
IC201	IC Counter/Decoder	3130-3193-502

TRANSISTORS

Item	Description	Part No.
Q201	Silicon, PNP (WT)	4801-0000-060
Q202	Silicon, PNP (WT)	4801-0000-060
Q203	Silicon, PNP (WT)	4801-0000-060
Q204	Silicon, PNP (WT)	4801-0000-060
Q205	Silicon, NPN MPS 5172	4801-0000-100
Q206	Silicon, NPN MPS 5172	4801-0000-100
Q207	Silicon, NPN MPS 5172	4801-0000-100
Q208	Silicon, NPN MPS 5172	4801-0000-100
Q209	Silicon, NPN MPS 5172	4801-0000-100

DIODES

Item	Description	Part No.
CR201	Silicon, IN4148	4805-1241-200
CR202	Silicon, IN 4148	4805-1241-200
CR203	Silicon, IN4148	4805-1241-200
CR204	Silicon, IN4148	4805-1241-200
CR205	Silicon, IN4148	4805-1241-200
CR206	Silicon, IN4148	4805-1241-200
CR207	Silicon, IN4148	4805-1241-200
CR208	Silicon, IN4148	4805-1241-200
CR209	Zener, 6V 5% MZ92-6.0B	4808-0000-012

LIGHT-EMITTING DIODES

Item	Description	Part No.
LD201	Led, Red	4810-1282-900
LD202	Led, Red	4810-1282-900
LD203	Led, Red	4810-1282-900
LD204	Led, Red	4810-1282-900

SWITCHES

Item	Description	Part No.
S201	SPDT, Slide	5113-3221-001
S202	SPDT, Slide	5113-3221-001
S203	SPDT, Slide	5113-3221-001
S204	SPDT, Slide	5113-3221-001
S205	1 Pole, 3 Pos., Slide	5113-3221-101

MAIN BOARD 600-386

RESISTORS

All Resistors are † 10%, ¼W, unless otherwise noted

Item	Description	Part No.
R101	1K	
R102	680 ohm	
R103	10K	
R104	10K	
R105	100 ohm	
R106	680 ohm	
R107	470 ohm	
R108	68 ohm	

Item	Description	Part No.
R109	22K	
R110	10K	
R111	10K	
R112	10K	
R113	100 ohm	
R114	680 ohm	
R115	68 ohm	
R116	22K	
R117	10K	
R118	220 ohm	
R119	22K	
R120	10K	
R121	220 ohm	
R122	680 ohm	
R123	100 ohm	
R124	68 ohm	
R125	1.2K	
R126	1.2K	
R127	1.2K	
R128	1.2K	
R129	1.2K	
R130	1.2K	
R131	1.2K	
R132	1.2K	
R133	1.2K	
R134	1.2K	
R135	10K	
R136	39K	
R137	470 ohm	
R138	1K	
R139	82K	
R140	4.7K	
R141	39K	
R142	39K	
R143	82K	
R144	10K	
R145	1K	
R146	1K	
R147	1K, Trimmer	4751-0102-005
R148	820 ohm	
R149	1K	
R150	3.3K	
R151	150 ohm	
R152	1K	
R153	3.3K	
R154	4.7K	
R155	47K	
R156	10K	
R157	4.7K	
R158	10K	
R159	4.7K	
R160	10K	
R161	4.7K	
R162	180 ohm	
R163	22K	
R164	180 ohm	
R165	12K	
R166	1.8K	
R167	1.5K	
R168	150 ohm	

Item	Description	Part No.
R169	33 ohm	
R170	47K	
R171	270K	
R172	2.2K	
R173	4.7K	
R174	8.2K	
R175	39K	
R176	10K	
R177	5.6K	
R178	2.2K	
R179	100 ohm	
R180	10K	
R181	10K	
R182	10K	
R183	2.2K	
R184	27K	
R185	27K	
R186	47K	
R187	22K	
R188	47K	
R189	2.2 ohm	
R190	33 ohm	
R191	4.7K	
R192	82 ohm, 2W, 5%	4707-0820-031
R193	22K	
R194	47K	
R195	33K	
R196	4.7K	
R197	18K	
R198	220 ohm	
R199	470 ohm	
R200	2.2K	
R201	68 ohm, 2W, 5%	4707-0680-031
	CAPACITORS	
C101	27pf 10% 50V NPO (Disc.)	1500-0270-650
C102	82pf 5%, 50V NPO (Disc.)	1524-0820-002
C103	120pf 5%, 50V (Mica)	1506-0121-550
C104	82pf 5%, 50V NPO (Disc.)	1524-0820-002
C105	.005mf +80%-20% 500V, Z5U (Disc.)	1503-0502-002
C106	.001mf +80%-20% 500V, Z5U (Disc.)	1503-0102-001
C107	68pf 5%, 50V NPO (Disc.)	1524-0680-002
C108	.005mf +80%-20% 500V, Z5U (Disc.)	1503-0502-002
C109	5.6pf 10%, 500V NPO (Disc.)	1500-0569-905
C110	27pf 10%, 50V NPO (Disc.)	1500-0270-650
C111	2.7pf 10%, 500V NPO (Disc.)	1500-0279-905
C112	8.2pf 10%, 500V NPO (Disc.)	1500-0829-905
C113	.47pf 10% (Composition)	1510-0478-900
C114	5.6pf 10%, 500V NPO (Disc.)	1500-0569-905
C115	3.9pf 10%, 500V NPO (Disc.)	1500-0399-905
C116	.001mf +80%-20% 500V, Z5U (Disc.)	1503-0102-001
C117	.005mf +80%-20% 500V, Z5U (Disc.)	1503-0502-002
C118	5.6pf 10% 500V NPO (Disc.)	1500-0569-905
C119	.005mf +80%-20% 500V, Z5U (Disc.)	1502-0502-002
C120	1.0pf 10% (Composition)	1510-0010-900
C121	1.5-5.5pf, Trimmer	1517-0000-011
C122	1.5-5.5pf, Trimmer	1517-0000-011
C123	.005mf +80%-20% 500V, Z5U (Disc.)	1503-0502-002
C124	.005mf +80%-20% 500V, Z5U (Disc.)	1503-0502-002
C125	.47pf 10% (Composition)	1510-0478-900
C126	.005mf +80%-20% 500V, Z5U (Disc.)	1503-0502-002

Item	Description	Part No.
C127	390pf 5% 50V (Mica)	1506-0391-550
C128	.005mf +80%-20% 500V, Z5U (Disc.)	1503-0502-002
C129	.005mf +80%-20% 500V, Z5U (Disc.)	1503-0502-002
C130	.01mf +80%-20% 500V, Z5U (Disc.)	1503-0103-001
C131	.001mf +80%-20% 500V, Z5U (Disc.)	1503-0102-001
C132	82pf 5% 50V NPO (Disc.)	1524-0820-002
C133	27pf 10% 50V NPO (Disc.)	1500-0270-650
C134	27pf 10% 50V NPO (Disc.)	1500-0270-650
C135	.001mf +80%-20% 500V, Z5U (Disc.)	1503-0102-001
C136	.001mf +80%-20% 500V, Z5U (Disc.)	1503-0102-001
C137	.2mf +80%-20% 12V (Disc.)	1502-0204-006
C138	150pf 20% 50V, Z5F (Disc.)	1523-0151-002
C139	1.5-5.5pf Trimmer	1517-0000-011
C140	470pf 20% 50V, Z5F (Disc.)	1523-0471-002
C141	1mf 85° 16V (Electrolytic)	1513-0010-002
C142	3.9pf 10% (Composition)	1510-0399-900
C143	390pf 5% 50V (Mica)	1506-0391-550
C144	3.9pf 10% 500V NPO (Disc.)	1500-0399-905
C145	680pf 5% 50V (Mica)	1506-0681-550
C146	390pf 5% 50V (Mica)	1506-0391-550
C147	.01mf +80%-20% 25V (Disc.)	1502-0103-004
C148	.01mf +80%-20% 25V (Disc.)	1502-0103-004
C149	68pf 5% 50V NPO (Disc.)	1524-0680-002
C150	180pf 5% 50V (Mica)	1506-0181-550
C151	390pf 5% 50V (Mica)	1506-0391-550
C152	.2mf +80%-20% 12V (Disc.)	1502-0204-006
C153	.2mf +80%-20% 12V (Disc.)	1502-0204-006
C154	270pf 5% 50V (Mica)	1506-0271-550
C155	250pf 5% 50V (Mica)	1506-0251-550
C156	.2mf +80%-20% 12V (Disc.)	1502-0204-006
C157	.2mf +80%-20% 12V (Disc.)	1502-0204-006
C158	.2mf +80%-20% 12V (Disc.)	1502-0204-006
C159	.002pf 20% 50V, Z5F (Disc.)	1523-0202-002
C160	39pf 10% 50V NPO (Disc.)	1500-0390-650
C161	.047mf 10% 100V (Mylar film)	1508-0473-610
C162	.1mf 20% 12V (Disc.)	1502-0104-005
C163	.1mf 20% 12V (Disc.)	1502-0104-005
C164	.01mf 10% 100V (Mylar film)	1508-0103-610
C165	470pf 20% 50V, Z5F (Disc.)	1523-0471-002
C166	.015mf 10% 100V (Mylar film)	1508-0153-610
C167	.047mf 10% 100V (Mylar Film)	1508-0473-610
C168	.002pf 20% 50V, Z5F (Disc.)	1523-0202-002
C169	1mf 85° C 16V (Electrolytic)	1513-0010-002
C170	1mf 85° C 16V (Electrolytic)	1513-0010-002
C171	.2mf +80%-20% 12V (Disc.)	1502-0204-006
C172	.015mf 10% 100V (Mylar film)	1508-0153-610
C173	.0033mf 10% 100V (Mylar film)	1508-0332-610
C174	5mf 85° C 10V (Electrolytic)	1513-0050-001
C175	.05mf +80%-20% 25V (Disc.)	1502-0503-004
C176	.2mf +80%-20% 12V (Disc.)	1502-0204-006
C177	100mf 85° C 16V (Electrolytic)	1513-0101-002
C178	.047mf 10% 100V (Mular film)	1508-0473-610
C179	10mf 85° C 10V (Electrolytic)	1513-0100-001
C180	3.3mf 20% 85° C 10V (Electrolytic)	1513-0339-005
C181	.015mf 10% 100V (Mylar film)	1508-0153-610
C182	250mf 85° C 10V (Electrolytic)	1513-0251-001
C183	100mf 16V (Electrolytic)	1511-0102-002

COILS

Item	Description	Part No.
L101	Choke .68 UHY	1802-0688-003
L102	Coil, RF input (Green)	1800-3152-005
L103	Coil, RF (Yellow)	1800-3152-004
L104	Coil, RF input (Red)	1800-3152-002
L105	Coil, RF input (Red)	1800-3152-002
L106	Coil, RF (Orange)	1800-3152-003
L107	Coil, RF 450 MHZ	1800-3160-001
L108	Coil, RF AMP out	1800-3160-002
L109	Choke 15 MHY	1802-0152-004
L110	Choke 15 MHY	1802-0152-004
L111	Choke 15 MHY	1802-0152-004
L112	Choke 15 MHY	1802-0152-004
L113	Choke 15 MHY	1802-0152-004
L114	Choke 15 MHY	1802-0152-004
L115	Choke 15 MHY	1802-0152-004
L116	Choke 15MHY	1802-0152-004
L117	Choke 15MHY	1802-0152-004
L118	Choke 15MHY	1802-0152-004
L119	Coil, RF Inj. (White)	1800-3152-009
L120	Coil, OSC.	1801-1236-900
L121	Coil OSC. 450 MHZ	1800-3160-003
L122	Coil RF (White)	1800-3191-401
L123	Coil RF (White)	1800-3191-401
L124	Coil RF (Yellow)	1800-3191-402
L125	Choke 6.8 uh 10%	1802-0689-003
L126	Choke	1802-0000-002
L127	Choke	1802-0000-002
L128	Coil, Quadrature	1800-3151-700

INTEGRATED CIRCUITS

Item	Description	Part No.
IC101	IC, 10.7 IF	3130-3167-901
IC102	IC, 455 KHZ Detector	3130-3157-603
IC103	IC, Audio Amplifier	3130-3157-614
IC104	IC, Counter	3130-3157-608
IC105	IC, Decoder Driver	3130-3193-501

TRANSISTORS

Item	Description	Part No.
Q101	Silicon NPN (RT)	4801-0000-035
Q102	Silicon NPN (RT)	4801-0000-035
Q103	Silicon NPN (RT)	4801-0000-035
Q104	Silicon NPN (RT)	4801-0000-035
Q105	Junct. FET	4811-0000-015
Q106	Silicon NPN (RT)	4801-0000-035
Q107	Silicon NPN, MPS5172	4801-0000-100
Q108	Silicon NPN (RT)	4801-0000-035
Q109	Silicon PNP (WT)	4801-0000-060
Q110	Silicon NPN, MPS5172	4801-0000-010
Q111	Silicon NPN, MPS5172	4801-0000-010
Q112	Silicon NPN, MPS 5172	4801-0000-010
Q113	Silicon PNP (W T)	4801-0000-060
Q114	Silicon PNP (WT)	4801-0000-060
Q115	Silicon PNP (WT)	4801-0000-060
Q116	Silicon NPN, MPS 5172	4801-0000-060
Q117	Silicon NPN, MPS 5172	4801-0000-010
Q118	Silicon PNP (WT)	4801-0000-060
Q119	Silicon NPN, MPS 5172	4801-0000-010
Q120	Silicon PNP (WT)	4801-0000-060
Q121	Silicon NPN, MPS 5172	4801-0000-010
Q122	Silicon NPN, MPS 5172	4801-0000-010
Q123	PN Silicon, Unijunction	4813-0000-001
Q124	Silicon NPN, MPS 5172	4801-0000-010

(WT) = White Top (RT) = Red Top

Item	Description	Part No.
	DIODES	
CR101	Germanium-junction, signal	4807-1233-900
CR102	Germanium-junction, signal	4807-1233-900
CR103	Germanium-junction, signal	4807-1233-900
CR104	Germanium-junction, signal	4807-1233-900
CR105	Germanium-junction, signal	4807-1233-900
CR106	Germanium-junction, signal	4807-1233-900
CR107	Germanium-junction, signal	4807-1233-900
CR108	Germanium-junction, signal	4807-1233-900
CR109	Germanium-junction, signal	4807-1233-900
CR110	Germanium-junction, signal	4807-1233-900
CR111	VARICAP	4809-0000-001
CR112	Silicon, IN4148	4805-1241-200
CR113	Silicon, IN4148	4805-1241-200
CR114	Zener 5% 5.1V	4808-0000-007
CR115	Silicon, IN4148	4805-1241-200
CR116	Silicon, IN4148	4805-1241-200
CR117	Silicon, IN4148	4805-1241-200
CR118	Silicon, IN4148	4805-1241-200
CR119	Silicon, IN4148	4805-1241-200
CR120	Silicon, IN4148	4805-1241-200
CR121	Silicon, IN4148	4805-1241-200
CR122	Silicon, IN4148	4805-1241-200
CR123	Silicon, IN4148	4805-1241-200
CR124	Silicon, IN4148	4805-1241-200
CR125	Silicon, IN4148	4805-1241-200
CR126	Silicon, IN4148	4805-1241-200
CR127	Zener 8.2V	4808-0000-009
CR128	Silicon Rectifier	4806-0000-004
CR129	Silicon Rectifier	4806-0000-004
	FILTER	
CF101	Ceramic	2700-0000-007
	CRYSTAL	
Y111	10.245 MHZ	2301-3151-601
	MISCELLANEOUS	
SW101-	Switch (10POS)	5112-5038-401
SW110	Shield Assembly (301-602	2508-1256-300
		2508-1256-400

LIGHT BOARD 501-027

Item	Description	Part No.
	LAMPS	
M301-	Lamp, incandescent, 14.4V, 80MA (WB # 73)	
M310		3901-0000-007

Item	Description	Part No.
	MECHANICAL	
	Bulb Cup	2830-1276-900
	Connector (12PIN)	2105-0000-012

CHASSIS ASSEMBLY—ACT-R10H/L/U

Item	Description	Part No.
	ELECTRICAL COMPONENTS	
R1	Res. Var. 10K/SW	4750-3212-101
R2	Res. Var. 7.5K	4750-3212-102
R3	Res. Comp. 1MEG., 10%, ½W	4701-0105-044
C1	.005MF, +80-20%, 1.4KV Z5U (Disc.)	1501-0502-002
SPK-1	Speaker 2 x 6 in. 8 ohm	1301-5101-000
ANT-1	Antenna, H/L Telescoping	1201-0000-002
ANT-2	Antenna, UHF Telescoping	1201-0000-003
T1	Power Transformer	5604-5100-600
J1, J2	Antenna Connector	2105-0000-005
P1	AC Connector	2105-1279-100
P2	DC Connector	2105-1277-900
Y100	Crystal, Receive (Low Band)	2303-0000-000
Y100	Crystal, Receive (Highband)	2302-0000-000
Y100	Crystal, Receive (450-470MHZ)	2304-0000-000
Y100	Crystal, Receive (470-500MHZ)	2320-0000-000
MA-16	AC Power Cord	6041-3215-900
MA-17	DC Power Cord Assembly	7011-1047-800
SW1-2	Switch	5112-6035-820
	MECHANICAL COMPONENTS	
	AC Shield	2508-3215-800
	Speaker, MTG. Speed Nut	2853-0000-003
TB-1	Terminal Board (3Lug)	2103-3007-907
	Front Panel (Bezel)	1405-6034-302
	Face Plate	2403-3212-600
	Knob, Volume & Squelch	2402-1276-202
	Lens, 10 Station	3900-5100-800
	Cabinet/Wrap Assembly	1408-6038-500
	Bushing, Mounting Bolt	2501-3214-000
	Bushing, UHF Antenna	2501-0000-006
	Bushing, H/L Antenna	2501-0000-007
	Feet, Rubber	1402-0000-001
	Mounting Bolt (¼-20 x 5/8	2806-0625-002
	Bracket, Mobile, Mounting	1400-3214-200
	Manual, Owner's Instruciton	7001-1047-500
	Manual, Service ($5.00 Prepaid)	

Chapter 5

Midland Monitors

At this writing, Midland is selling four models, which are really only three basic designs. The difference between the 13-930 and 13-940 is largely in the type of cabinet used. Midland is one of the largest marketers of private label Japanese equipment in this country, with the discount stores as its primary outlets. This means that most of the Midland units that you will service were sold by people who know little or nothing about what they were selling. You may therefore find that you will, as an independent service shop operator, find it worth your while to stock and install crystals for these units.

In this chapter you will find crystal data provided by Midland. I have also included all service data that is available; it should be adequate for service and alignment of the units covered.

You might also find that other import units can be serviced by application (carefully) of the Midland data because of the many similarities—major and minor—between competing products.

UNITS COVERED

The index above will refer you to sections listing specifications, locations of necessary diagrams and illustrations, and alignment data for units listed. Doublecheck to see that you are using the proper data for the model under repair. Parts lists are at the end of the chapter.

Follow the alignment steps (A, B, C, etc.) exactly and in order, at least until you are familiar with the unit. Note that some units are almost identical, with only one or two changes in alignment procedure.

CRYSTAL DATA FOR ALL MIDLAND UNITS

Once you have determined the frequencies you want to monitor, crystals may be ordered from your Midland dealer or by writing directly to a crystal manufacturer.

The following information may be required by the crystal supplier to properly prepare the crystals.

Low Band (30–50 MHz)

Crystal frequency	Channel frequency + 10.7 MHz
Frequency tolerance	+2000, −1500 Hz
Crystal type	HC-25/U plug-in type
Series resonance	−450 Hz, third overtone
Maximum impedance	35Ω
Drive level	2 mW
Shunt capacitance	Less than 5 pF

High Band (144–173 MHz)

Crystal frequency	Channel frequency — 10.7 MHz
Frequency tolerance	+1000, −800 Hz

Crystal type	HC-25/U plug-in type
Series resonance	−450 Hz, third overtone
Maximum impedance	35Ω
Drive level	25 mW
Shunt capacitance	Less than 6 pF

UHF Band (450–470 MHz)

Crystal frequency	$\frac{\text{Channel frequency} + 10.7\text{ MHz}}{9}$
Frequency tolerance	+800, -600 Hz
Crystal type	HC-25/U (plug-in type)
Series resonance	Third overtone, load capacity of 18 pF

MIDLAND 13-904

Specifications

Circuitry	27 transistors, 20 diodes, 1 integrated circuit
Frequency	150–170 MHz
Channels	Four
Sensitivity, 20 dB quieting	0.5 μV
Adjacent-channel rejection	More than 60 dB
Scan rate	10 channels per sec
Power supply	6V DC
Audio output	0.3W
Intermediate frequencies	First, 10.7 MHz Second, 455 kHz
Accessories	Telescopic antenna Antenna wire Owner's guide

Service Data

The 13-904 is a 4-channel, single-band unit requiring no crystal programming.

Crystal layout	Fig. 5-1
Schematic	Schematic 5-1
Internal layout	Figs. 5-1, 5-3
Alignment data	Steps A, B

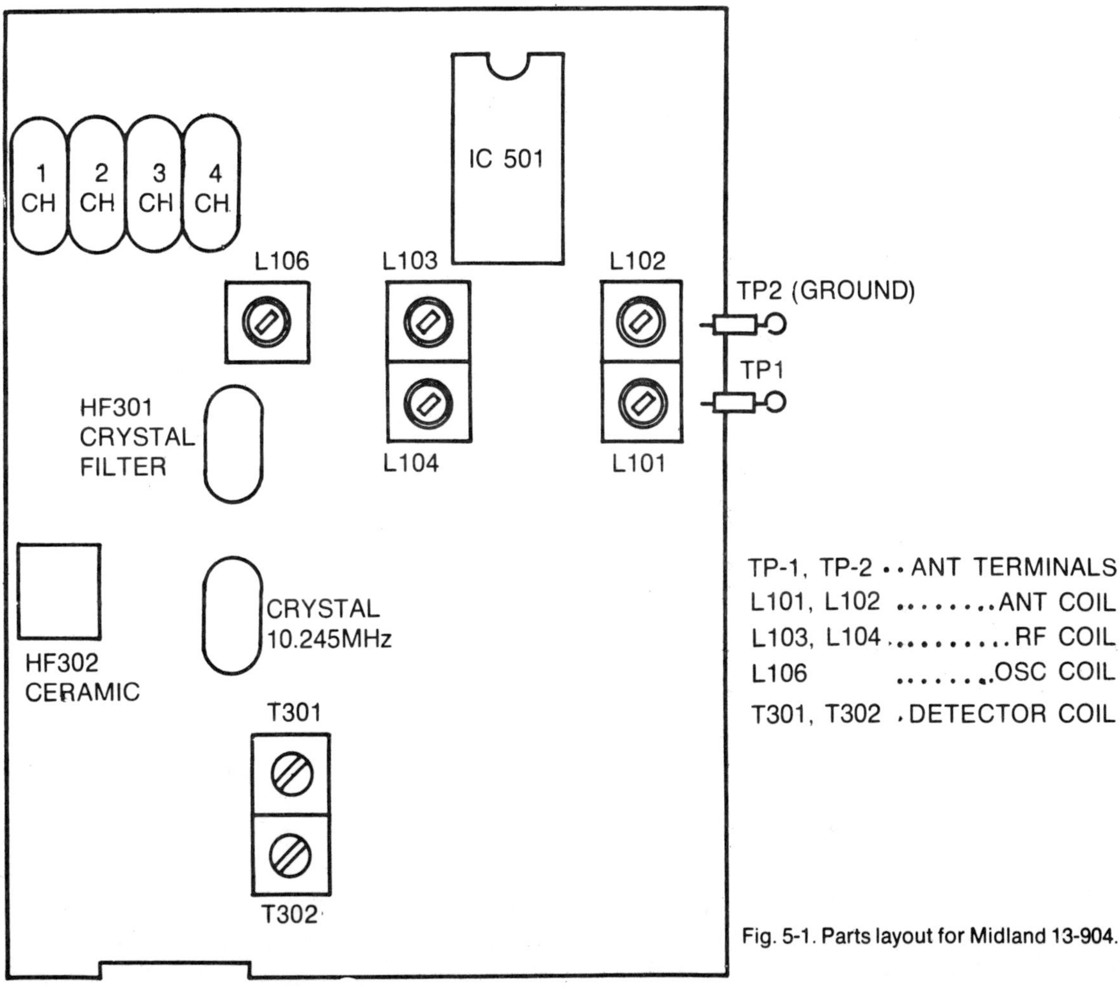

Fig. 5-1. Parts layout for Midland 13-904.

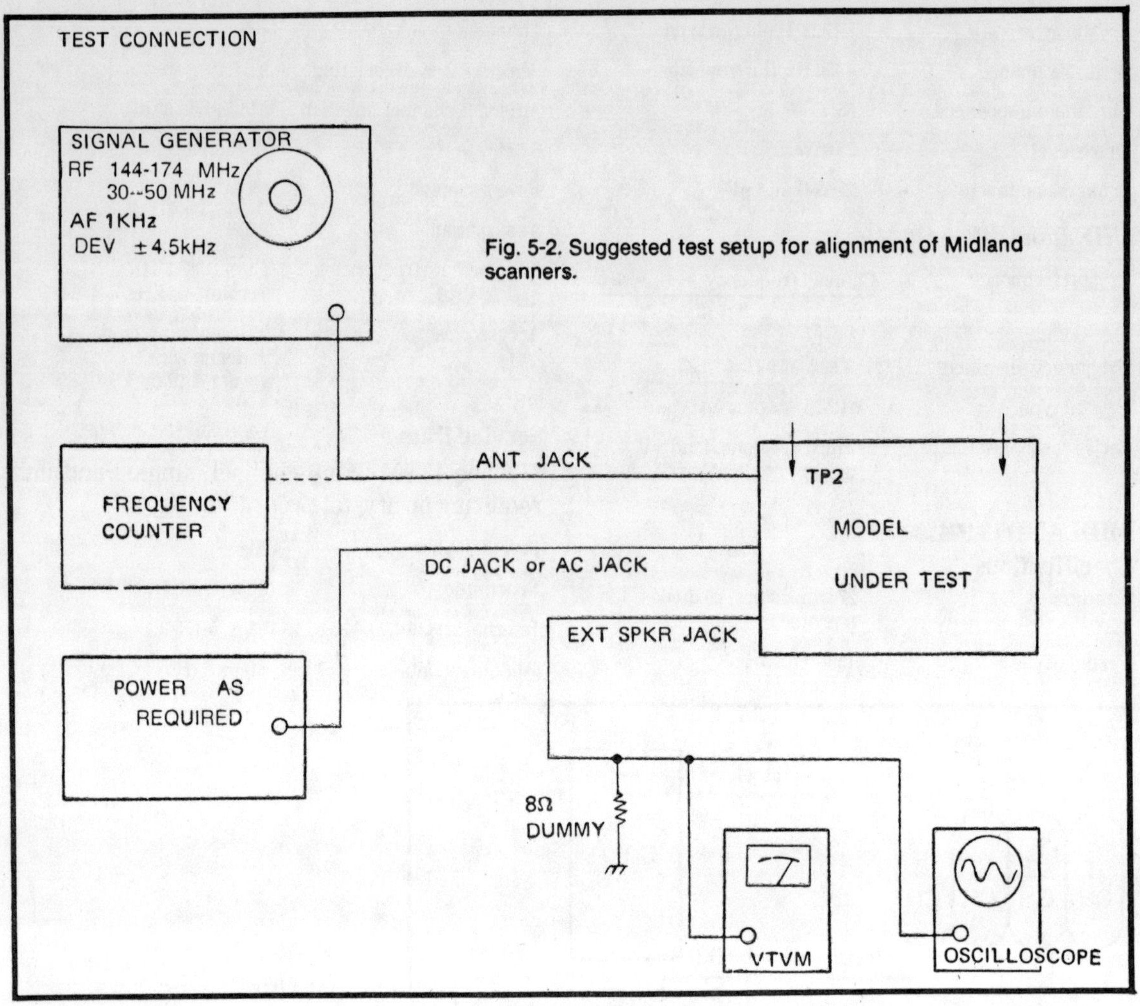

Fig. 5-2. Suggested test setup for alignment of Midland scanners.

MIDLAND 13-930, 13-940

Specifications

Frequency	Low band. 33–47 MHz High band. 144–173 MHz
Channels	Eight in any sequence high or low
Sensitivity. 20 dB quieting	More than 60 dB
Adjacent-channel rejection	Less than 1V
Scan rate	8–24 channels per sec
Power supply	117V AC . 12V DC
Audio output	1W or more
Intermediate frequency	First. 10.7 MHz Second. 455 kHz
Accessories	AC power cord unit DC power cord unit Spare fuse Spare lamp Mounting bracket with screws Owner's guide Telescopic antenna

Table 5-1. Oscillator Voltage Data for Midland 13-940, 13-950, and 13-930.

Frequency	Voltage at Collector of Q302 and Q204 on High and Low Bands, Respectively
144.0 MHz	1.0 ±0.2V
150.0 MHz	2.0 ±0.5V
155.16 MHz	3.0 ±0.5V
162.55 MHz	5.0 ±0.5V
173.0 MHz	6.0 or over
33 MHz	1.0 ±0.2V
36 MHz	1.5 ±0.5V
39 MHz	2.0 ±0.5V
41 MHz	3.0 ±0.5V
44 MHz	5.0 ±0.5V
47 MHz	6.0 or over

Service Data

The 13-930 and 13-940 are programmed for high or low band simply by plugging the crystal in either the *high* or *low* side of the crystal board.

Crystal layout	Fig. 5-4
Voltage date	Table 5-1 (also see schematic)

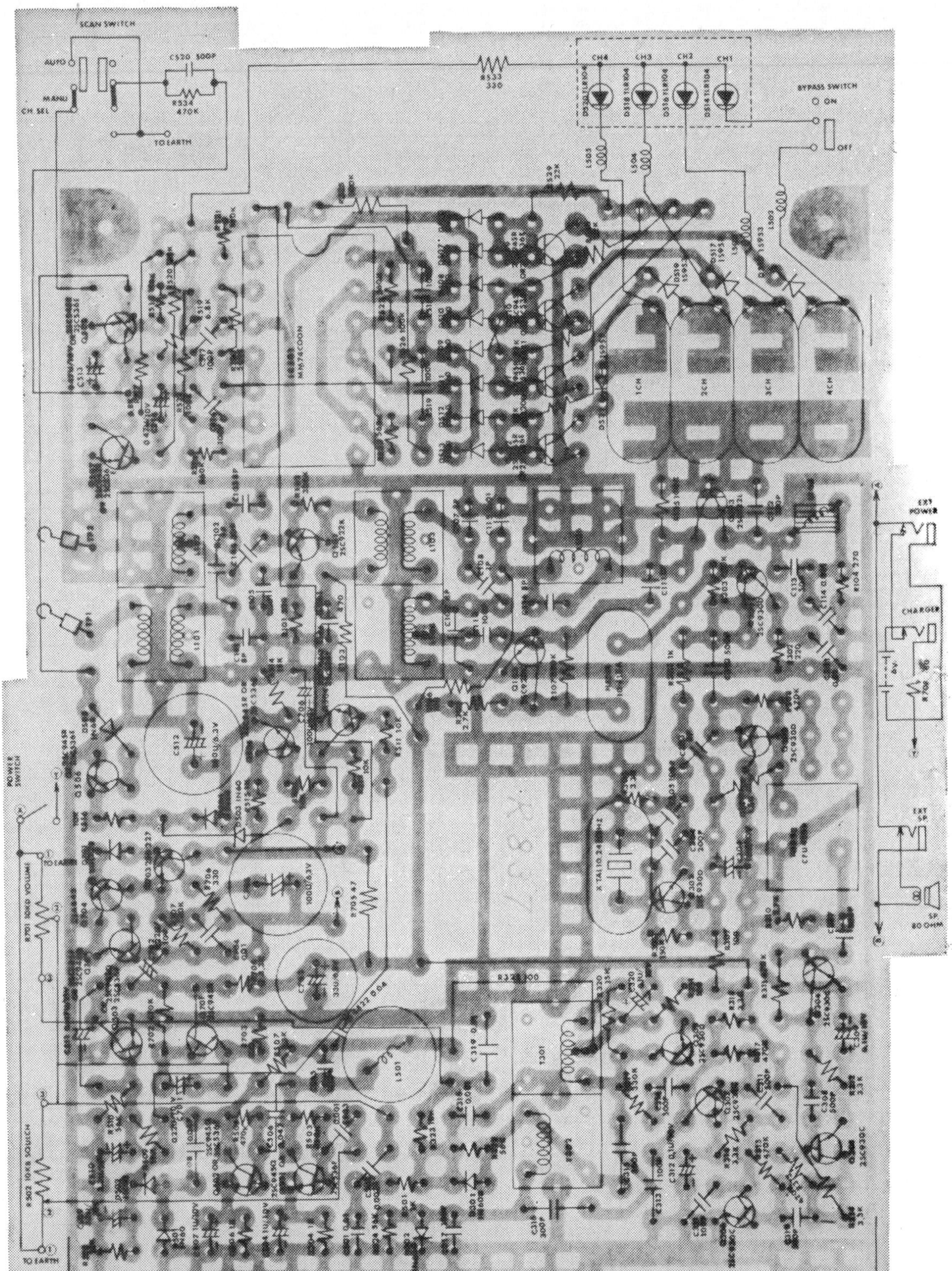

Fig. 5-3. Midland 13-904 circuit board (bottom view).

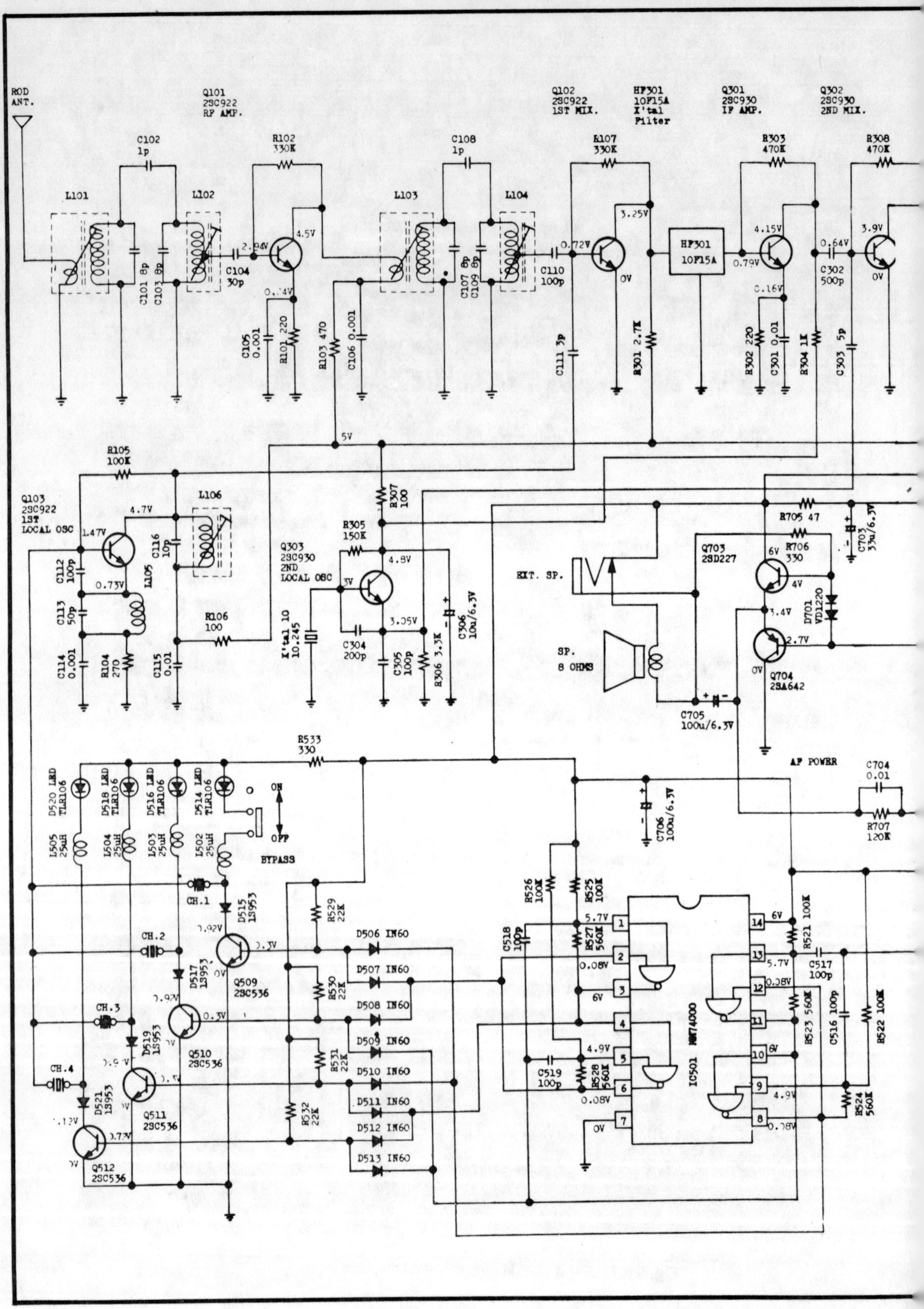
ROD ANT.
Q101 2SC922 RF AMP.
Q102 2SC922 1ST MIX.
HF301 10F15A X'tal Filter
Q301 2SC930 IF AMP.
Q302 2SC930 2ND MIX.
Q103 2SC922 1ST LOCAL OSC
Q303 2SC930 2ND LOCAL OSC
X'tal 10 10.245
EXT. SP.
SP. 8 OHMS
Q703 2SD227
Q704 2SA642
D701 VD1220
AF POWER
BYPASS
ON
OFF
CH.1
CH.2
CH.3
CH.4
D520 LED TLR106
D518 LED TLR106
D516 LED TLR106
D514 LED TLR106
Q509 2SC536
Q510 2SC536
Q511 2SC536
Q512 2SC536
IC501 MN74000

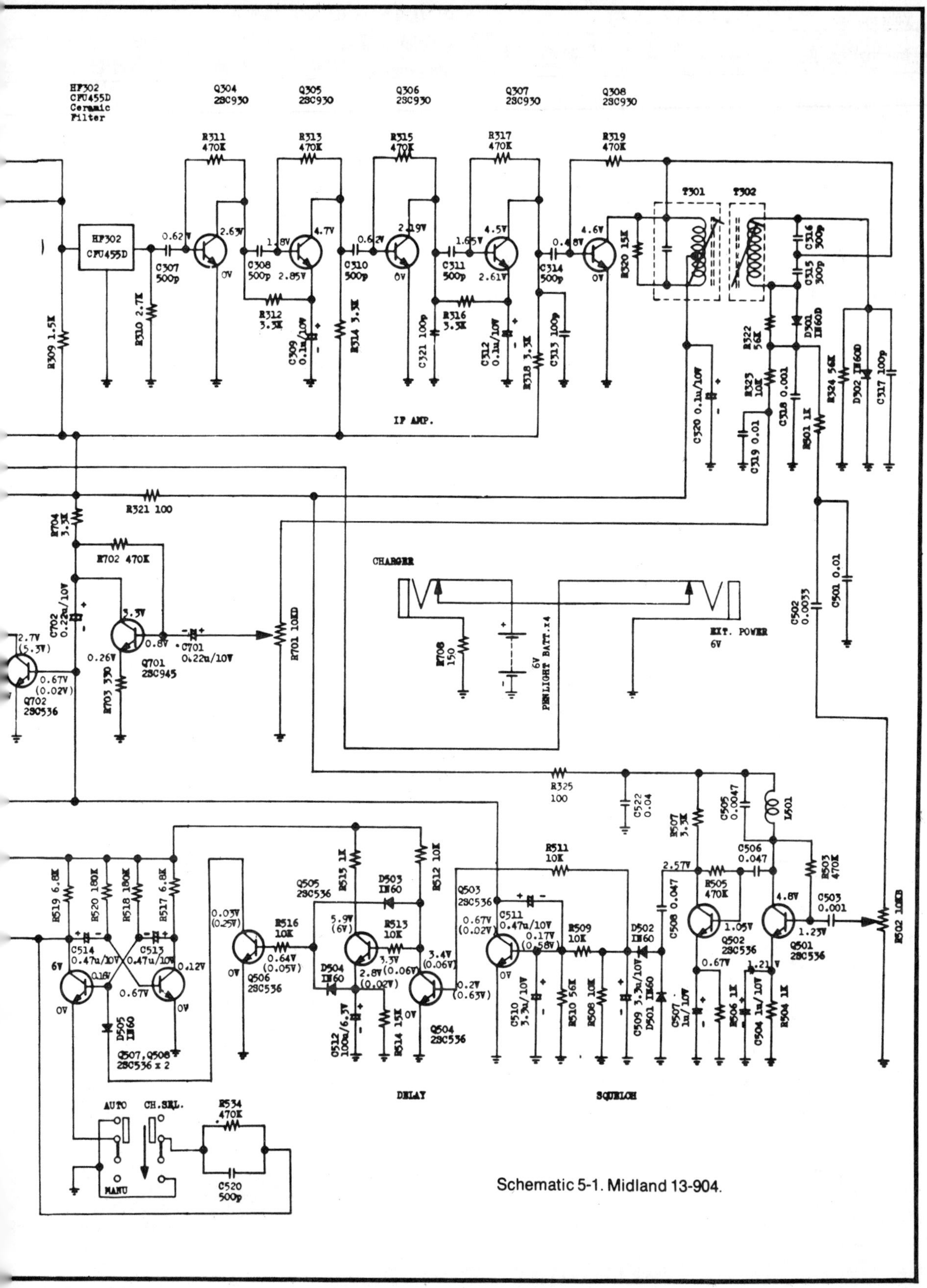

Schematic 5-1. Midland 13-904.

Schematic	Schematic 5-2
Internal layout	Figs. 5-4, 5-5 (13-940), 5-6 (13-930)
Alignment data	Steps A, C, D, E, F, G, H, I

MIDLAND MODEL 13-950

Specifications

Frequency	Low band, 33–47 MHz High band, 144–173 MHz UHF band, 450–470 MHz

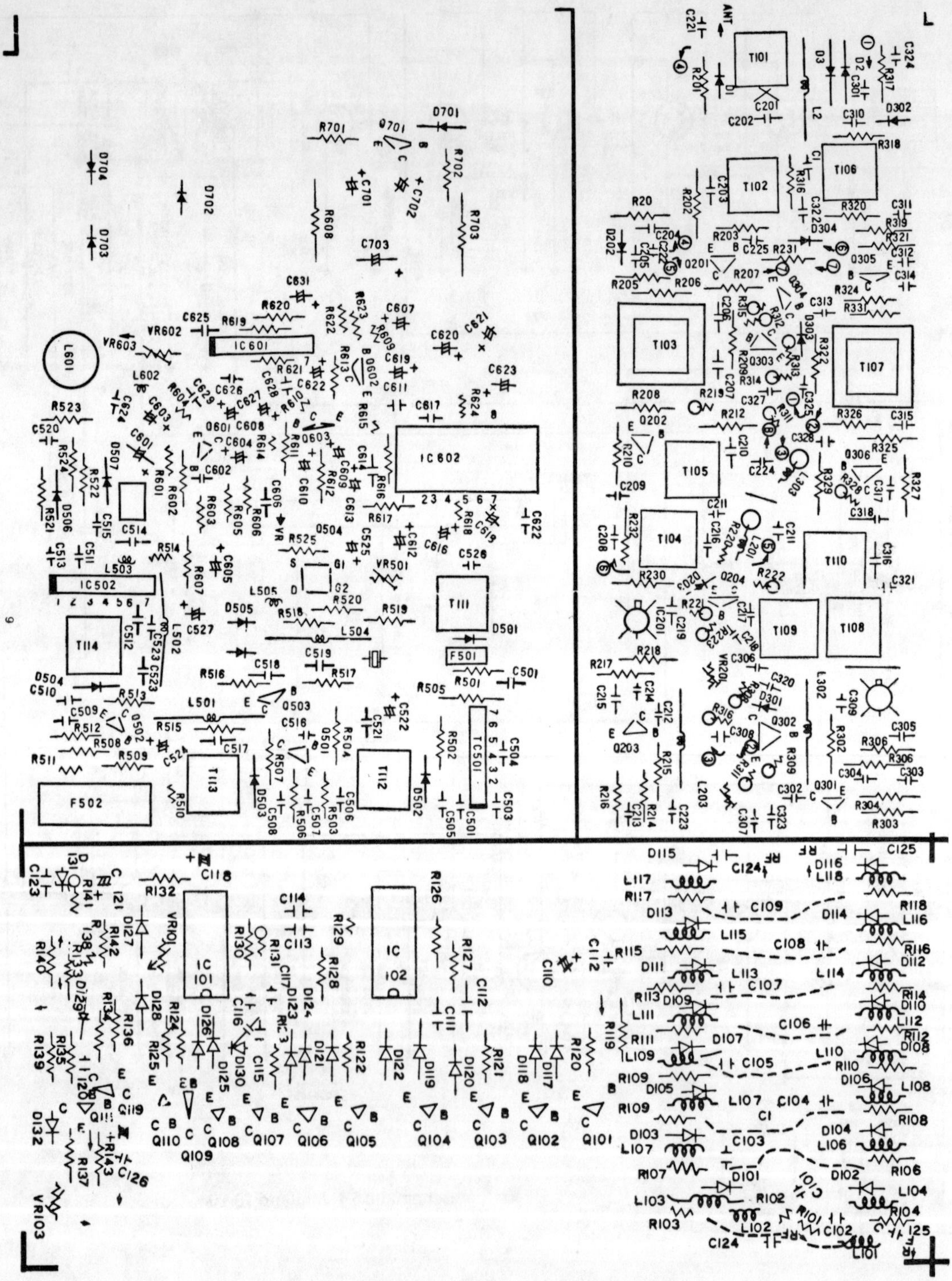

Fig. 5-4. Circuit board layout—13-930 and 13-940.

Channels	Eight of any sequence, high or low; eight of any sequence, UHF
Sensitivity, 20 dB quieting	Less than 1.0 μV
Adjacent-channel rejection	More than 60 dB
Scan rate	8–24 channels per sec
Power supply	117V AC, 12V DC
Audio output	1W or more
Intermediate frequency	First, 10.7 MHz Second, 455 kHz
Accessories	AC power cord unit DC power cord unit Spare fuse Spare lamp Owner's guide Two telescopic antennas

Service Data Band programming on the 13-950 is only necessary in the VHF high–low section and is accomplished by simply plugging the crystal into the *high* or *low* side of the crystal board.

Crystal layout	Fig. 5-7
Voltage data	Table 5-1 (also see schematic)
Schematic	Schematics 5-3, 5-8
Internal layout	Figs. 5-7, 5-8
Alignment data	Steps A, C, D, E, I, J

A—Equipment Connection for Alignment

Connect FM signal generator to antenna terminal as in Fig. 5-2 and tune to exact frequency of desired crystal. Use of channel near center of band is suggested.

B—Alignment, RF and IF

1. Turn BYPASS switch off.
2. SCAN switch at MAN.
3. Set SQUELCH control at minimum (unsquelched).
4. Adjust L101, L102, L103, and L104 for maximum output, measured at 8Ω dummy load.
5. Adjust L106 for maximum output, reducing generator output as sensitivity increases.
6. Adjust T301 and T302 for maximum.
7. Repeat steps B-4, B-5, B-6 until maximum output with minimum distortion is achieved.

C—Alignment, Crystal Oscillator

1. Check for specified frequency of second local oscillator (10.245 MHz) at the base of Q501.
2. Short out resistor R523.
3. Adjust VR501 to obtain the second-local-oscillator frequency of 10.245 MHz ± 1 kHz.
4. Adjust T112, T113, and T114 for maximum noise level on the oscilloscope or voltmeter at speaker.
5. Restore R523 to operation.
6. Recheck the frequency of the second local oscillator. If the frequency is 10.245 MHz ±2 kHz, the oscillator is functioning properly. Repeat alignment steps C-1 through C-4 if the oscillator frequency is below or above the specified frequency.

The emitter voltage of Q301 (high band) or Q203 (low band) should increase slightly when a crystal is placed in an operative channel. An absence of any increase in voltage indicates either a defective crystal or a malfunctioning oscillator.

D—Self-Tuning Control

Check automatic self-tuning control for RF stages. The checkpoint on high band is the collector of Q302, for low band the collector of Q204.

E—Alignment, Crystal Oscillator

Place a 144 MHz crystal in an operative channel. Adjust VR301 to obtain a 1V or less indication at the collector of Q302. Place a 33 MHz crystal in an operative channel. Adjust VR201 to obtain a 1V or less indication at the collector of Q203. Place a 162.55 MHz crystal in an operative channel. Adjust T109 for 5V at the collector of Q302. Place a 44 MHz crystal in an operative channel. Adjust T104 for 5V at the collector of Q204.

Caution:

- Adjust high band first.
- Do not try to obtain peak point in voltage at 162.55 MHz on high band and at 44 MHz on low band.

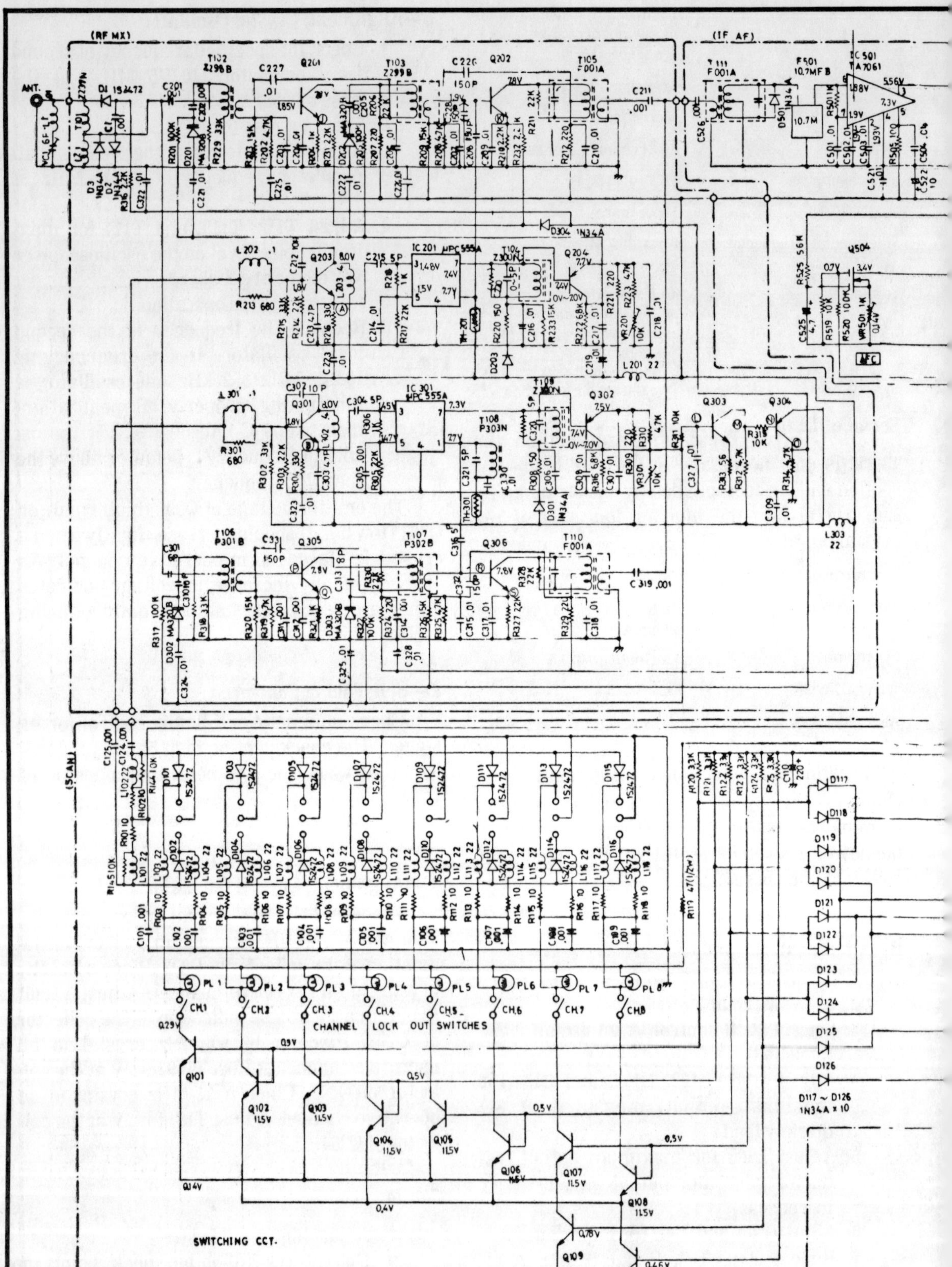

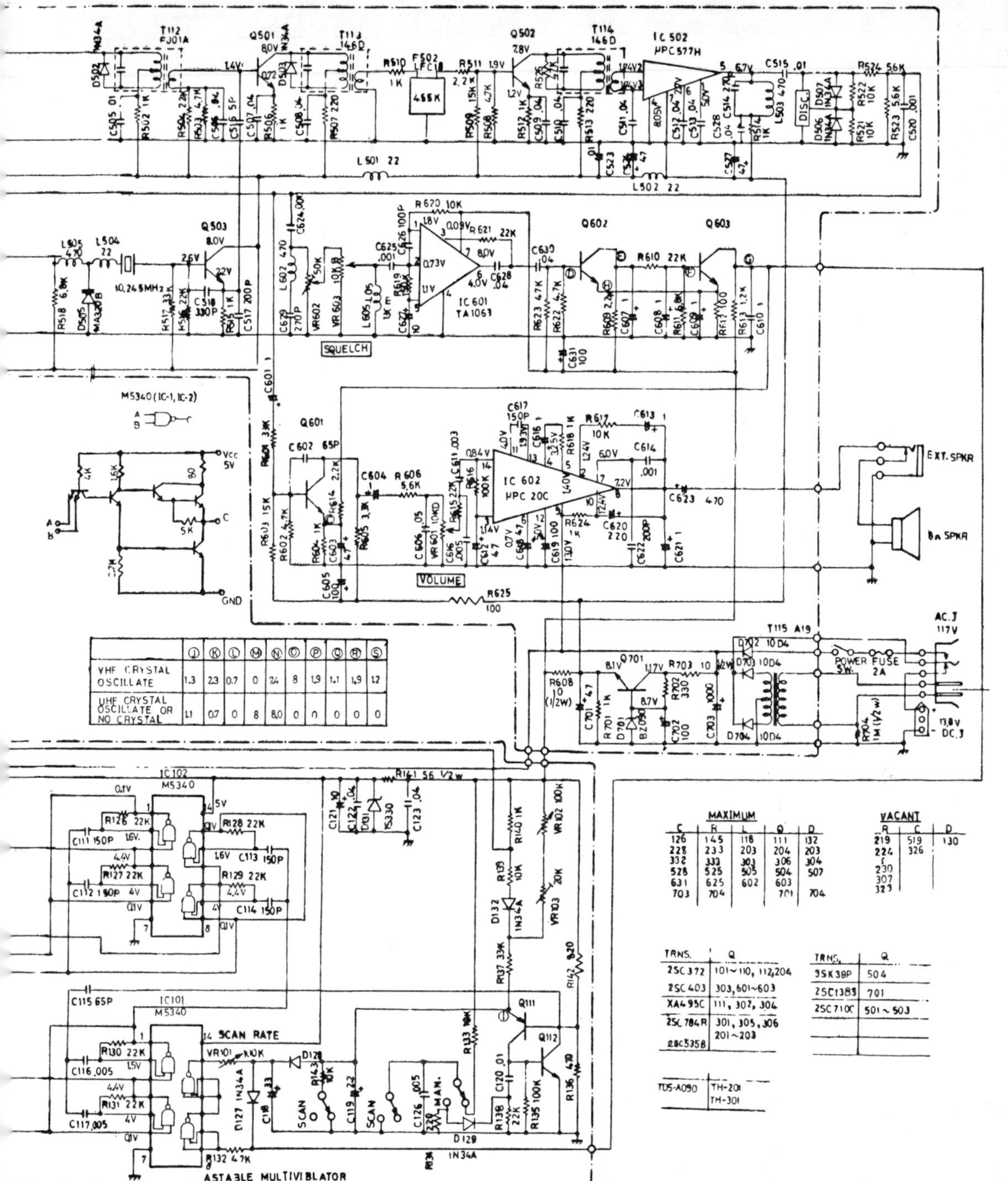

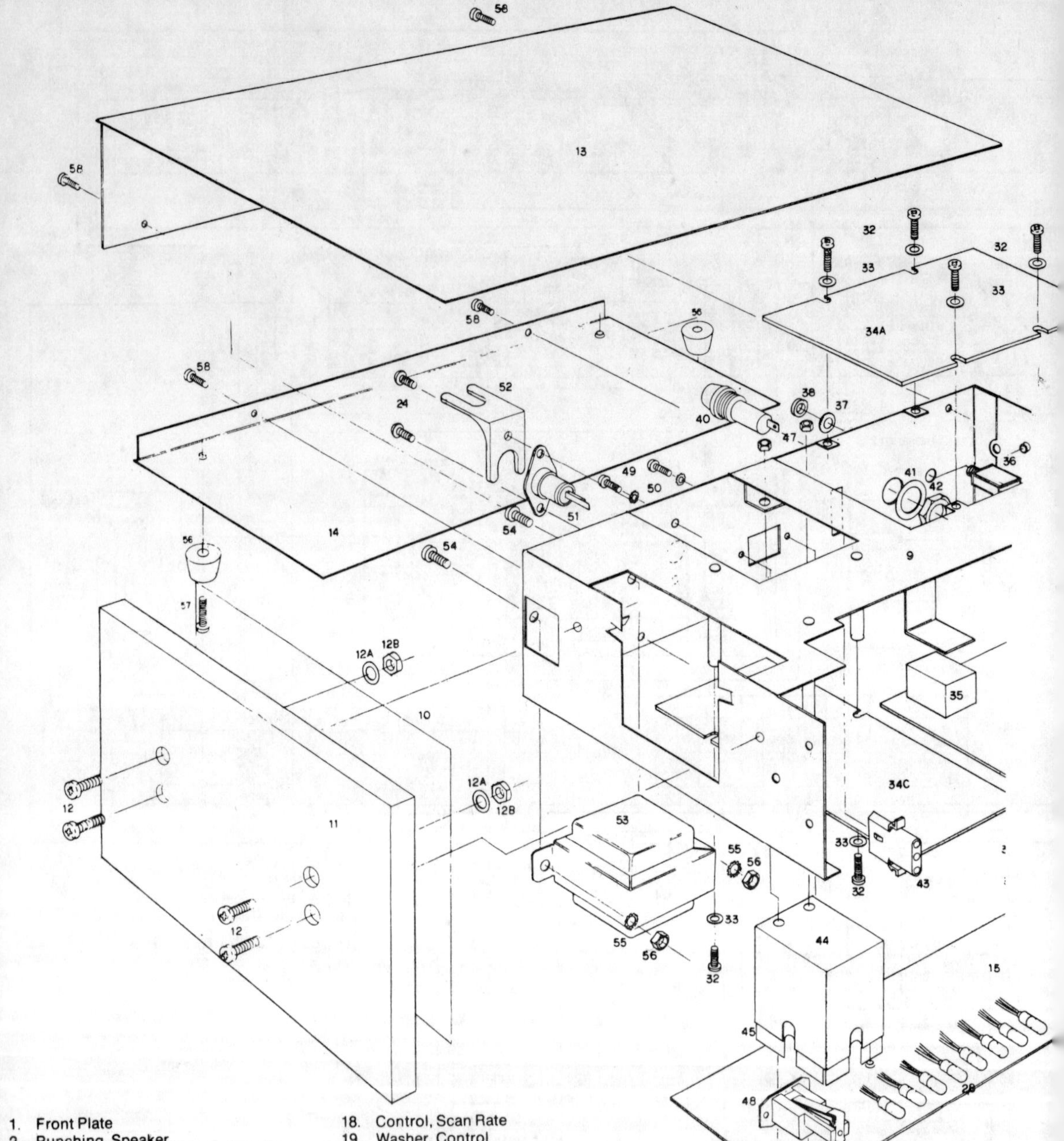

1. Front Plate
2. Punching, Speaker
3. Extrusion, Aluminum
4. Card Frame
5. Frequency Card
6. Card Cover
7. Plate for Brand
8. Chassis, Front
9. Chassis, Main
10. Punching, Metal
11. Wooded Side Cabinet
12. Screw 4×15 mm
13. Case, Top
14. Case, Bottom
15. Door, Crystal Access
16. Control, Volume with Power Switch
17. Control, Squelch
18. Control, Scan Rate
19. Washer, Control
20. Nut, Control
21. Knob, Volume, Squelch, Scan Rate
22. Spin for knob
23. 2-Pushbutton Switch
24. Screw, Self-Tapping 3×6 mm
25. Speaker
26. Chassis, Switch Mounting
27. 8-Pushbutton Switch
28. Lamp, Channel Indication
29. Rubber, Lamp Assemble
30. Lamp Assemble
31. Grommet, Channel Lamp
32. Screw, Self-Tapping 3×10 mm
33. Washer

34A. Printed Cricuit Board—AF, IF

34B. Printed Circuit Board, Scan
34C. Printed Circuit Board, RF
35. Rubber Cushion
36. Jack, External Speaker
37. Washer, External Speaker Jack
38. Nut, External Speaker Jack
39. Locker Button G-2-1
40. Holder, Fuse
41. Washer, Fuse Holder
42. Nut, Fuse Holder
43. Socket, DC Power
44. Inside Cover
45. Insulating Paper for Inside Cover
46. Screw 3×6 mm
47. Nut 3 mm
48. Socket, AC Power
49. Screw 2.6×6 mm
50. Nut 2.6 mm
51. Jack, Antenna
52. Holder, Antenna
53. Power Transformer
54. Screw 4×8 mm
55. Washer 4ϕ
56. Rubber Foot
57. Screw 2.6×4 mm
58. Screw Self-Tapping 3×6 mm

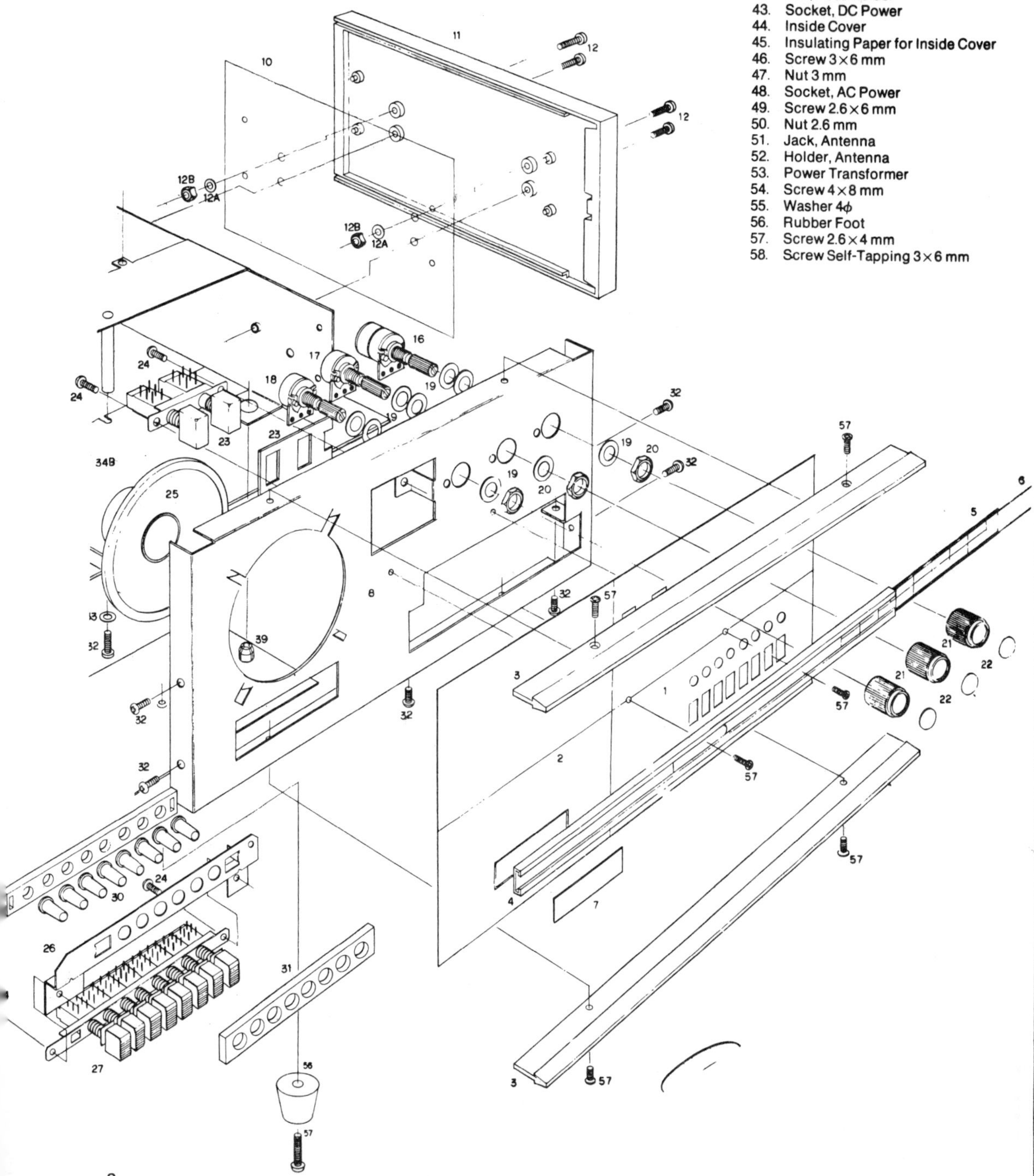

Fig. 5-5. Chassis layout—13-940.

1. CHASSIS, MAIN
2. CASE, TOP
3. CASE BOTTOM
4. MOUNTING BRACKET, AUTO
5. CHASSIS, FRONT
6. GRILLE, FRONT
7. PLATE, FRONT
8. KNOB, VOLUME, SQUELCH, SCAN RATE
9. GROMMET, CHANNEL LAMP
10. LAMP, CHANNEL LAMP
12. TRANSFORMER
13. CONTROL, SCAN RATE
14. CONTROL—VOLUME, SWITCH
15. CONT, SQUELCH
16. SWITCH, CHANNEL PUSHBUTTON ASSEMBLY
17. SWITCH, PUSHBUTTON ASSEMBLY FOR SCAN, MAN
18. JACK, ANTENNA
19. HOLDER, FUSE
20. JACK, EXTERNAL SPEAKER
21. SOCKET, AC POWER
22. CORD, AC POWER
23. COVER, AC SOCKET
24. SPEAKER, 8Ω 0.5W
25. BRACKET, SPEAKER
26. BUTTON, CABINET LOCKING
27. CORD, DC POWER
28. SOCKET, DC POWER
29. HOLDER, ANTENNA

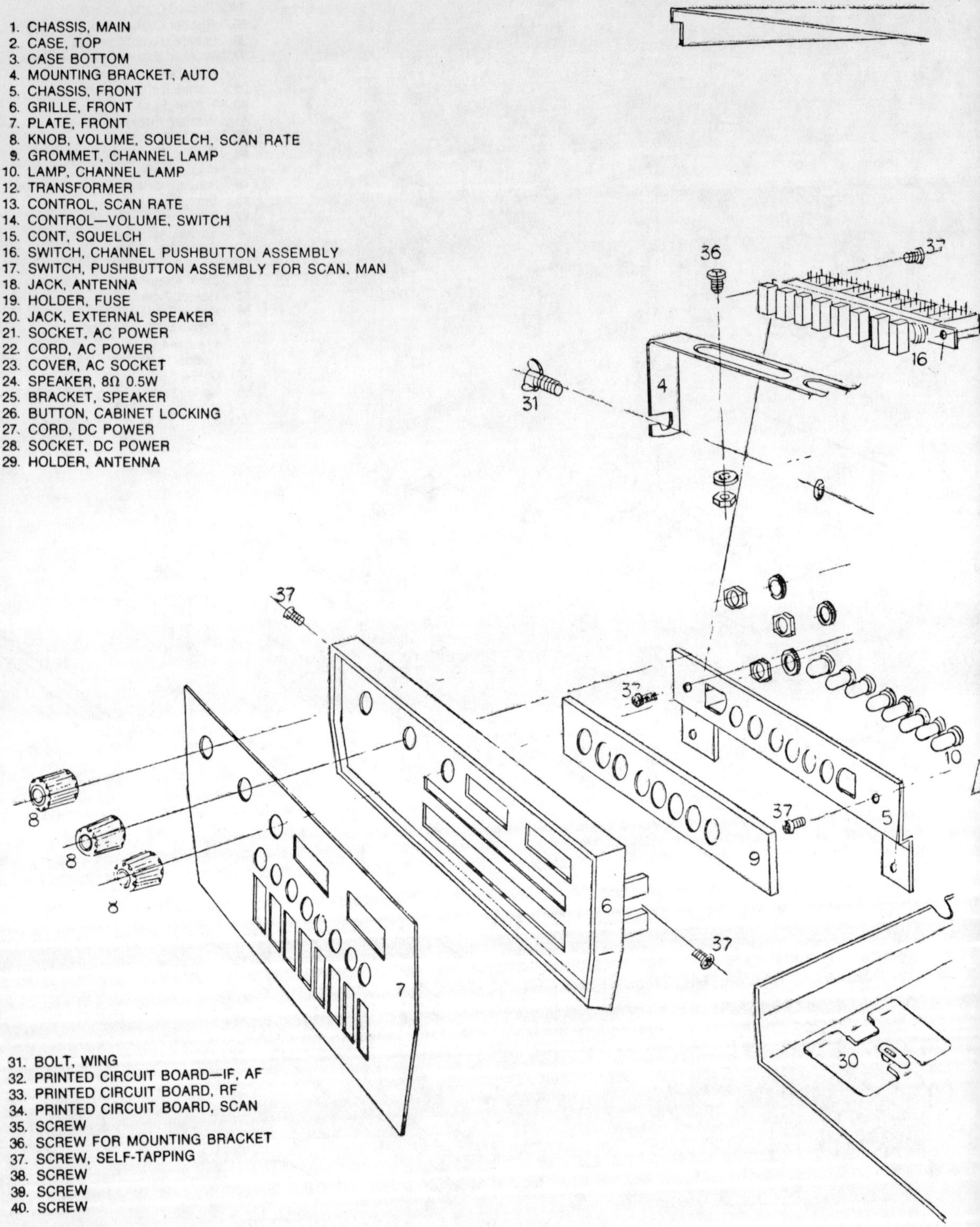

31. BOLT, WING
32. PRINTED CIRCUIT BOARD—IF, AF
33. PRINTED CIRCUIT BOARD, RF
34. PRINTED CIRCUIT BOARD, SCAN
35. SCREW
36. SCREW FOR MOUNTING BRACKET
37. SCREW, SELF-TAPPING
38. SCREW
39. SCREW
40. SCREW

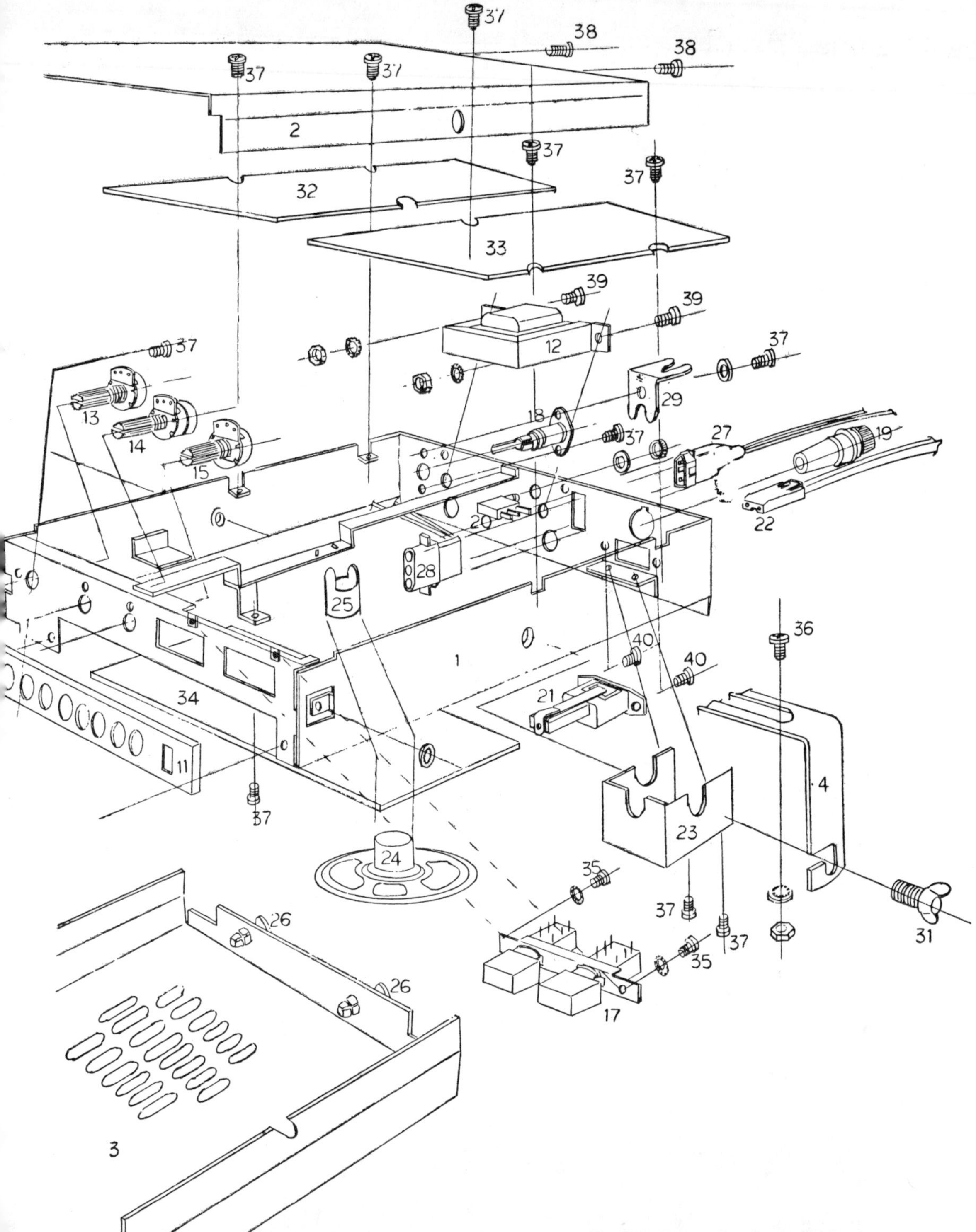

Fig. 5-6. Chassis layout 13-930.

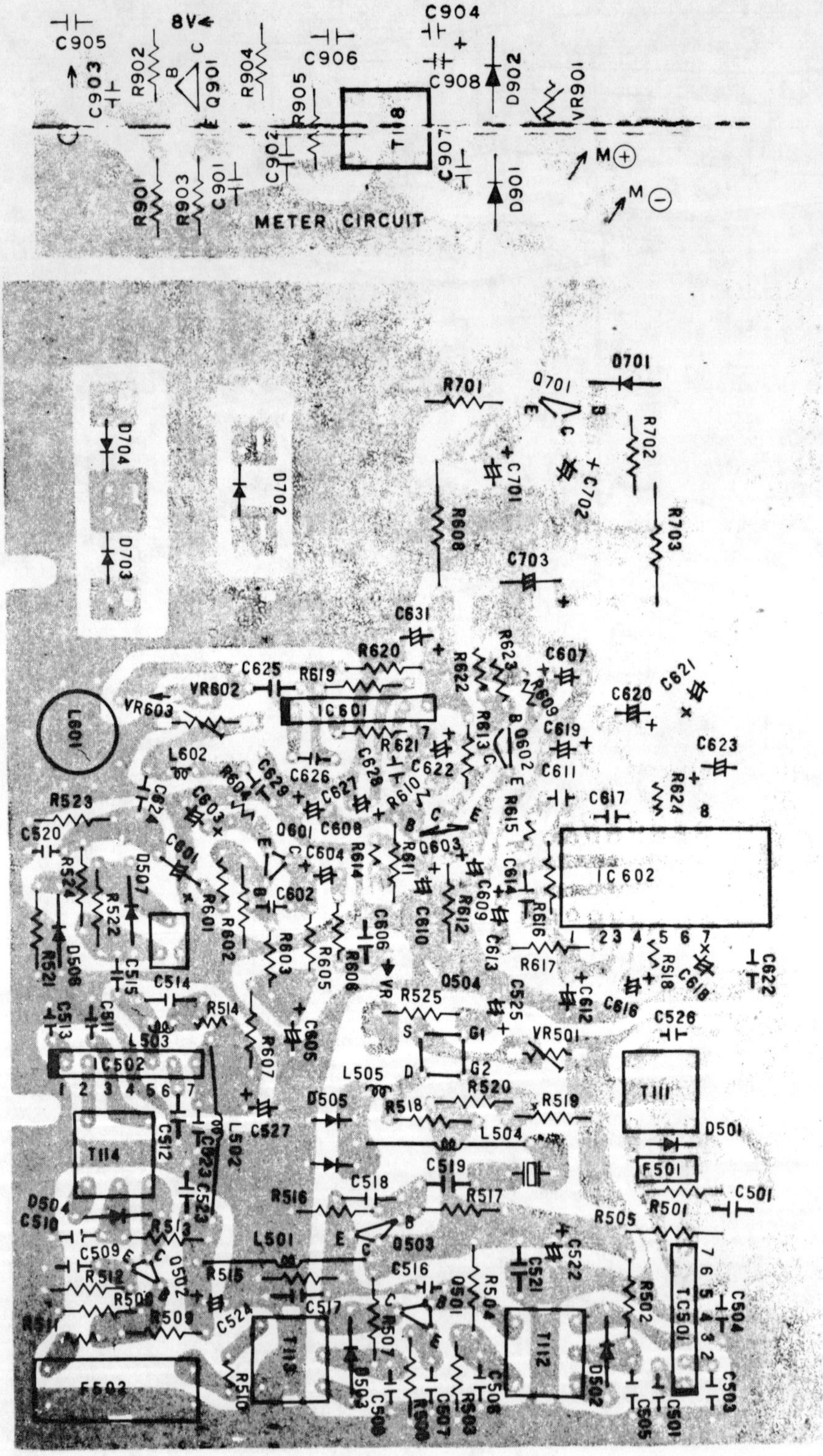
METER CIRCUIT

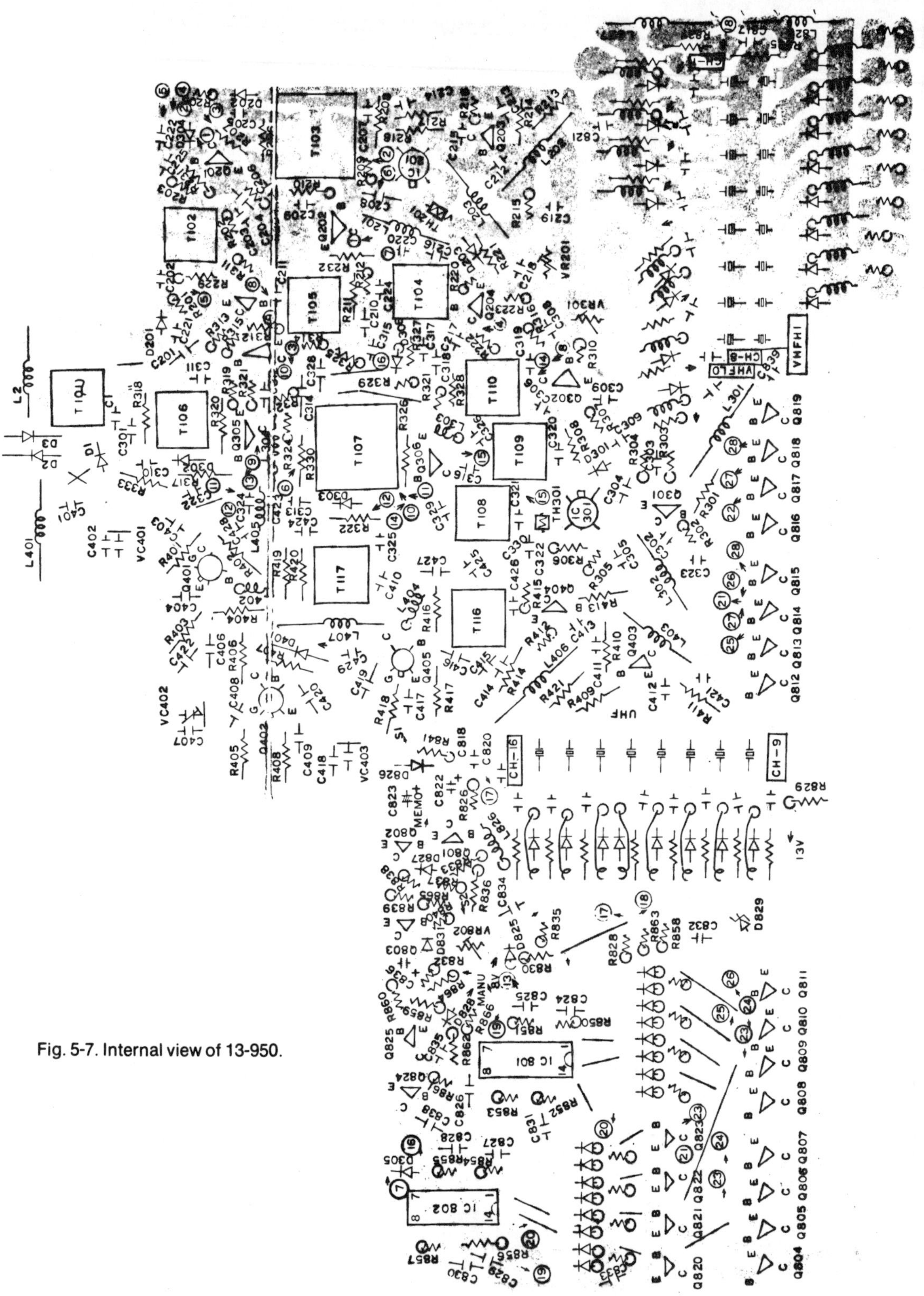

Fig. 5-7. Internal view of 13-950.

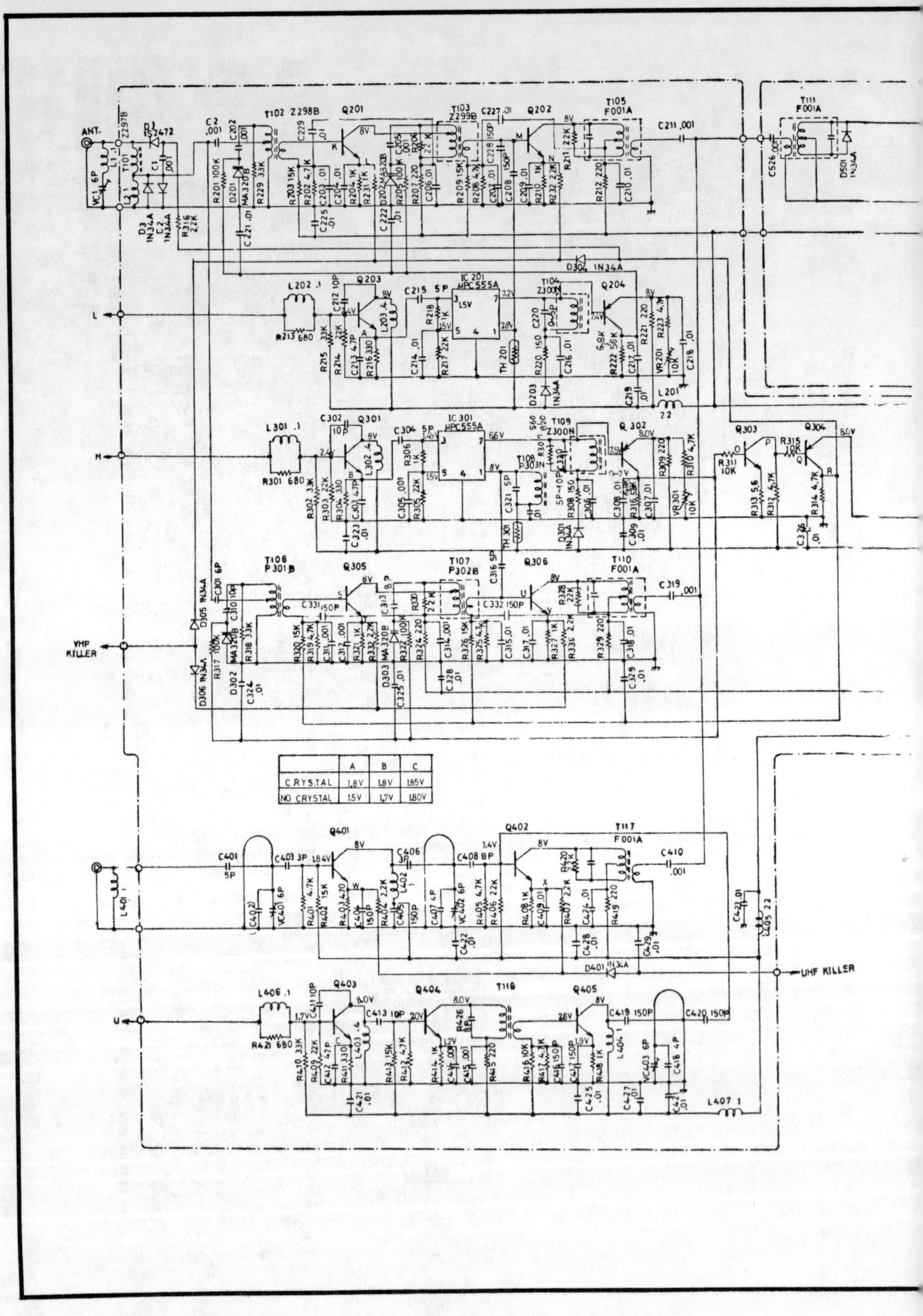

	A	B	C
CRYSTAL	1.8V	1.8V	1.85V
NO CRYSTAL	1.5V	1.7V	1.80V

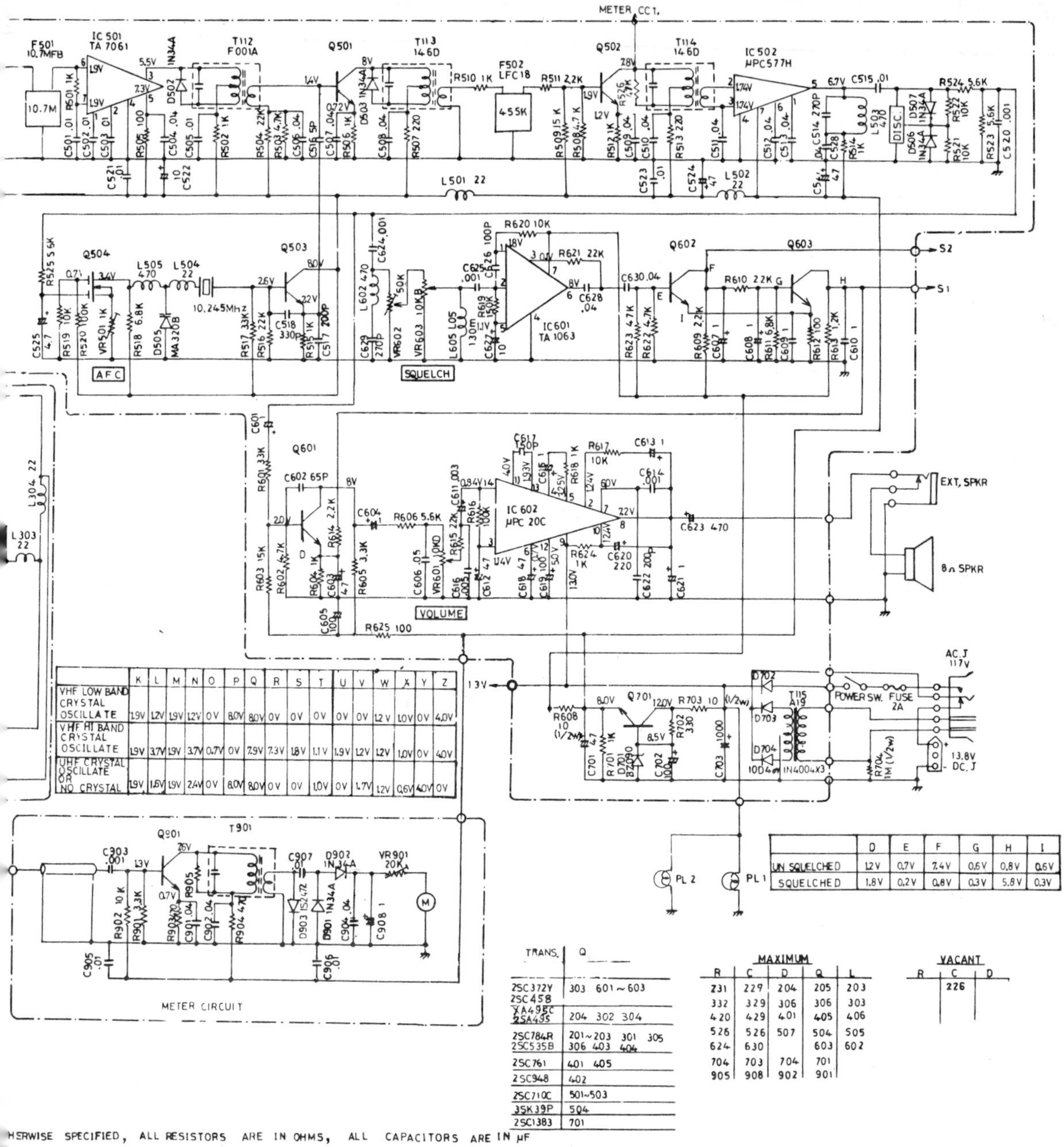

	K	L	M	N	O	P	Q	R	S	T	U	V	W	X	Y	Z
VHF LOW BAND CRYSTAL OSCILLATE	1.9V	1.2V	1.9V	1.2V	0V	8.0V	8.0V	0V	0V	0V	0V	0V	1.2V	1.0V	0V	4.0V
VHF HI BAND CRYSTAL OSCILLATE	1.9V	3.7V	1.9V	3.7V	0.7V	0V	7.9V	7.3V	1.8V	1.1V	1.9V	1.2V	1.2V	1.0V	0V	4.0V
UHF CRYSTAL OSCILLATE OR NO CRYSTAL	1.9V	1.6V	1.9V	2.4V	0V	8.0V	8.0V	0V	0V	1.0V	0V	1.7V	1.2V	0.6V	4.0V	0V

	D	E	F	G	H	I
UN SQUELCHED	1.2V	0.7V	7.4V	0.6V	0.8V	0.6V
SQUELCHED	1.8V	0.2V	0.8V	0.3V	5.8V	0.3V

TRANS.	Q
2SC372Y 2SC458	303 601~603
XA495C 2SA495	204 302 304
2SC784R 2SC535B	201~203 301 305 306 403 404
2SC761	401 405
2SC948	402
2SC710C	501~503
3SK39P	504
2SC1383	701

MAXIMUM

R	C	D	Q	L
231	229	204	205	203
332	329	306	306	303
420	429	401	405	406
526	526	507	504	505
624	630		603	602
704	703	704	701	
905	908	902	901	

VACANT

R	C	D
	226	

HERWISE SPECIFIED, ALL RESISTORS ARE IN OHMS, ALL CAPACITORS ARE IN μF
NDUCTORS ARE IN μH.
Y BE CHANGED FOR QUALITY CONTROL.

Schematic 5-3. The RF, IF, and audio sections of 13-950.

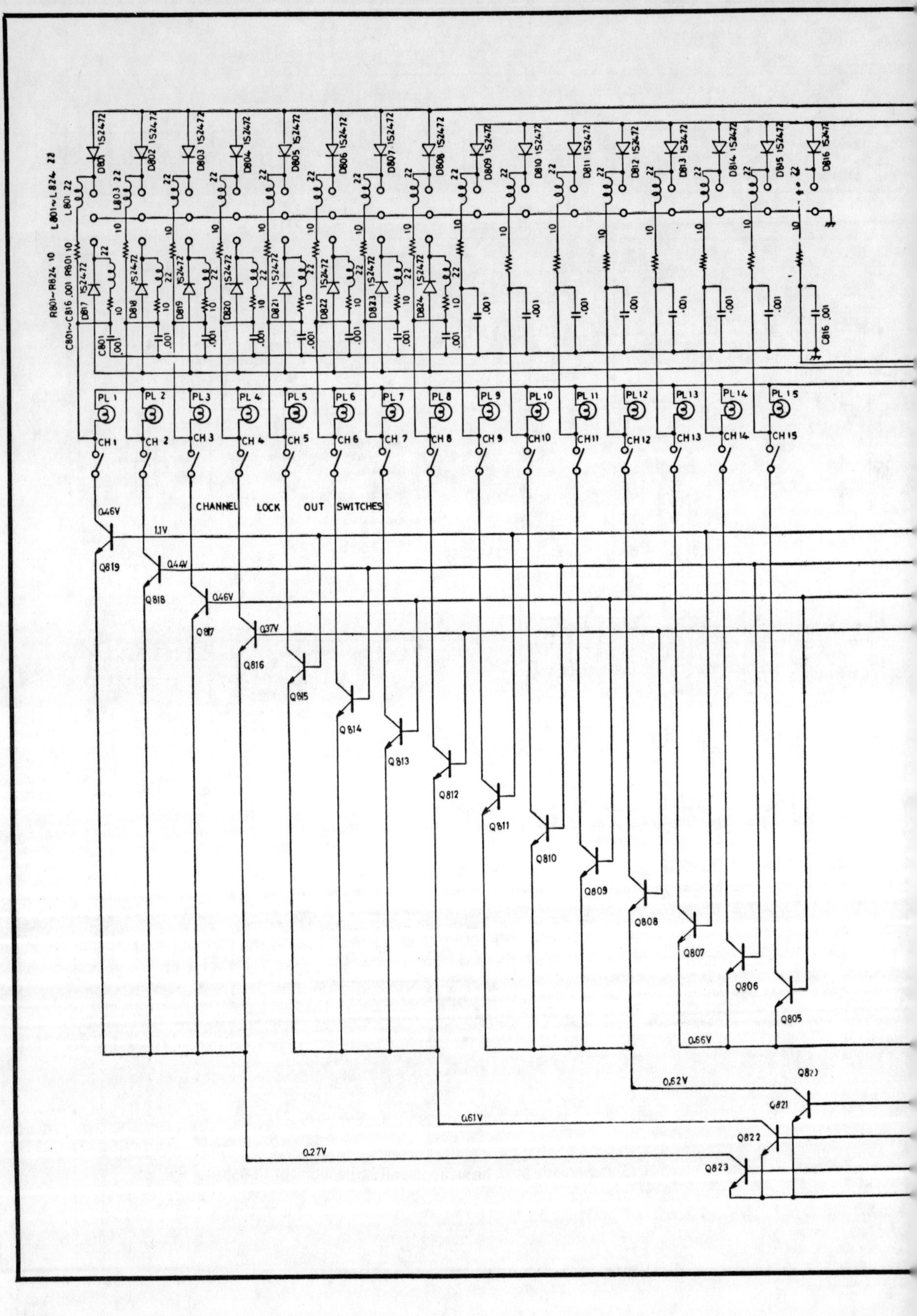
L801~L824 22
R801~R824 10
C801~C816 .001
CHANNEL LOCK OUT SWITCHES
PL 1
PL 2
PL 3
PL 4
PL 5
PL 6
PL 7
PL 8
PL 9
PL 10
PL 11
PL 12
PL 13
PL 14
PL 15
CH 1
CH 2
CH 3
CH 4
CH 5
CH 6
CH 7
CH 8
CH 9
CH 10
CH 11
CH 12
CH 13
CH 14
CH 15
Q819
Q818
Q817
Q816
Q815
Q814
Q813
Q812
Q811
Q810
Q809
Q808
Q807
Q806
Q805
Q821
Q822
Q823
0.46V
1.1V
0.44V
0.37V
0.66V
0.62V
0.61V
0.27V

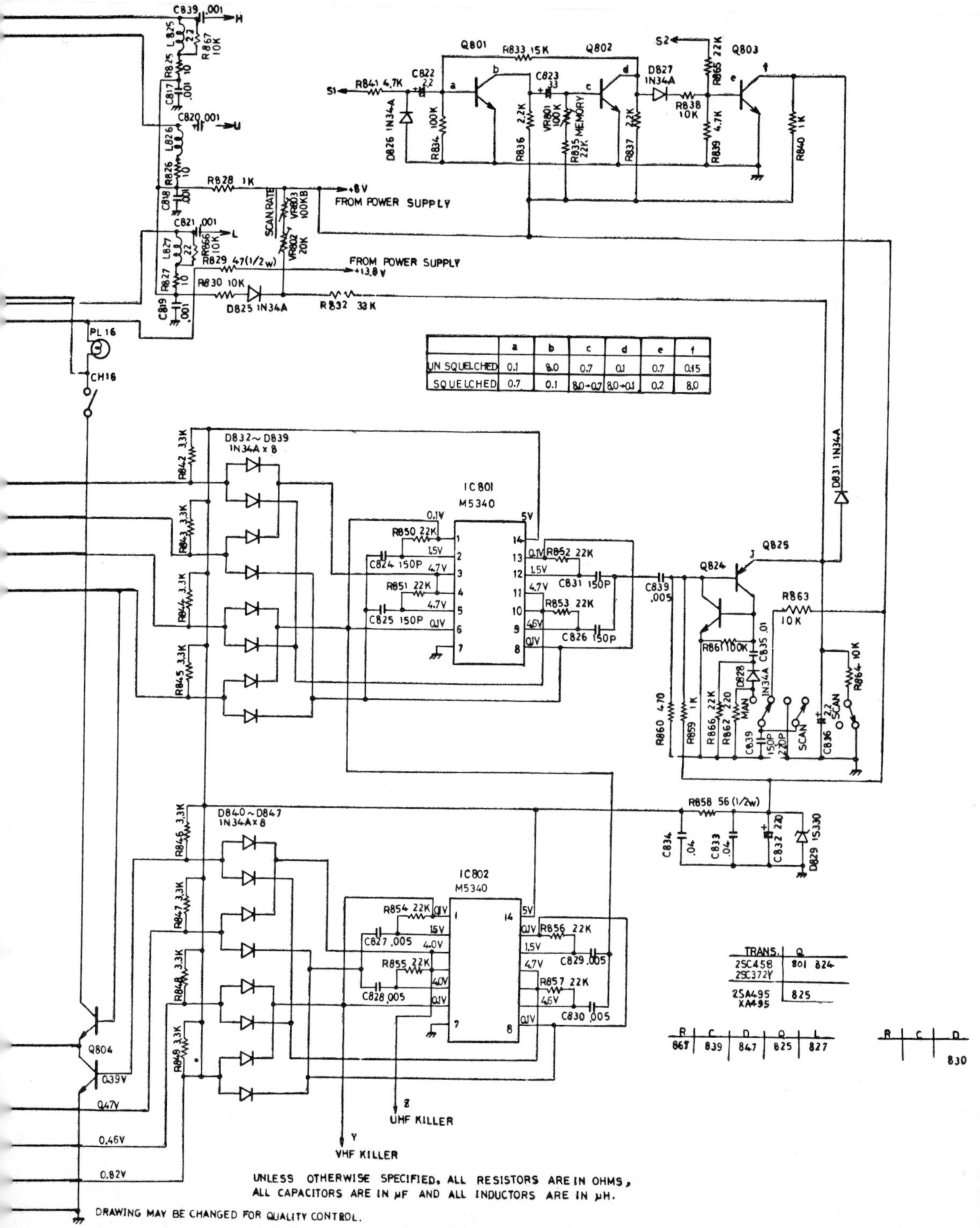

	a	b	c	d	e	f
UN SQUELCHED	0.1	8.0	0.7	0.1	0.7	0.15
SQUELCHED	0.7	0.1	8.0→0.7	8.0→0.1	0.2	8.0

Schematic 5-4. Scan section of 13-950.

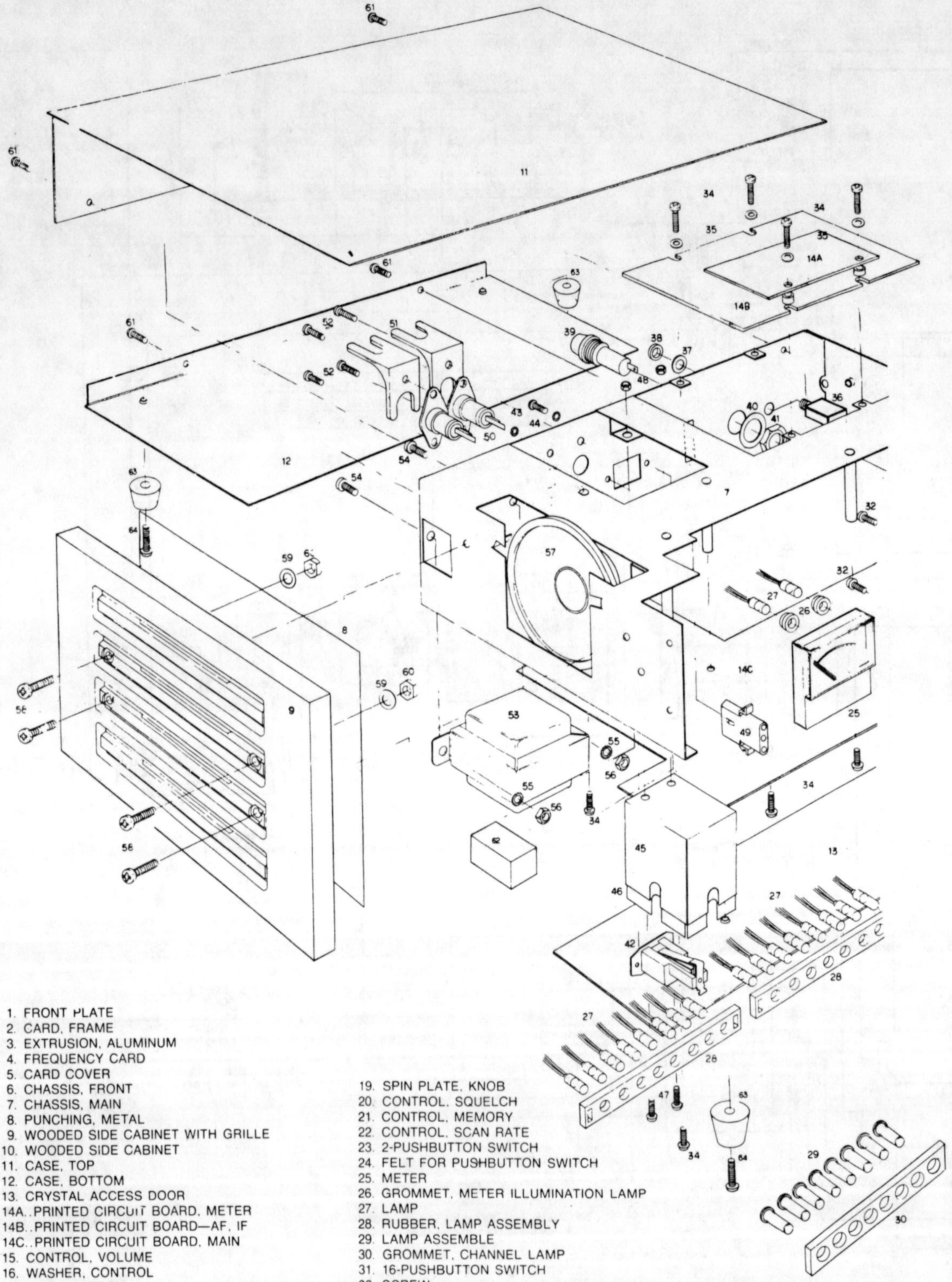

1. FRONT PLATE
2. CARD, FRAME
3. EXTRUSION, ALUMINUM
4. FREQUENCY CARD
5. CARD COVER
6. CHASSIS, FRONT
7. CHASSIS, MAIN
8. PUNCHING, METAL
9. WOODED SIDE CABINET WITH GRILLE
10. WOODED SIDE CABINET
11. CASE, TOP
12. CASE, BOTTOM
13. CRYSTAL ACCESS DOOR

14A..PRINTED CIRCUIT BOARD, METER

14B..PRINTED CIRCUIT BOARD—AF, IF

14C..PRINTED CIRCUIT BOARD, MAIN

15. CONTROL, VOLUME
16. WASHER, CONTROL
17. NUT, CONTROL
18. KNOB, VOLUME, SCAN RATE, SQUELCH, MEMORY
19. SPIN PLATE, KNOB
20. CONTROL, SQUELCH
21. CONTROL, MEMORY
22. CONTROL, SCAN RATE
23. 2-PUSHBUTTON SWITCH
24. FELT FOR PUSHBUTTON SWITCH
25. METER
26. GROMMET, METER ILLUMINATION LAMP
27. LAMP
28. RUBBER, LAMP ASSEMBLY
29. LAMP ASSEMBLE
30. GROMMET, CHANNEL LAMP
31. 16-PUSHBUTTON SWITCH
32. SCREW
33. SCREW

34. SCREW
35. WASHER
36. JACK, EXTERNAL SPEAKER
37. WASHER, EXTERNAL SPEAKER JACK
38. NUT, EXTERNAL SPEAKER JACK
39. HOLDER, FUSE
40. WASHER, FUSE HOLDER
41. NUT, FUSE HOLDER
42. SOCKET, AC POWER
43. SCREW
44. WASHER
45. INSIDE COVER
46. INSULATING PAPER FOR INSIDE COVER
47. SCREW
48. NUT
49. SOCKET, DC POWER
50. JACK, ANTENNA
51. HOLDER, ANTENNA
52. SCREW
53. POWER TRANSFORMER
54. SCREW
55. WASHER
56. NUT
57. SPEAKER
58. SCREW
59. WASHER
60. NUT
61. SCREW
62. CUSHION
63. RUBBER FOOT
64. SCREW

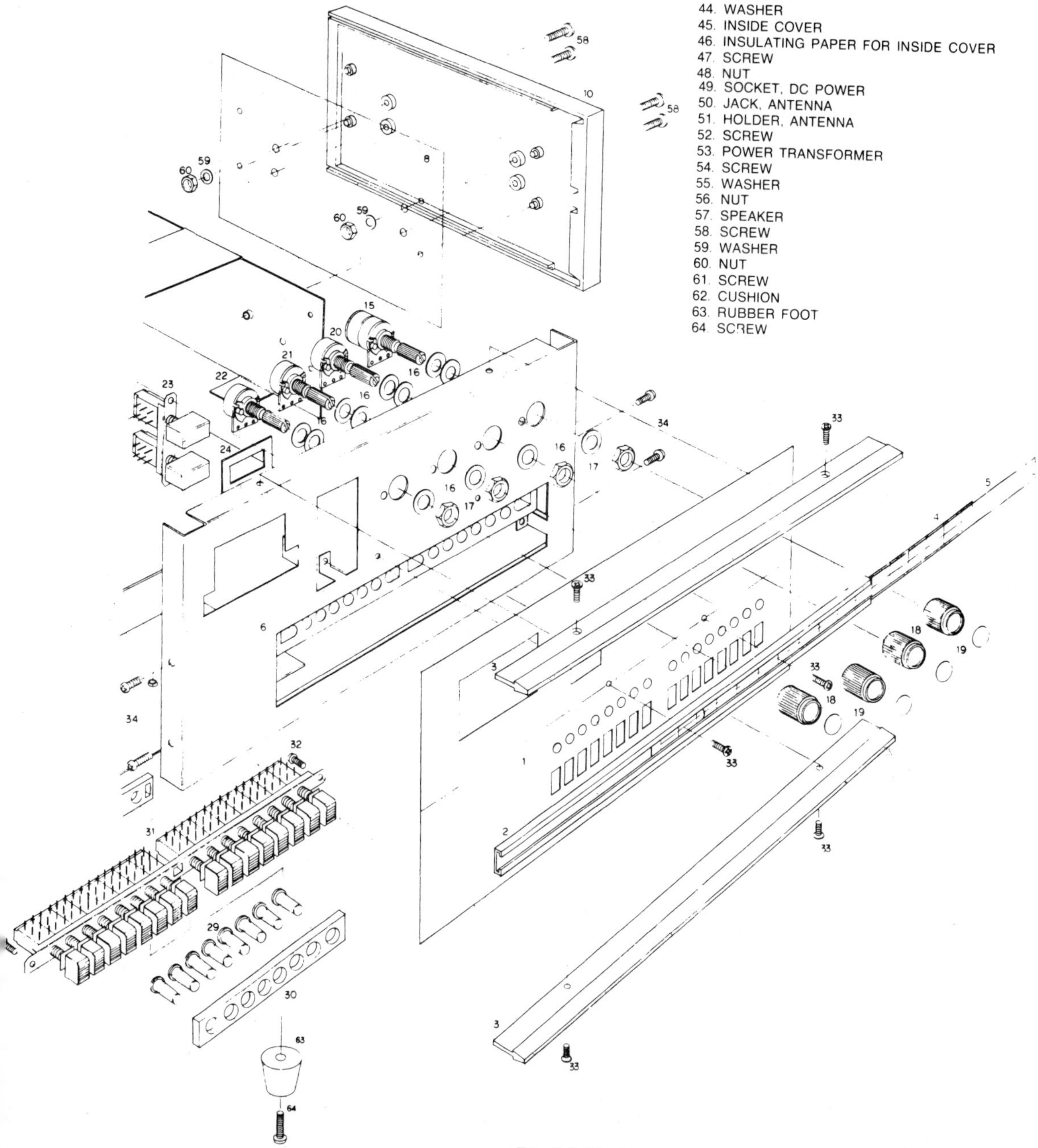

Fig. 5-8. Exploded view of 13-950.

F—Alignment, Second IF

1. Turn VOLUME control to obtain proper audio level (around 0.5W).
2. Turn SQUELCH control fully counterclockwise.
3. Apply a 455 kHz unmodulated signal to the base of Q501 (second mixer) through a 0.001 μF capacitor.
4. Adjust T113 and T114 for maximum quieting (minimum noise) while decreasing the output of the signal generator. With the output of the signal generator at about 300 μV, quieting sensitivity should be around 20 dB (monitored at speaker).

G—Alignment, First IF Section

5. Apply a 10.7 MHz unmodulated signal to the base of Q202 or Q306 through a 0.001 μF capacitor.
6. Adjust T105 (T110), T111, and T112 for maximum noise quieting while decreasing the output of the signal generator. The 100 μV quieting sensitivity should be around 20 dB (monitored at speaker).

H—Alignment, RF Section

1. Short out T101.
2. Apply a channel frequency in high band from signal generator to antenna connector of the unit. Modulation of signal generator should be 1 kHz with 2 kHz deviation.
3. Adjust T106, T107, T108, and T110 on high band for best audio waveform in terms of sine wave, while decreasing output of signal generator.
4. Change the frequency of signal generator to low-band channel frequency.
5. Adjust T102, T103, and T105 for low-band RF section alignment.
6. Adjust T110 (T105), T111, and T112 to obtain best sinusoidal audio wave on oscilloscope while decreasing the output of signal generator.
7. Check the difference between noise level and signal level on VTVM. If it is less than 3 dB, your adjustment is correct.
8. T101 should be replaced in the circuit.

I—Squelch Adjustment

1. With signal generator, apply any frequency for which a crystal is installed, modulated at 1 kHz with $\pm$4.5 kHz deviation, to antenna connector.
2. Turn SQUELCH control fully clockwise.
3. Increase the signal generator output until the audio output is silenced.
4. Slowly decrease the generator output until audio is heard or seen on the oscilloscope. Generator output level should be 1V to 0.5V to open squelch.
5. Adjust V602 to vary the tight squelch level.

J—Alignment, RF Section

1. Short out T101.
2. Apply a channel frequency in high band from signal generator to antenna connector of the unit. Modulation of signal generator should be 1 kHz with 2 kHz deviation.
3. Adjust T106, T107, T108, and T110 on high band for best audio waveform in terms of sine wave, while decreasing output of signal generator.
4. Change the frequency of signal generator to low-band channel frequency.
5. Adjust T102, T103, and T105 for low-band RF section alignment.
6. Change the frequency of signal generator to UHF-band channel frequency.
7. Adjust VC401, VC402, VC403, T116, and T117 for UHF-band RF section alignment.
8. Adjust T110 (T105), T111, and T112 to obtain best sinusoidal audio wave on oscilloscope while decreasing the output of signal generator.
9. Check the difference between noise level and signal level on VTVM. If it is less than 3 dB, your adjustment is correct.
10. T101 should be replaced in the circuit.

Parts List

MIDLAND 13-940

REF. NO.	DESCRIPTION	PART NO.
	CASE PARTS	
	Cabinet	13-010245
	Panel, Front	13-010243
	Lid, Rear	13-013108
	Lid, Crystal Compartment	13-018021
	Platė, Marking	13-020578
	Plate, Marking	13-020575
	Plate, Rating	13-020579
	Plate, Name	13-020574
	Knob, Volume/Squelch	13-110166
	Label	13-020577
	Cover	13-022001
	Cover, By-Pass Switch	13-022002
	Battery Holder Ass'y	13-030120
	Knob, Lever	13-155007
	Cap, Rotary Knob	13-110165
	Cap, Lever Knob	13-110167
	Knob, Push Button	13-115101
	HARDWARE	
	Bracket, By-Pass Switch	13-158332
	Bracket, PCB Mounting	13-158329
	Bracket, PCB Mounting	13-158330
	Bracket	13-158331
	Screw (Special), Rod Antenna	13-151406
	Lever	13-155008
	Spring, Lever	13-152008
	Terminal	13-155015
	Terminal	13-155013
	Terminal	13-155014
	MISCELLANEOUS	
SP	Speaker 8 OHM o.3W	13-068003
	Board, PC LED	13-070116
	Board, PC	13-070117
	Plate, Shield	13-155009
R502	Control, Trimmer 10KB	13-161003
R701	Control, Trimmer 10KD	13-161004
	Antenna, Telescopic	13-040078
	Switch, Slide	13-180084
	Switch, Auto-Man/Ch. Select	13-180085
	Socket, Crystal	13-159119
	Socket	13-159181
	Socket	13-159180
	Crystal 10.245 MHZ HC-18/U	13-128317
	Plug	13-159182
	COILS & TRANSFORMERS	
L101, 103	Coil, RF	13-176429
L102, 104	Coil, RF	13-176430
L105	Coil, VHF	13-090317
L106	Coil, RF	13-176428
L501	Choke 8.2 mH	13-178183
L502, 503 504, 505	Choke 25 uH	13-178127

REF. NO.	DESCRIPTION	PART NO.
T301	IFT	13-090320
T302	IFT	13-090323
	FILTERS	
HF301	Crystal 10F15A	13-179042
HF302	Ceramic CFU-455D	13-179041
	TRANSISTORS	
Q101, 102,103	2SC 922	09-309072
Q301,2,3,4,5, 6,7,8	2SC 930	09-302037
Q501,2,3,4,5, 6,7,8,9,10, 11,12 & 701 & 2	2SC 945	09-302125
Q703	2SD 227	09-303025
Q704	2SA 642	09-305134
	DIODES	
D301,302 501, 513	IN 60	09-306336
D514,16,18,20	TLR 104 LED	09-306371
D515,17,19,21	IS 953	09-306313
D701	VD 1220	09-306372
	INTEGRATED CIRCUIT	
IC 501	CMOS MM74COON	09-308074

RESISTORS

ALL RESISTORS NOT SHOWN ON THIS PARTS LIST ARE CARBON, 1/4W. SE SPECIFIC VALUES.

REF. NO.	DESCRIPTION	PART NO.
R708	150 OHM ½W	77-102151

CAPACITORS

ALL CAPACITORS NOT SHOWN ON THIS PARTS LIST ARE CERAMIC DISC, 50 WV. SEE SCHEMATIC FOR SPECIFIC VALUES.

REF. NO.	DESCRIPTION	PART NO.
	CERAMIC, 25V.	
C115,301	0.01 uF	78-121103
C522	0.04 uF	78-121403
	MYLAR, 50V.	
C318,503	0.001 uF	78-651102
C319,501,704	0.01 uF	78-651103
C502	0.0033 uF	78-651332
C505	0.0047 uF	78-651472
C506,508	0.047 uF	78-651473

REF. NO.	DESCRIPTION	PART NO.	REF. NO.	DESCRIPTION	PARTS NO.

MIDLAND 13-940—Continued

REF. NO.	DESCRIPTION	PART NO.	REF. NO.	DESCRIPTION	PARTS NO.
	ELECTROLYTICS				
C306	10 uF 10V	77-336106	C511,513,514	0.47 uF 10V	77-336474
C309,312,320	0.1 uF 10V	77-336104	C512,705,706	100 uF 6.3V	77-332107
C504,507	1 uF 10V	77-336105	C701,702	0.22 uF 10V	77-336224
C509,510	3.3 uF 10V	77-336335	C703	33 uF 6.3V	77-332336

MIDLAND 13-950

REF. NO.	DESCRIPTION	PARTS NO.
	CASE PARTS	
	Case Top	13-010208
	Case Bottom	13-010209
	Door, Crystal Access	13-018017
	Front Plate	13-020521
	Extrusion, Aluminum	13-158283
	Card Frame	13-158284
	Frequency Card	13-033123
	Label FCC	13-023154
	Label Caution	13-033119
	Label Caution (B)	13-033120
	TRANSISTORS	
Q201-203, 301, 305, 306, 403, 404	2SC784R (2SC535B)	09-302095
Q303, 601-603, 801-824	2SC372Y (2SC403C)	09-302039
Q204, 302, 304, 825	XA495C	09-304012
Q401, 405	2SC761	09-302149
Q402	2SC948	09-302148
Q501-503	2SC710C	09-302012
Q701	2SC1383	09-302155
Q504	3SK39P (FET)	09-305040
	DIODE	
D2, 3, 203, 304, 305, 306, 825-828, 831-847	1N34A	09-306020
D1, 801-824	1S2472	09-306219
D201, 202, 302, 303, 505	MA320B	09-306240
D829	1S330	09-306191
D701	BZ090	09-306180
D702-704	10D4 (1N4004)	09-306149
TH301, 210	TD5-A090	09-307072

REF. NO.	DESCRIPTION	PARTS NO.
	INTEGRATED CIRCUITS	
IC801, 802	M5340P	09-308015
IC201, 301	uPC555A	09-308013
IC501	TA7061	09-308041
IC502	uPC577H	09-308052
IC601	TA7063	09-308043
IC602	uPC20C	09-308053
	FILTERS	
F501	10.7MHz Ceramic Filter	13-179021
F502	455MHz Crystal Filter	13-179022
Disc.	455D Ceramic Disc.	13-179023
	RESISTORS	
	All resistors not shown on this parts list are carbon 1/8 watt. For specific values consult Schematic Diagram.	
R703, 608	Solid 10 ohm 1/2W	77-202100
R829	Solid 47 ohm 1/2W	77-202470
	ELECTROLYTIC CAPACITOR	
	All capacitors not shown on this parts list are available at local parts supply houses. Only electrolytic capacitors are shown on this list.	
C525	4.7μF/16V	77-337475
C620, 110	220μF/16V	77-337227
C703	1000μF/16V	77-337108
C612, 618, 527, 524, 701, 603	47μF/16V	77-337476
C522, 121	10μF/16V	77-337107
C631, 619, 702	100μF/16V	77-337107
C623	470μF/16V	77-337477
	MISCELLANEOUS	
	Knob, Volume/Squelch/Scan rate/Memory	13-110149
	Inside cover	13-158285
	Grommet, Channel Lamp	13-157166
	Cover, Channel Lamp	13-020509
	Lamp, Channel indicator and meter	13-201038
	Lamp assemble	13-201049
	Meter	13-200047
VR601	Control Volume	13-160087
VR602	Control Squelch	13-166048
VR803	Control Scan Rate	13-166039
VR801	Control Memory	13-166039
	16 Push Button Switch	13-188005
	2 Push Button Switch	13-183164

REF. NO.	DESCRIPTION	PARTS NO.
MIDLAND 13-950—Continued		
	Jack Antenna	13-153088
	Socket AC Power	13-159110
	Socket DC Power	13-159162
	Jack, External Speaker	13-153089
	Cord AC Power	13-034049
	Cord DC Power	13-034062
	Holder, Fuse	13-159154
	Socket, Crystal	13-159155
	Speaker 8 ohm 0.5W	13-060073
	Label Crystal	13-033124
	Fuse 2A	Purchase locally
	Fuse 1A	Purchase locally
	Telescopic Antenna/Motorola plug for VHF	13-040067
	Telescopic Antenna/Motorola plug for UHV	13-159163

COIL AND TRANSFORMERS

REF. NO.	DESCRIPTION	PARTS NO.
T101	Coil Z297N	13-094072
T102	Coil Z298B	13-094073
T103	Coil Z299B	13-094074
T104, 109	Coil Z300N	13-094071
T105, 110, 111, 112, 117	Transformer 10.7MHz	13-094047
T106	Coil P301B	13-094068
T107	Coil P302B	13-094069
T108	Coil P303N	13-094070
T113, 114	Transformer 455KHz	13-090282
T116	Coil P122DS	13-094004
L201, 303, 304, 405, 501, 502, 504, 801-827	Choke 22μH	13-178151
L503, 505, 602	Micro inductor 470μH	13-178152
L503, 505, 602	Micro inductor 470μH	13-178152
L203, 302, 403	Choke 0.4H	13-178153
L1, 202, 301, 406	Choke 0.1H	13-178154
L2, 401, 402, 404, 407	Choke 1.0H	13-178155
L605	Choke L05	13-178110

MIDLAND 13-930

CASE PARTS

REF. NO.	DESCRIPTION	PARTS NO.
2	Case, Top	13-010203
3	Case, Bottom	13-010204
30	Door, Crystal Accessory	13-018015
6	Grille, Front	13-020507
7	Plate, Front	13-020508
	Label, FCC	13-023150
	Label, Crystal Accessory	13-033116
8	Knob, Volume, Squelch, Scan Rate	13-033116
29	Holder, Antenna	13-158280

REF. NO.		DESCRIPTION	PARTS NO.
		MISCELLANEOUS	
	23	Cover, AC Socket	13-157165
	4	Mounting Bracket, Auto	13-158205
	9	Grommet, Channel Lamp	13-157166
		Cover, Channel Lamp	13-020509
		Lamp, Channel Indicator	13-201038
VR601	14	Control, Volume W/Swich	13-160087
VR602	15	Control, Squelch, VR602	13-166048
	16	Switch, Channel Push Button Assemble	13-188005
	17	Switch, Push Button Assembly for Scan, Man	13-183164
VR102	13	Control, Scan Rate	13-166039
	18	Jack, Antenna	13-153088
	21	Socket, AC Power	13-159110
	28	Socket, DC Power	13-159153
	20	Jack, External Speaker	13-153089
	22	Cord, AC Power	13-034049
	27	Cord, DC Power	13-034062
	19	Holder, Fuse	13-159154
		Socket, Crystal	13-159155
	31	Screw, Auto Mount Bracket	13-151378
	24	Speaker, 8ohm 0.5W	13-060073
		Label, Caution	13-033117
		Label, Crystal	13-033118
		Fuse, 2A	Purchase locally
		Fuse, 1A	Purchase locally
		Antenna, Telescopic/Motorala Plug	13-040067
		COILS AND TRANSFORMERS	
T101		Coils Z297N	13-094043
T102		Coil Z298B	13-094944
T103		Coil Z299B	13-094956
T104, 109		Coil Z300N	13-094946
T105, 110, 111, 112		Transformer 10.7MHz	13-094047
T106		Coil P301B	13-094048
T107		Coil P302B	13-094049
T108		Coil P303N	13-094050
T113, 114		Transformer 455KHz	13-094051
L101, 118, 504, 201, 303, 501, 502, 504		Choke Coil 22μH	13-178151
L503, 505, 602		Micro inductor 470μH	13-178152
L203, 302		Choke Coil 0.4μH	13-178153
L301, 202, L1		Choke Coil 0.1μH	13-178154
L2		Choke Coil 1.0μH	13-178155
L05, 605		AF Coil	13-178156

TRANSISTORS

REF. NO.	DESCRIPTION	PARTS NO.
Q301, 305, 306, 201-203	2SC784R(2SC535B)	09-302095
Q101-110, 112, 204, 303, 601-603	2SC372(2SC403)	09-302039

REF. NO.	DESCRIPTION	PARTS NO.
MIDLAND 13-930—Continued		
Q111, 302, 304, 204	XA495	09-304012
Q504	FET3SK39P	
Q701	2SC1383	09-302155
Q501, 502, 503	2SC710	09-302012
	DIODES	
D101-106, 1	1S247Z	09-306219
D201, 202, 302, 305, 505	MA320B	09-306240
D117-129, 132, 2,3 301, 501-504, 507	1N34A	09-306020
D131	1S330	09-306191
D701	BZ090	09-306180
D702, 703, 704	10D4(1N4004)	09-306149
TH301, 201	TD5-A090	09-307072
	INTEGRATED CIRCUITS	
IC101, 102	M5340P	09-308015
IC201, 301	μPC555A	09-308013
IC501	TA7061	09-308041
IC502	μPC577H	09-308052
IC601	TA7063	09-308043
IC602	μPC20C	09-308053
	FILTERS	
F501	10.7MHz Ceramic Filter	13-179021
F502	455KHz Ceramic Filter	13-179022
	455D Ceramic Disc.	13-179023
	RESISTORS	
	All resistros not shown on this parts list are carbon 1/8 watt. For specific values consult Schematic Diagram.	
R703, 608	Solid 10ohm 1/2W	77-202100
R119	Solid 47ohm 1/2W	77-202470
	CAPACITORS	
	All capacitors not shown on this parts list are available at local parts supply houses. Consult Schematic Diagram for specific values electrolytic capacitors are shown on this list	
C525	4.7μF/16V	77-337475
C620, 110	220μF/16V	77-337227
C703	1000μF/16V	77-337108
C612, 618, 527, 524 701, 603	47μF/16V	77-337476
C522, 121	10μF/16V	77-337106
C631, 619, 702	100μF/16V	77-337107
C623	470μF/16V	77-337477

REF. NO.	DESCRIPTION	PARTS NO.
MIDLAND 13-904		
	CASE PARTS	
	Case,Top	13-010208
	Case Bottom	13-010209
	Door, Crystal Access	13-018017
	Grille for Speaker	13-020516
	Front Plate	13-020517
	Extrusion, Aluminum	13-158283
	Card Frame	13-158284
	Plate for Brand	13-020518
	Frequency Card	13-033121
	Label FCC	13-025152
	Label Caution	13-033119
	Label Caution (B)	13-033120
	Knob, Volume/Squelch/Scan Rate	13-110149
	Inside Cover	13-158285
	Grommet, Channel Lamp	13-157166
	Cover, Channel Lamp	13-020509
	Lamp, Channel Indication	13-201038
	Lamp Assemble	13-201049
VR602	Control Squelch	13-166048
VR601	Control Volume	13-160087
VR102	Control Scan Rate	13-166039
	8-Push Button Switch	13-188005
	2 Push Button Switch	13-183614
	Jack Antenna	13-153088
	Socket AC Power	13-159110
	Socket DC Power	13-159162
	Jack External Speaker	13-153089
	Cord AC Power	13-034049
	Cord DC Power	13-034062
	Holder Fuse	13-159154
	Socket, Crystal	13-159155
	Speaker, 8ohm 0.5W	13-060073
	Label Crystal	13-033118
	Fuse 2A	Purchase locally
	Fuse 1A	Purchase locally
	Telescopic Antenn/Motrala Plug	13-040067
	COILS AND TRANSFORMER	
T101	Coil Z297N	13-094043
T102	Coil Z298B	13-094044
T103	Coil Z299B	13-094045
T104, 109	Coil Z300N	13-094046
T105, 110, 111, 112	Transformer, 10.7MHz	13-094047
T106	Coil P301B	13-094048
T107	Coil P302B	13-094049
T108	Coil P303	13-094045
T113, 114	Transformer	13-094051
L101-118, 504, 201 L303	Choke Coil 22μH	13-178151

REF. NO.	DESCRIPTION	PARTS NO.
MIDLAND 13-904—Continued		
L503, 505, 602	Micro inductor 470μH	13-178151
L203, 302	Choke Coil 0.4μH	13-178153
L301, 202, 1	Choke Coil 0.1μH	13-178154
L2	Choke Coil 1.0μH	13-178155
L605	AF Coil L05	13-178156
	TRANSISTORS	
Q301, 306, 201-203	2SC784R (2SC535B)	09-302095
Q101-110, 112, 204 Q303 601-603	2SC372 (2SC403)	09-302039
Q111, 302, 304, 204	XA495	09-304012
Q504	FET 3SK39P	09-305040
Q701	2SC1383	09-302155
Q501, 502, 503	2SC710	09-302012
	DIODE	
D101-106, 1	1S2472	09-306219
D201, 202, 302, 303 D505	MA320B	09-306240
D117-129, 132, 213	1N34A	09-306020
D203, 301, 501-504 D506, 507	1N34A	09-306020
D131	1S330	09-306191
D701	BZ090	09-306180
D702, 703, 704	10D4 (1N4004)	09-306149
TH301, 201	TD5-A090	09-307072
	INTEGRATED CIRCUITS	
IC101, 102	M4340P	09-308015
IC201, 301	μPC555A	09-308013
IC501	TA7061	09-308041
IC502	μPC577H	09-308052
IC601	TA7063	09-308043
IC602	μPC20C	09-308053
	FILTERS	
F501	10.7MHz Ceramic Filter	13-179021
F502	455KHz Ceramic Filter	13-179022
Disc.	455D Ceramic Disc.	13-179023

RESISTORS

All resistors not shown on this parts list are carbon 1/8 watt. For specific values consult Schematic Diagram.

R703, 608	Solid 10ohm 1/2W	77-202100
R119	Solid 47ohm 1/2W	77-202470

REF. NO.	DESCRIPTION	PARTS NO.

MIDLAND 13-904—Continued

ELECTROLYTIC CAPACITORS

All capacitors not shown on this parts list are available at local parts supply houses. Only electrolytic capacitors are shown on this list.
Consult Schematic Diagram for specific values. Value and voltage are given for electrolytic.

REF. NO.	DESCRIPTION	PARTS NO.
C525	4.7μF/16V	77-337475
C620, 110	220μF/16V	77-337227
C703	1000μF/16V	77-337108
C612, 618, 527, 524 C701, 603	47μF/16V	77-337476
C522, 121	10μF/16V	77-337106
C631, 619, 702	100μF/16V	77-337107
C623	470μF/16V	77-337477

Chapter 6

Servicing Unimetrics' Dura Scan 8

Unimetrics' *Dura-Scan 8* is a high-quality import line and is actually three different monitors under a single model designation. It is designed to monitor any one of the three public service bands, depending on whether the low-band, VHF, or UHF module is installed.

Data is included for crystal ordering, as is a chart for modifying the low-band module for optimum reception of a specific segment of the band. If modification is deemed necessary, do it all. Half-way modification may cause degraded performance on all channels.

If reception of a wider range of frequencies than called for is desired, you may have to advise the customer to use two or more units. Your own experience will be your best guide here.

The *Dura-Scan 8* may be employed for fixed (home or office) applications by using the internal power supply, which supplies 12V DC from the 117V AC house current.

Specifications

Frequency range	Any eight crystal-controlled FM channels in the following frequency bands: • VHF low band: 30–50 MHz, total spread 8 MHz • VHF high band: 150–174 MHz, total spread 8 MHz • UHF band: 450–470 MHz, total spread 8 MHz
Sensitivity (20 dB quieting)	0.5 μV
Selectivity	60 dB down at ±25 kHz
Antenna	50Ω
Signal-to-noise ratio	30 dB
Bandwidth (−6 dB)	15 kHz
Intermediate frequencies	First, 10.7 MHz Second, 455 kHz
Spurious response rejection	More than 70 dB
Scanning rate	Approx 15 channels per sec
Current drain	Less than 500 mA
Power source	11.5V to 14.5V DC, negative ground; 117V, 50 or 60 Hz
Dimensions	8⅜ × 8¾ × 2⅞ in.
Net weight	5 lb

Operation on any of the three bands is possible through the use of interchangeable RF modules in the front end.

Crystal Data

Crystals used in *Dura-Scan 8* are type HC-25/U. Crystal frequencies are determined as follows:

Frequency for low-band VHF, 30–50 MHz:

$$\text{Crystal fundamental frequency} = \text{channel frequency} + 10.7\text{ MHz}$$

Frequency for high-band VHF, 150–174 MHz:

$$\text{Crystal 3rd overtone frequency} = \frac{\text{Channel frequency} - 10.7\text{ MHz}}{3}$$

Frequency for UHF band, 450–470 MHz:

$$\text{Crystal 9th overtone frequency} = \frac{\text{Channel frequency} - 10.7\ \text{MHz}}{9}$$

Frequency deviation should be less than 20 ppm in VHF bands, less than 10 ppm in UHF band.

Crystal Correlation

Resonance	Parallel
Overtone	Fundamental, third, ninth
Load capacity[1]	32 pF (UHF, 22 pF)
Maximum series resistance	Less than 40Ω
Frequency tolerance	±0.001% (at 25°C)

Service Data

Crystal layout	Fig. 6-1
Schematic	Schematic 6-1
Internal layout	Figs. 6-2, 6-3
Alignment data	Steps A, B (low band only), C (high band only), D (UHF only), E (all), A (repeated)

A—Alignment Procedure

1. Stop scanner on a channel as near to the center frequency of crystals installed as possible.
2. Feed a 10.7 MHz signal modulated with ±4.5 kHz deviation through a 0.001 μF capacitor to point *3* on the IF board.
3. Reducing the generator's output as necessary, set the VOLUME control to measure 0.5V across the speaker.
4. Adjust L301, L302, and L303 for maximum output with minimum distortion.
5. Repeat above until no further improvement can be made.
6. Transformer L301 may need touching up slightly when RF section is tuned.

B—Alignment, Low-Band Tuner

1. Oscillator circuit alignment: Connect a frequency counter to the collector of Q102 through a capacitor of 1 pF. Turn the core of L106 clockwise until oscillation stops; then turn the core one turn in the counterclockwise direction from the position at which the oscillation starts.
2. Tracking alignment: Set signal generator output (with no modulation) to provide a level which will give 20 dB noise-quieting sensitivity. Adjust L104 for maximum 20 dB noise-quieting sensitivity at the low end of the band. Adjust L103 and L105 for maximum 20 dB noise-quieting sensitivity at the center and high end of the band, respectively.
3. Changing center frequency (low band only): Since the *Dura-Scan 8* has been tuned to the standard center frequency (35 MHz) and its specific frequency spread is about ±4 MHz, to receive the signals higher or lower the center frequency must be shifted to the proper frequency by changing some capacitors according to the following chart. In this case the entire circuit must be readjusted as stated in steps B-1, B-2, and E.

[1] This particular number could vary depending upon crystal manufacturer, but crystal trimmer will take care of any error.

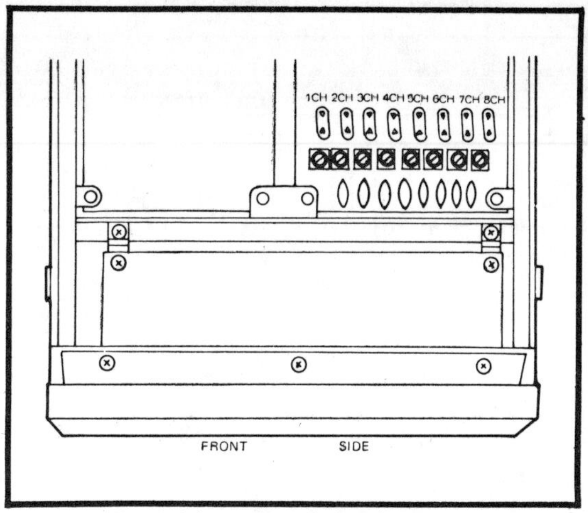

Fig. 6-1. Location of crystal sockets.

Frequency	30–37 MHz	37–44 MHz	44–50 MHz
	C103, 56 pF	C103, 56 pF	C103, 56 pF
Preselector	C104, 47 pF	C104, 47 pF	Remove C104
	C105, 56 pF	Remove C105	Remove C105
	C108, 47 pF	C108, 47 pF	C108, 47 pF

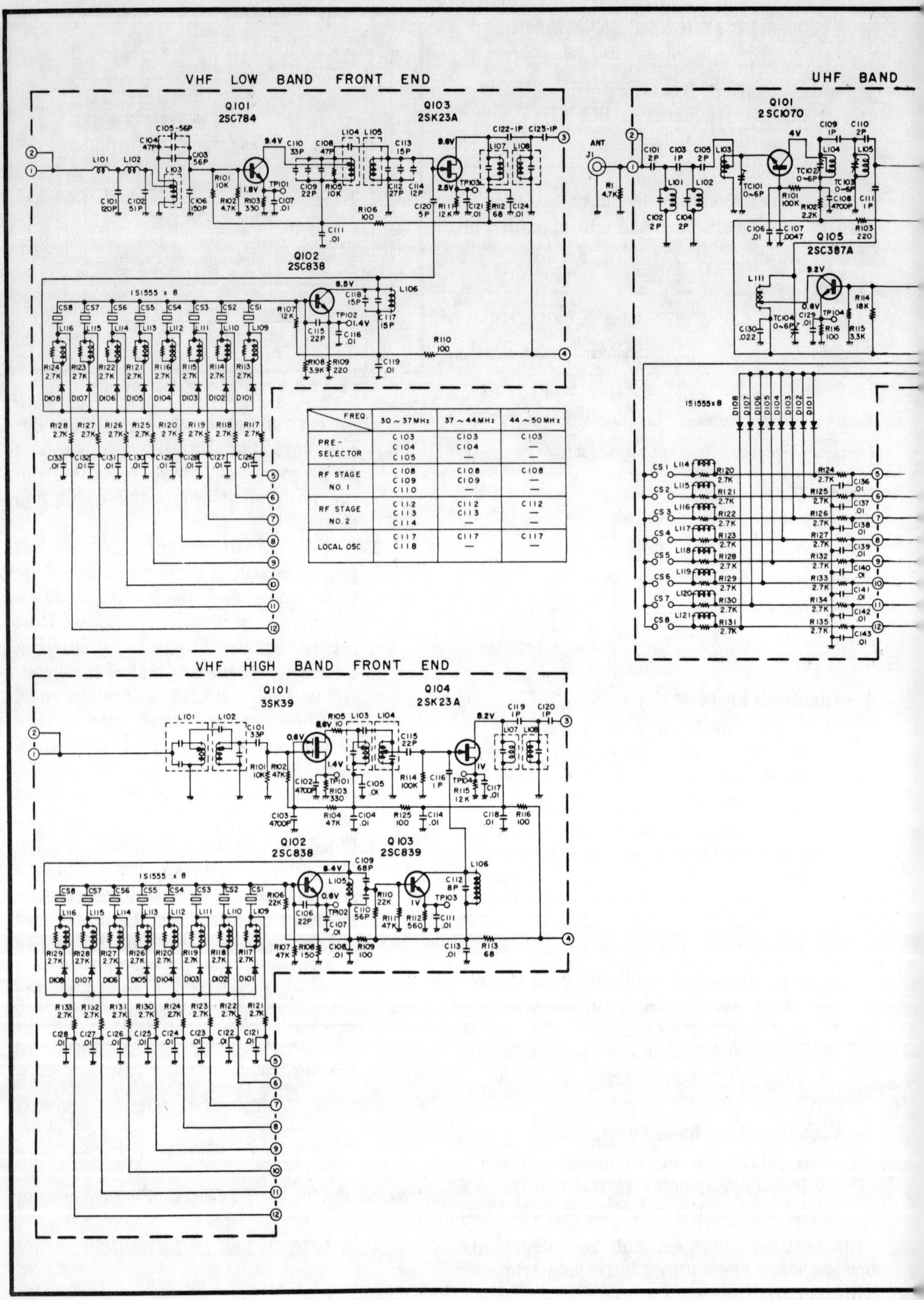

FREQ.	30 ~ 37MHz	37 ~ 44MHz	44 ~ 50MHz
PRE-SELECTOR	C103 C104 C105	C103 C104 —	C103 — —
RF STAGE NO. 1	C108 C109 C110	C108 C109 —	C108 — —
RF STAGE NO. 2	C11.2 C113 C114	C112 C113 —	C112 — —
LOCAL OSC	C117 C118	C117 —	C117 —

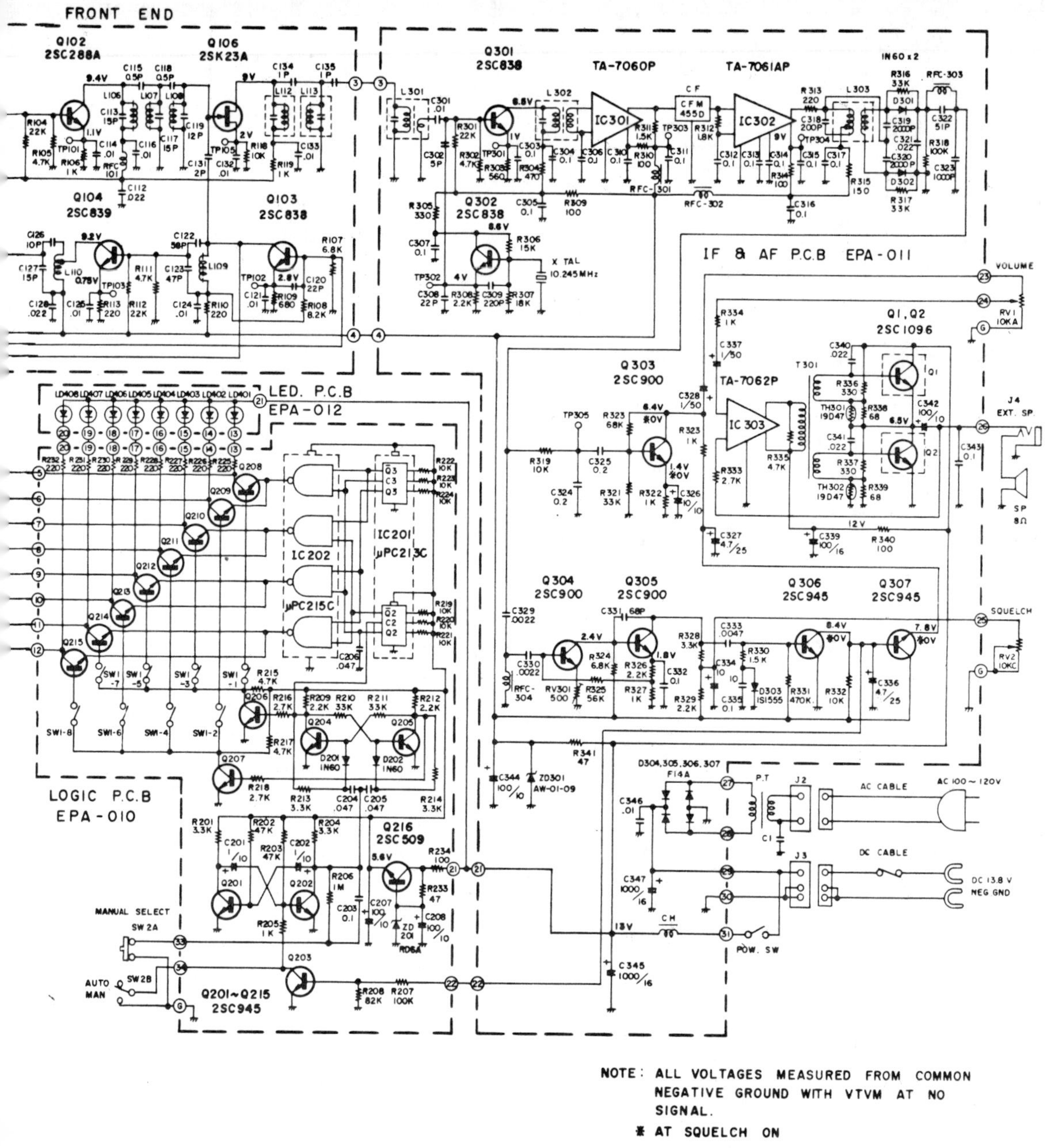

Schematic 6-1. Unimetrics Dura-Scan 8, showing high, low, and UHF boards.

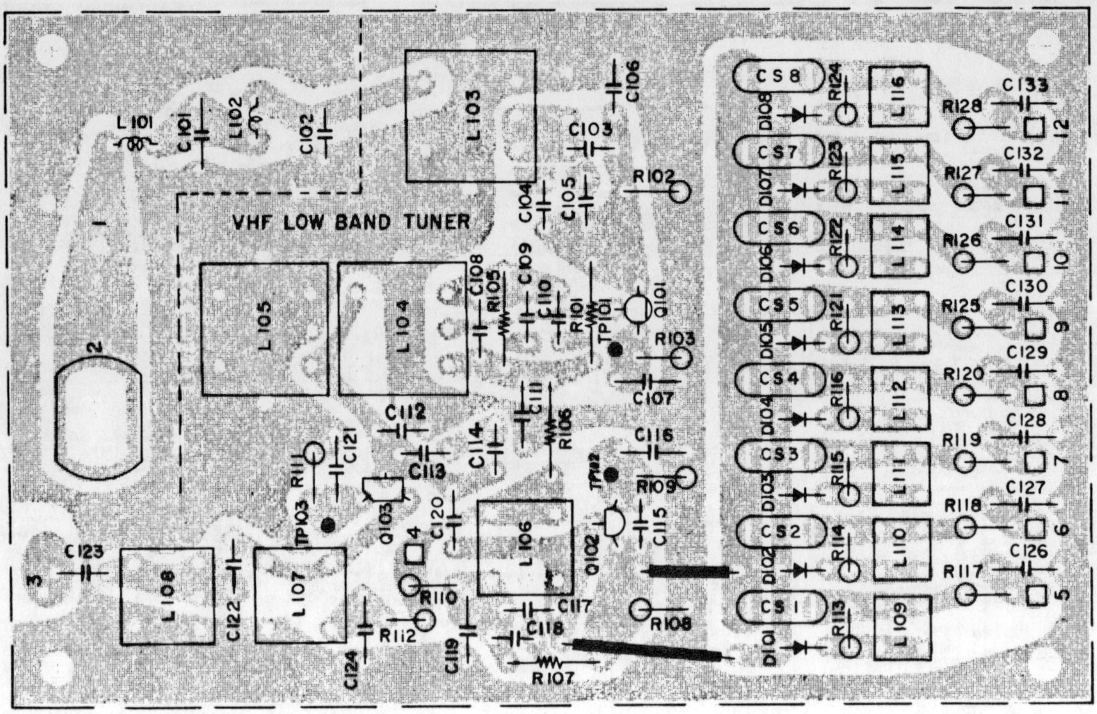

EPO-557L VHF LOW-BAND TUNER

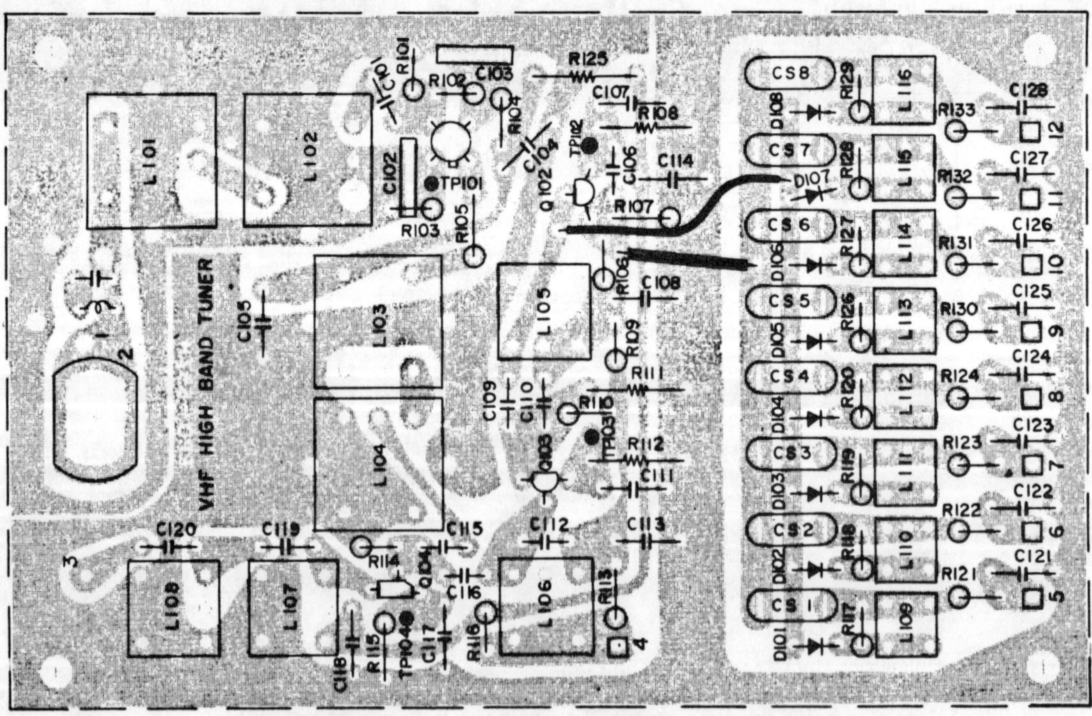

EPO-557H VHF HI-BAND TUNER

Fig. 6-2. Printed circuit board parts location diagram.

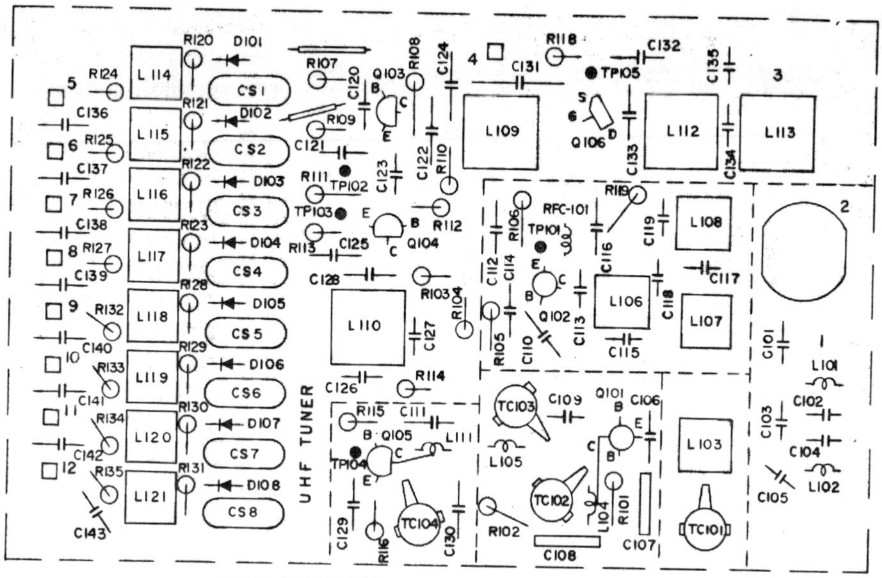

EPO-577U/2 UHF TUNER

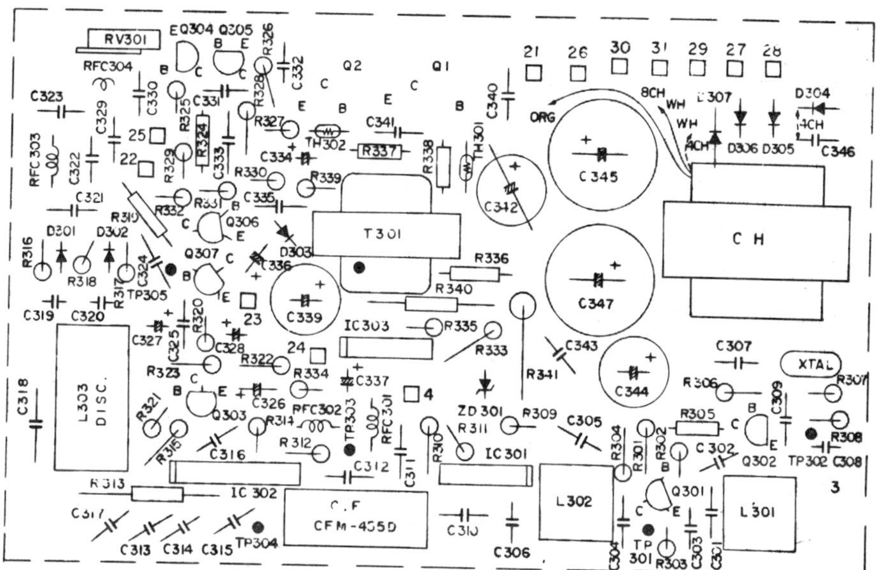

EPO-557A/2 IF AMP/AUDIO AMP

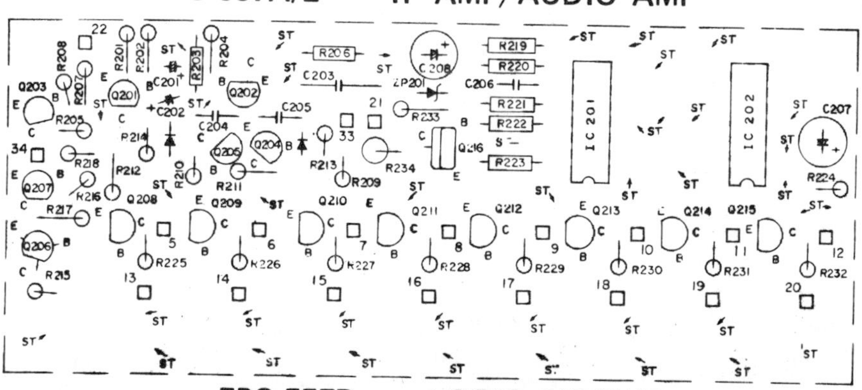

EPO-557D LOGIC CIRCUIT

Fig. 6-3. Internal layout Dura-Scan 8.

RF stage 1	C109. 22 pF	C109. 22 pF	'Remove C109
	C110. 33 pF	Remove C110	Remove 110
	C112. 27 pF	C112. 27 pF	C112. 27 pF
RF stage 2	C113. 15 pF	C113. 15 pF	Remove C113
	C114. 12 pF	Remove C114	Remove C114
Local oscillator	C117. 10 pF	C117. 10 pF	C117. 10 pF
	C118. 15 pF	Remove C118	Remove C118

C—Alignment, High-Band Tuner

1. Oscillator circuit alignment: Connect a frequency counter to the base of Q103 through a capacitor of 1 pF. Turn the core of L105 clockwise until oscillator stops its oscillation; then turn the core again three-fourths turn in the counterclockwise direction from the position at which the oscillation starts. Adjust L106 for the third overtone.
2. Tracking alignment: Set signal generator output (with no modulation) to provide about the level which will give 20 dB noise-quieting sensitivity. Adjust L102 and L104 for maximum 20 dB noise-quieting sensitivity at low end of the band. Adjust L101 and L103 for maximum 20 dB noise-quieting sensitivity at high end of the band.
3. Repeat the above steps until no further improvement is obtained.

D—Alignment, UHF-Band Tuner

1. Oscillator circuit alignment: Connect a frequency counter to the base of Q103 through a capacitor of 1 pF. Turn the core of L109 clockwise until oscillator stops its oscillation; then turn the core three-fourths turn in the counterclockwise direction from the position at which the oscillation starts. Adjust L110 for third overtone. Adjust L111 for ninth overtone.
2. Tracking alignment: Set a signal generator (with no modulation) to provide about the level which will give 20 dB noise-quieting sensitivity. Adjust TC101, TC102, TC103, L106, L107, and L108 for maximum 20 dB noise-quieting sensitivity.

E—Alignment, Crystal Frequency

The crystal frequency for each of eight channels may be adjusted slightly by turning the core of L109 (channel 1), L110 (2), L111 (3), L112 (4), L113 (5), L114 (6), L115 (7), and L116 (8).

Parts List

Description	Designation	Unimetrics' Part #
UNIMETRICS DURA-SCAN 8		
GENERAL		
P.C. Board	EPO-557D	60-114
Tie Point	MYO-9	60-115
P.C. Board	EPO-557A	60-116
Ceramic Filter	CF	60-117
Crystal Oscillator	EXT=14	60-118
Test Point	TP301, 302, 303, 304, 305	60-119
Tie Point	MYO-9	60-120
Shield Cover (IF)	MYO-119	60-121
P.C. Board	EPO-557F	60-122
Tie Point	MYO-9	60-123
CR Module	C1	60-124
DC Cord Ass.	ENO-9A	60-125

Description	Designation	Unimetrics' Part #
GENERAL		
AC Cord Ass.	ENO-33	60-126
Push Switch	SW1-1 -- 8	60-127
Push Switch	SW2A, B	60-128
Ant. Jack	J1	60-129
Speaker Jack	J4	60-130
DC Power Jack	J3	60-131
AC Power Jack	J2	60-132
Speaker	SP	60-133
Escutcheon Ass.	MM15P1	60-134
Chassis	MM15P2	60-135
Front Panel	MM15P3	60-136
Rear Panel	MM15P4	60-137
Cage	MM15P5	60-138
Mobile Mtg. Brckt.	MM15P7	60-139
Inner Carton Box	MM15P101	60-140
Partitioner	MM15P102	60-141
Partitioner	MM15P103	60-142
Master Carton Box	MM15P104	60-143
Poly-Bag (Manual)	MWO-16	60-144
" " (Accessory)	MWO-200	60-145
" " (Unit)	MZ0iP104	60-146
FCC Label	EBP-41	60-147
Antenna Catalog	EBP-149	60-148
Warranty Card	EBP-147U	60-149
Instruction Manual	EBP-153	60-150
RESISTORS		
Solid Resistor 100	R234	70-160
" " 220	R225, 226, 227, 228, 229, 230, 231, 232	70-161
" " 47	R233	70-162
" " 10K	R219, 220, 221, 222, 223, 224	70-163
" " 1K	R205	70-164
" " 100K	R207	70-165
" " 1M	R206	70-166
" " 2.2K	R209, 212	70-167
" " 2.7K	R216, 218	70-168
" " 3.3K	R214, 201, 204, 213	70-169
" " 4.7KΩ	R217, 202, 203, 215	70-170

Description	Designation	Unimetrics' Part #
RESISTORS		
" " 82KΩ	R208	70-171
Thermistor	TH301, 302	70-172
Potentiometer	RV301	70-173
Solid Resistor 100Ω	R309, 310, 314, 340	70-174
" " 1KΩ	R322, 323, 327, 334	70-175
" " 10KΩ	R332	70-176
" " 150KΩ	R315	70-177
" " 1.5KΩ	R311, 330	70-178
" " 15KΩ	R306	70-179
" " 1.8KΩ	R312	70-180
" " 18KΩ	R307	70-181
" " 220Ω	R313	70-182
" " 2.2KΩ	R308, 326, 329	70-183
" " 22KΩ	R301	70-184
" " 330Ω	R305, 336, 337	70-185
" " 3.3KΩ	R328	70-186
Solid Resistor 47Ω	R341	70-187
" " 470Ω	R304	70-188
" " 4.7KΩ	R302, 335	70-189
" " 560Ω	R303	70-190
" " 68Ω	R338, 339	70-191
" " 6.8KΩ	R324	70-192
Carbon Resistor 10KΩ	R319	70-193
" " 100KΩ	R318	70-194
" " 33KΩ	R316, 317, 321	70-195
" " 470KΩ	R331	70-196
" " 56KΩ	R325	70-197
" " 68KΩ	R320	70-198
Variable Resistor	RV1	70-199
" "	RV2	70-200
Solid Resistor 4.7KΩ	R1	70-201
CAPACITORS		
Ceramic Capacitor .047	C204, 205, 206	80-145
Elyt. Capacitor 1/10	C201, 202	80-146
" " 100/10	C207, 208	80-147
Mylar " .1	C203	80-148
Ceramic Capacitor .01	C301, 338, 347	80-149
" " 51P	C322	80-150

Description	Designation	Unimetrics' Part #
CAPACITORS		
” ” 5P	C302	80-151
” ” 22P	C308	80-152
” ” 68P	C331	80-153
” ” .1	C303, 304, 305, 306, 307, 332, 310, 311, 312, 313, 314, 315, 316, 317, 335, 343	80-154
” ” .2	C324, 325	80-155
” ” .022	C321, 340, 341	80-156
Elyt. Capacitor 10/10	C326, 334	80-157
” ” 100/10	C342, 344	80-158
” ” 100/16	C339	80-159
Elyt. Capacitor 1000/16	C345, 346	80-160
” ” 4.7/25	C327, 336	80-161
” ” 1/50	C328, 337	80-162
Styroflex Capacitor 2000PF	C319, 320	80-163
” ” 200PF	C318	80-164
” ” 220PF	C309	80-165
Mylar Capacitor 1000PF	C323	80-166
” ” 2200PF	C329, 330	80-167
” ” 4700PF	C333	80-168
TRANSISTORS		
Transistor	Q201, 202, 203, 204, 205, 206, 207, 208, 209, 210, 211, 212, 213, 214, 215	90-65
”	Q216	90-66
I.C.	IC201	90-67
”	IC202	90-68
Transistor	Q306, 307	90-69
”	Q303, 304, 305	90-70
”	Q301, 302	90-71
I.C.	IC303	90-72
”	IC302	90-73
”	IC301	90-74
Transistor	Q1, 2	90-75
DIODES		
Zener Diode	ZD201	100-135
Diode	D301, 302	100-136
”	D303	100-137

Description	Designation	Unimetrics' Part #
"	D304, 305, 306, 307	100-138
Zener Diode	ZD301	100-139
L.E.D.	LD401, 402, 403, 404, 405, 406, 407, 408	100-140
UHF TUNER ASSEMBLY		
HARDWARE		
Crystal Socket	CS1, 2, 3, 4	40-15
GENERAL		
P. C. Board	PT	60-95
Tie Point	MYO-9	60-96
L-Type Shield Plate (L)	MYO-112	60-97
L-Type Shield Plate (S)	MYO-113	60-98
Shield Plate (Ant.)	MYO-114	60-99
Shield Plate (IF)	MYO-115	60-100
Shield Plate (PF)	MYO-116	60-101
Shield Cover (RF)	MYO-117	60-102
RESISTORS		
Solid Resistor	R116	70-115
" "	R106, 119	70-116
" "	R118	70-117
" "	R101	70-118
" "	R115	70-119
" "	R103, 110, 113	70-120
" "	R102	70-121
" "	R104, 112	70-122
" "	R120, 121, 122, 123, 124, 125, 126, 127	70-123
" "	R114	70-124
" "	R105, 111, 117	70-125
" "	R109	70-126
" "	R107	70-127
" "	R108	70-128
CAPACITORS		
Ceramic Capacitor	C106, 114, 116, 121, 124, 125, 136, 137, 138, 129, 132, 133, 139	80-105

Description	Designation	Unimetrics' Part #
CAPACITORS		
Ceramic Capacitor	C112, 128, 130	80-106
" " "	C115, 118	80-107
" "	C122	80-108
" "	C103, 109, 111, 134, 135	80-109
" "	C101, 102, 104, 105, 110, 131	80-110
" "	C119	80-111
" "	C113, 117	80-112
" "	C126	80-113
" "	C120	80-114
" "	C127	80-115
" "	C123	80-116
" "	C107, 108	80-117
Trimmer Capacitor	TC101, 102, 103, 104	80-118
TRANSISTORS		
Transistor	Q101	90-45
"	Q102	90-46
"	Q103	90-47
"	Q104	90-48
"	Q105	90-49
"	Q106	90-50
DIODES		
Diode	D101, 102, 103, 104	100-120
COILS		
RF Coil	L109	50-55
RF "	L114, 115, 116, 117	50-56
RF "	L110	50-57
RF "	L103	50-58
RF "	L106, 107, 108	50-59
RF Choke Coil	RFC-101	50-60
RF " "	L101, 102	50-61
RF Choke Coil	L104, 105, 111	50-62
VHF HIGH TUNER ASSEMBLY		
HARDWARE		
Crystal Socket	EZS-27	40-40

Description	Designation	Unimetrics' Part #
COILS & TRANSFORMERS		
RF Coil	L101	50-65
RF "	L102	50-66
RF "	L103	50-67
RF "	L104	50-68
RF "	L105	50-69
RF "	L109, 110, 111, 112	50-70
RF "	L106	50-71
I.F.T.	L107, 108	50-72
GENERAL		
Tie Point	MYO-9	60-105
Test Point	MMO-77	60-106
P.C. Board	EPO-557H	60-107
RESISTORS		
Solid Resistor	R105	70-130
" "	R109, 116, 125	70-131
" "	R101	70-132
" "	R114	70-133
" "	R115	70-134
" "	R108	70-135
" "	R106, 110	70-136
" "	R117, 118, 119, 120, 121, 122, 123, 124	70-137
" "	R103	70-138
" "	R107, 111	70-139
" "	R102, 104	70-140
" "	R112	70-141
" "	R113	70-142
CAPACITORS		
Ceramic Capacitor	C104, 105, 107, 108, 111, 113, 114, 117, 118, 121, 122, 123, 124	80-120
" "	C116, 119, 120	80-121
" "	C109	80-122
" "	C112	80-123
" "	C106, 115	80-124
" "	C101	80-125
" "	C110	80-126
" "	C102, 103	80-127

Description	Designation	Unimetrics' Part #
TRANSISTORS		
Transistor (FET)	Q104	90-55
”	Q102	90-56
”	Q103	90-57
”	Q101	90-58
VHF LOW TUNER ASSEMBLY		
HARDWARE		
Crystal Socket	EZS-27	40-45
COILS & TRANSFORMERS		
RF Choke Coil	L101	50-75
RF ”	L102	50-76
RF ”	L103	50-77
RF ”	L104	50-78
RF ”	L105	50-79
RF ”	L106	50-80
RF ”	L109, 110, 111, 112	50-81
I.F.T.	L107, 108	50-82
GENERAL		
Tie Point	MYO-9	60-110
Test Point	MMO-66	60-111
P.C. Board	EPO-557L	60-112
Z-Type Shield	MYO-118	60-113
RESISTORS		
Solid Resistor	R106, 110	70-145
” ”	R103	70-146
” ”	R101, 105	70-147
” ”	R103, 111	70-148
” ”	R109	70-149
” ”	R113, 114, 115, 116, 117, 118, 119, 121	70-150
” ”	R108	70-151
” ”	R104	70-152
” ”	R102	70-153
” ”	R112	70-154

Description	Designation	Unimetrics' Part #
CAPACITORS		
Ceramic Capacitor	C122, 123	80-130
" "	C120	80-131
" "	C102, 111, 116, 119, 121, 124, 125, 126, 127, 128, 129	80-132
" "	C113, 118	80-133
" "	C101	80-134
" "	C106	80-135
" "	C117	80-136
" "	C114	80-137
" "	C109, 115	80-138
" "	C112	80-139
" "	C110	80-140
" "	C104, 108	80-141
" "	C102	80-142
" "	C103, 105	80-143
TRANSISTORS		
Transistor	Q101	90-60
"	Q108-2	90-61
"	Q103	90-62
DIODES		
Diode	D101, 102, 103, 104	100-130

Chapter 7

Heath GR-110 VHF Scanning Monitor

The Heath Model GR-110 VHF scanning monitor is an 8-channel, crystal-controlled, high-frequency receiver that can be tuned to any selected 9 MHz segment of the 146–174 MHz band.

The Heath GR-110 is assembled by purchasers, who are seldom electronics experts. Nothing is completely foolproof, and sooner or later you will probably be called upon to straighten out one of these units.

Heath was very cooperative in that it provided servicemen with a list of common problems and their solutions. Wouldn't it be nice if all manufacturers did that?

The GR-110 is a VHF-only monitor and has no options to make service more complex. Careful application of the data provided should make GR-110 servicing relatively simple.

If you encounter problems, a review of Chapter 1 may help.

SPECIFICATIONS

Frequency range	146 to 174 MHz
Scanning range	Any 9 MHz segment of frequency range
Sensitivity (20 dB quieting)	Less than 1 μV
Usable sensitivity	Less than 0.5 μV
Adjacent-channel rejection	-40 dB (±30 kHz spacing)
Intermediate frequency	10.7 MHz, crystal filter
IF rejection	Greater than -80 dB
Image rejection	Greater than -40 dB
Antenna input impedance	50Ω
Scan rate	Approx 17 channels per sec
Audio output	2W into a 4Ω load with less than 10% distortion
Semiconductors	30 transistors, 8 integrated circuits, 17 diodes
Readout device	3015A 7-segment indicator tube
Power requirements	
AC	120V, 50 or 60 Hz, 13W
DC	13.8V; squelched, 0.275A; unsquelched, with full audio, 0.8A
Overall dimensions	8¼ × 3 × 9¾ in.
Net weight	5½ lb

CRYSTAL DATA

Crystals may be ordered in the following manner:

Order a *crystal certificate* from the Heath Company for each crystal.
Model number: GRA-110-1
Cost: $4.95 each

When you receive your crystal certificates, fill one in for each crystal, cut off the receipt, place a stamp on the card, and mail it to CTS Knights, Inc. Your crystals will be quickly mailed to you.

Fig. 7-1. Circuit board X-ray views of Heath GR-110 (shown from component side).

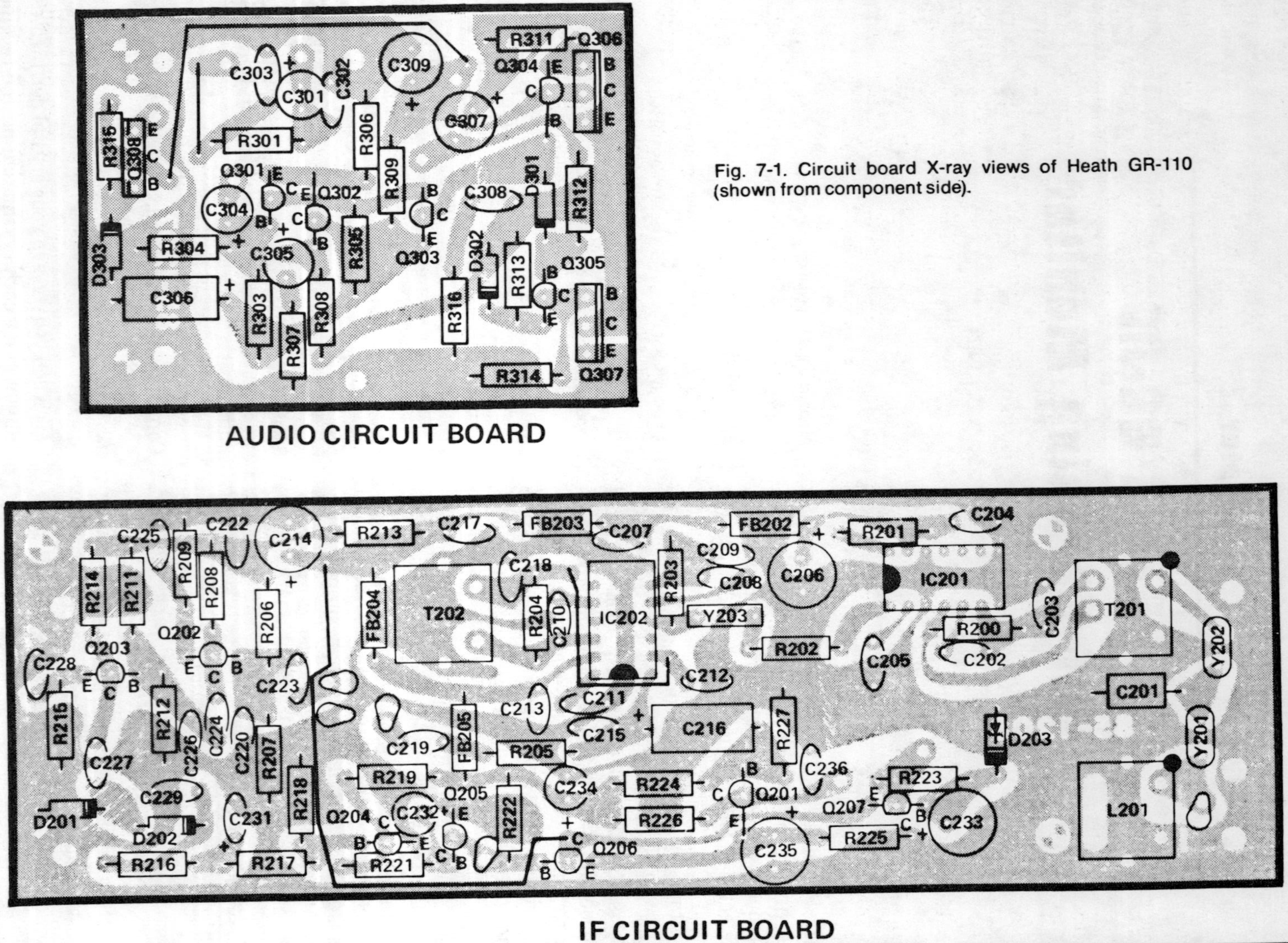

If you wish to obtain crystals from another source, you will have to provide the following information:

Resonance	Series
Case type	HE-025-0
Series resistance	Less than 25Ω
Shunt capacitance	7 pF
Tolerance	0.002%
Frequency (MHz)	Desired operating frequency (MHz) - 10.7 MHz / 3

Example

155.67 (desired frequency) − 10.7 (IF) = 144.97/3 = 48.3233 MHz (crystal frequency)

SERVICE DATA

Voltage data	Fig. 7-5. Table 7-2. Schematic 7-1
Schematic	Schematic 7-1
Internal layout	Figs. 7-1. 7-2. 7-3. 7-4
Alignment data. steps	A. B. C. D. E

Visual Tests

Refer to the circuit board X-ray views, Fig. 7-1, for the physical location of parts on the circuit board.

1. Recheck the wiring against the schematic.
2. About 90% of the kits that are returned to the Heath Company for repair do not function properly due to poor connections and soldering. Therefore, many troubles can be eliminated by reheating all connections to make sure that they are soldered.
3. Check to be sure that all transistors and diodes are in their proper locations. Make sure each lead is connected to the proper point.
4. Check that each of the ICs are properly installed in their sockets, and the pins are not bent out or under the IC. Also be sure the ICs are installed in their correct positions.
5. Check the values of the parts. Be sure in each step that the proper part has been wired into the circuit.
6. Check for bits of solder, wire ends, or other foreign matter that may be lodged in the wiring.
7. Look between each circuit board and the chassis to be sure all leads have been cut off and no solder bridges exist.

Precautions for Troubleshooting

1. Be cautious when testing IC and transistor circuits. Although they have almost unlimited life when used properly, they are much more vulnerable to damage from excessive voltage or current than tubes.
2. Be sure you do not short any terminals to ground when making voltage measurements. For example, if the probe should slip and short across components or voltage sources, it is very likely to cause damage to one or more ICs, transistors, or diodes.

A—Alignment Equipment

To perform the following alignment steps, you should have these test instruments: an 11M VTVM, or equivalent; an FM signal generator capable of covering the desired frequencies to be monitored plus 10.7 MHz, and modulated by a 5 kHz FM signal; and an AC VTVM.

B—Alignment, IF

Refer to Fig. 7-4 for the following steps:

1. Set the front-panel controls as follows:

   SQUELCH: clockwise
   AUDIO: counterclockwise to OFF
   Pushbuttons *0–7*: *0* depressed; all others out.
   MANUAL–AUTO: AUTO

2. Plug the AC line cord into a 120V AC wall receptacle. Do not plug in an antenna at this time.
3. Connect the alligator clip jumper to each lead of resistor R200 as shown in Fig. 7-4.
4. Turn the power switch to the ON position.
5. Set the controls to the following positions:

   SQUELCH control: fully clockwise
   AUDIO control: to a comfortable listening level

6. Adjust transformer T202 for maximum noise from the speaker. If measured, the DC voltage at pin 1 of IC202 should

Table 7-1. Troubleshooting Chart—GR-110

CONDITION	POSSIBLE CAUSE
No hum when you touch pin 4 of IC201.	1. IC201 incorrectly installed. 2. Incorrect IC installed at IC201. 3. Solder bridged on foil leads to IC201. 4. Defective IC201.
Crystal oscillator not working (less than 1V DC on scan circuit board TP; correct pushbutton out).	1. Defective crystal. 2. 8-wire cable incorrectly installed. 3. L107 and L109 interchanged. 4. Diode associated with selected channel. 5. Solder bridge in circuit. 6. IC105. 7. Pushbutton switch.
Monitor does not scan automatically.	1. SQUELCH control not properly set. 2. Incorrect voltage at pin A on IF circuit board (+2V DC or greater on collector of transistor Q205).
No +5V DC at ORG wire connection on scan circuit board.	1. Jumper wire on audio circuit board in wrong place or missing. 2. Regulator transistor Q308. 3. Regulator diode D303. 4. ICs incorrectly installed. 5. Transistors Q108, Q109, Q110, or Q111. 6. Solder bridges across scan circuit ICs.
One or more Display Indicator segments not lighting. (Check by manually scanning through all channels.)	1. Less than +4V DC at IC106 pin 16. 2. Defective Display Indicator. This can be determined by grounding each jumper wire, one at a time, between the indicator circuit board and the scan circuit board, except the power source. Each segment should light as its jumper is grounded. 3. Open foil leads or solder bridges at ICs on scan circuit board.
Unit scans but will not lock on transmitting station.	1. Associated crystal pushbutton is in (locking out reception). 2. Interchanged black and yellow wires on scan circuit board. 3. Incorrect wiring to MANUAL–AUTO switch. 4. Transistor Q111. 5. Defective crystal.

Table 7-1 (cont.)

CONDITION	POSSIBLE CAUSE
No sound, no visible sign of operation.	1. Fuse F1, if you are using 120V AC. 2. Fuse in 12V DC supply. 3. Power plug or socket; check for wiring errors. 4. Power switch S-1; check for wiring errors. 5. Speaker or speaker jack.
No sound, Display Indicator appears to scan properly.	1. Audio and squelch (shielded) cables incorrectly connected. 2. Squelch control incorrectly wired. 3. External SPKR jack incorrectly wired. 4. Spare phono plug left plugged into SPKR jack. 5. No B+ voltage to audio circuit board. Check "+12" pin (red lead) with VTVM.
No hum from audio amplifier. (Note: Lift the white wire from the audio circuit board connector. Then touch the WHT pin on the circuit board with a screwdriver blade. If you hear a loud hum, the audio amplifier section is okay.)	1. Faulty connection or component on audio circuit board.
No hum heard when you touch pin 14 of IC202.	1. IC202. 2. Transistor Q201.
No hum when you touch pin 4 of IC202.	1. IC202 incorrectly installed. 2. Solder bridge on foil leads to IC202. 3. Incorrect IC installed at IC202. 4. Defective IC202.
Audible distortion.	1. Crystal defective. 2. Crystal filter misaligned. 3. Defective quadrature coil.
Low Sensitivity.	1. Shorting spring not removed from Q101, Q102, or Q103. 2. Incorrect voltages at S (source leads) of Q101 Q102, or Q103; check each for correct installation. 3. RF coils L101-L105 interchanged, incorrectly installed, or misaligned.

Table 7-2. Scan Circuit IC Charts.

The following charts show the correct readings on each IC pin under certain conditions. To properly use the charts, some preliminary instructions are necessary:

1. Place the green wire on the scan circuit board OPTION pin.
2. Place the female end of the brown test wire on the scan circuit board GRN pin.
3. Place the MANUAL AUTO switch in the MANUAL position.
4. Manually scan to the desired channel. 0 through 7.
5. Check the pins on the IC with the end of the test wire. The key for the chart is as follows:

   0 – a segment or segments of the indicator are illuminated.

   5 = any illuminated segment or segments of the Display Indicator will go out.

   ⊗ – a segment or segments of the Display Indicator are illuminated when the MANUAL switch is activated.

If you use a VTVM to measure the voltages at these integrated circuit pins, the voltage values will be as follows:

0 = 0V DC
5 = +5V DC

When the tests are completed, disconnect the brown test wire from the GRN pin.

Note: In the DO NOT CHECK areas on these charts, no significant readings will be obtained.

A IC101

CHANNEL	IC PIN															
	1	2	3	4	5	6	7	8	9	10	11	12	13	14	15	16
0		0	0	0	5	0	0	0	0	0	0	0	0	⊗		
1		0	0	0	5	0	0	0	0	0	0	5	0	⊗		
2		0	0	0	5	0	0	0	5	0	0	0	0	⊗		
3		0	0	0	5	0	0	0	5	0	0	5	0	⊗		
4	DO NOT CHECK	0	0	0	5	0	0	0	0	0	0	0	0	⊗		
5		0	0	0	5	0	0	0	0	0	0	5	0	⊗		
6		0	0	0	5	0	0	0	5	0	0	0	0	⊗		
7		0	0	0	5	0	0	0	5	0	0	5	0	⊗		

B IC102

CHANNEL	IC PIN															
	1	2	3	4	5	6	7	8	9	10	11	12	13	14	15	16
0	0	5	0	5	5	5	5	5	0	5	5	5	5	5		
1	0	5	0	5	5	5	5	0	5	5	5	5				
2	0	5	5	5	5	0	5	5	0	5	5	5				
3	0	5	5	5	5	0	5	0	5	5	5	5				
4	0	5	0	5	5	5	5	5	0	5	5	5	DO NOT CHECK			
5	0	5	0	5	5	5	5	0	5	5	5	5				
6	0	5	5	5	5	0	5	5	0	5	5	5				
7	0	5	5	5	5	0	5	0	5	5	5	5				

C IC103

CHANNEL	IC PIN															
	1	2	3	4	5	6	7	8	9	10	11	12	13	14	15	16
0	5	5	5	0	0	0	5	5	0	5	0	5	5			
1	5	5	5	0	0	0	5	5	0	5	0	5	5			
2	5	5	5	0	0	0	5	5	0	5	0	5	5			
3	5	5	5	0	0	0	5	5	0	5	0	5	5			
4	5	5	5	0	0	0	0	5	5	5	0	5	5			
5	5	5	5	0	0	0	0	5	5	5	0	5	5	DO NOT CHECK		
6	5	5	5	0	0	0	0	5	5	5	0	5	5			
7	5	5	5	0	0	0	0	5	5	5	0	5	5			

D IC104

CHANNEL	IC PIN															
	1	2	3	4	5	6	7	8	9	10	11	12	13	14	15	16
0	5	0	5	0	5	0	0	0	5	0	5	0	5	5		
1	5	0	5	0	5	0	0	0	5	0	5	5	0	5		
2	5	0	5	0	5	0	0	5	0	0	5	0	5	5		
3	5	0	5	0	5	0	0	5	0	0	5	5	0	5		
4	5	0	5	0	0	5	0	0	5	0	5	0	5	5		
5	5	0	5	0	0	5	0	0	5	0	5	5	0	5		
6	5	0	5	0	0	5	0	5	0	0	5	0	5	5		
7	5	0	5	0	0	5	0	5	0	0	5	5	0	5		

E IC105

CHANNEL	IC PIN															
	1	2	3	4	5	6	7	8	9	10	11	12	13	14	15	16
0	0	0	0	0	5	0	0	5								0
1	0	5	5	0	5	0	0	5								
2	5	0	0	0	5	5	0	5								
3	0	5	5	0	5	5	0	5								
4	5	0	0	0	5	0	5	5				DO				
5	0	5	5	0	5	0	5	5				NOT				
6	5	0	0	0	5	5	5	5				CHECK				
7	0	5	5	0	5	5	5	5								

F IC106

CHANNEL	IC PIN															
	1	2	3	4	5	6	7	8	9	10	11	12	13	14	15	16
0	0	0				0	0	0								
1	0	0				0	5	0								
2	5	0				0	0	0								
3	5	0				0	5	0								
4	0	5		DO		0	0	0				DO				
5	0	5		NOT		0	5	0				NOT				
6	5	5		CHECK		0	0	0				CHECK				
7	5	5				0	5	0								

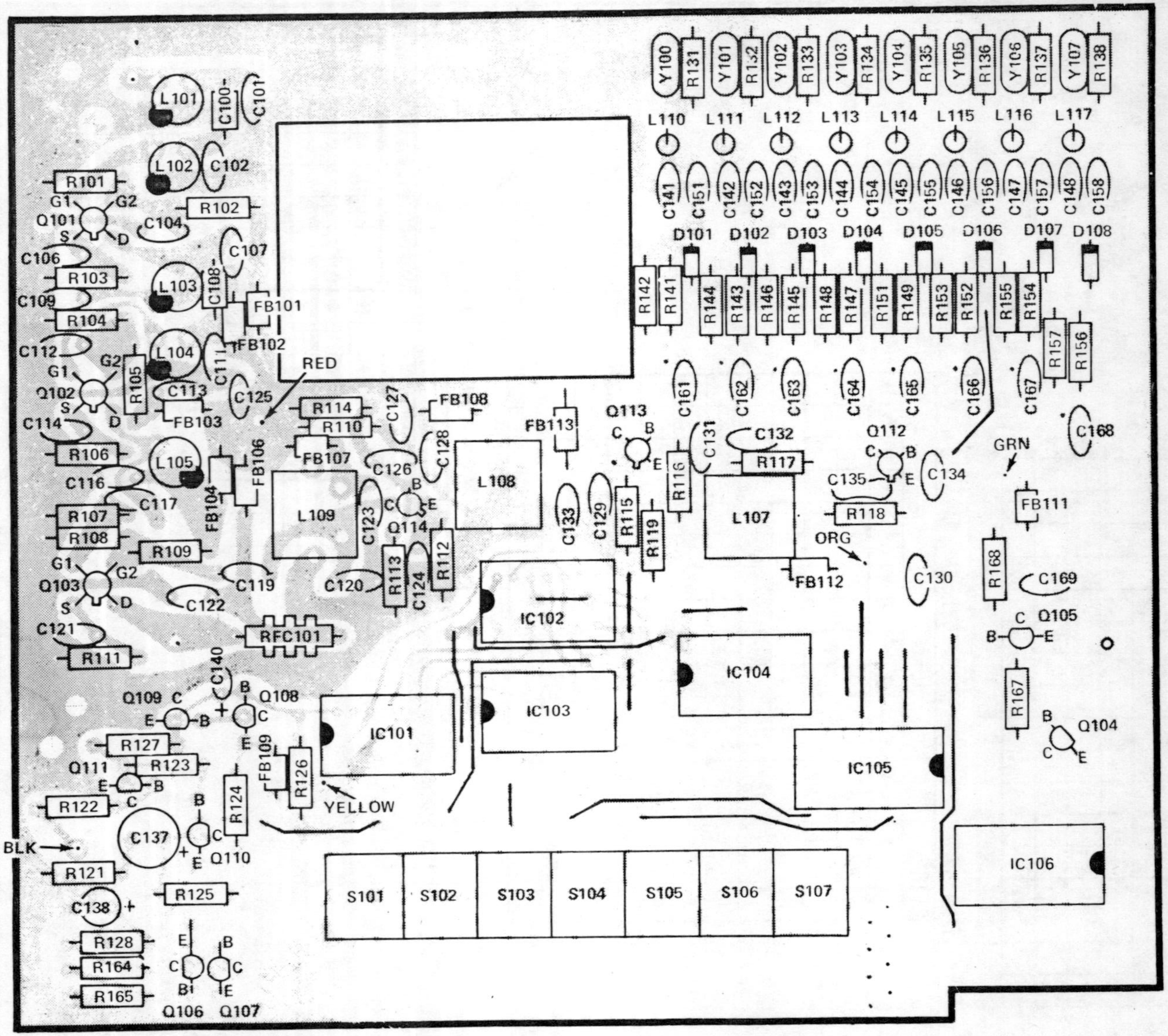

Fig. 7-2. Scan circuit board of Heath GR-110 (shown from component side).

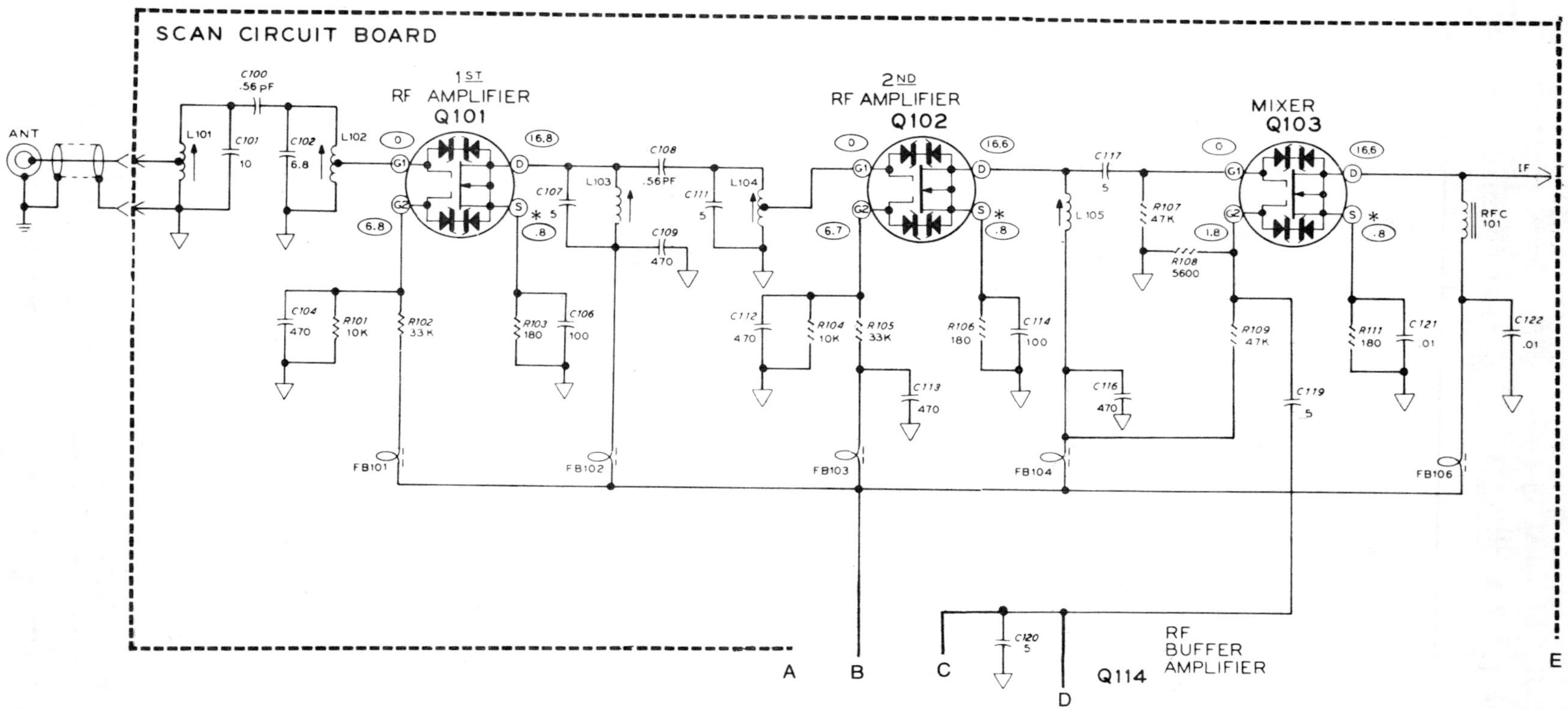

Schematic 7-1. Heathkit Model GR-110 VHF scanning monitor.

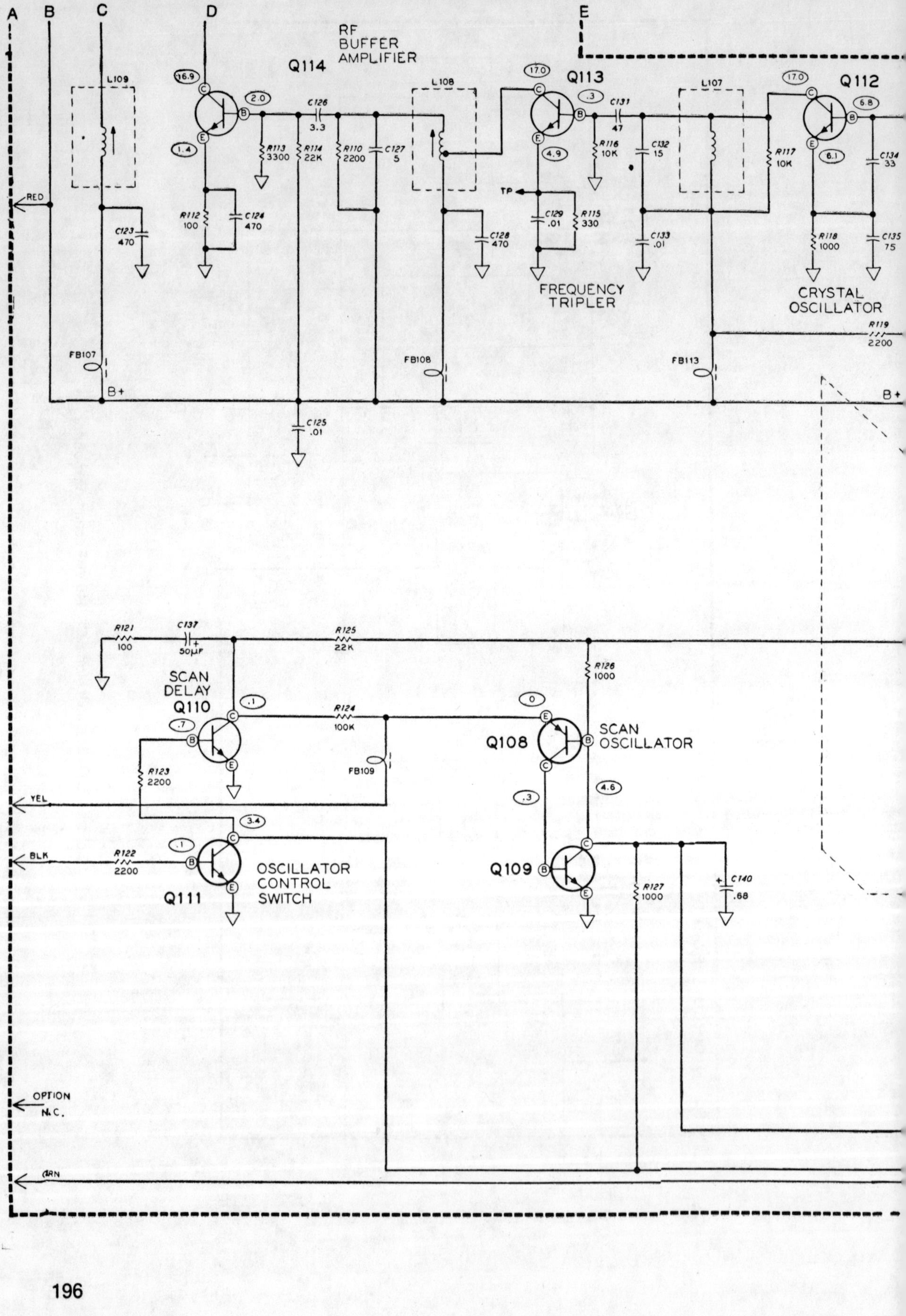
A
B
C
D
E
RF
BUFFER
AMPLIFIER
Q114
Q113
Q112
L109
L108
L107
16.9
2.0
1.4
17.0
.3
4.9
6.8
6.1
C126
3.3
C131
47
R113
3300
R114
22K
R110
2200
C127
5
R116
10K
C132
15
R117
10K
C134
33
RED
R112
100
C124
470
C123
470
TP
C129
.01
R115
330
C128
470
C133
.01
R118
1000
C135
75
FREQUENCY
TRIPLER
CRYSTAL
OSCILLATOR
R119
2200
FB107
FB108
FB113
B+
C125
.01
R121
100
C137
50µF
R125
22K
R126
1000
SCAN
DELAY
Q110
.1
.7
R124
100K
.0
SCAN
OSCILLATOR
Q108
R123
2200
FB109
.3
4.6
YEL
3.4
.1
BLK
R122
2200
OSCILLATOR
CONTROL
SWITCH
Q111
Q109
R127
1000
C140
.68
OPTION
N.C.
GRN

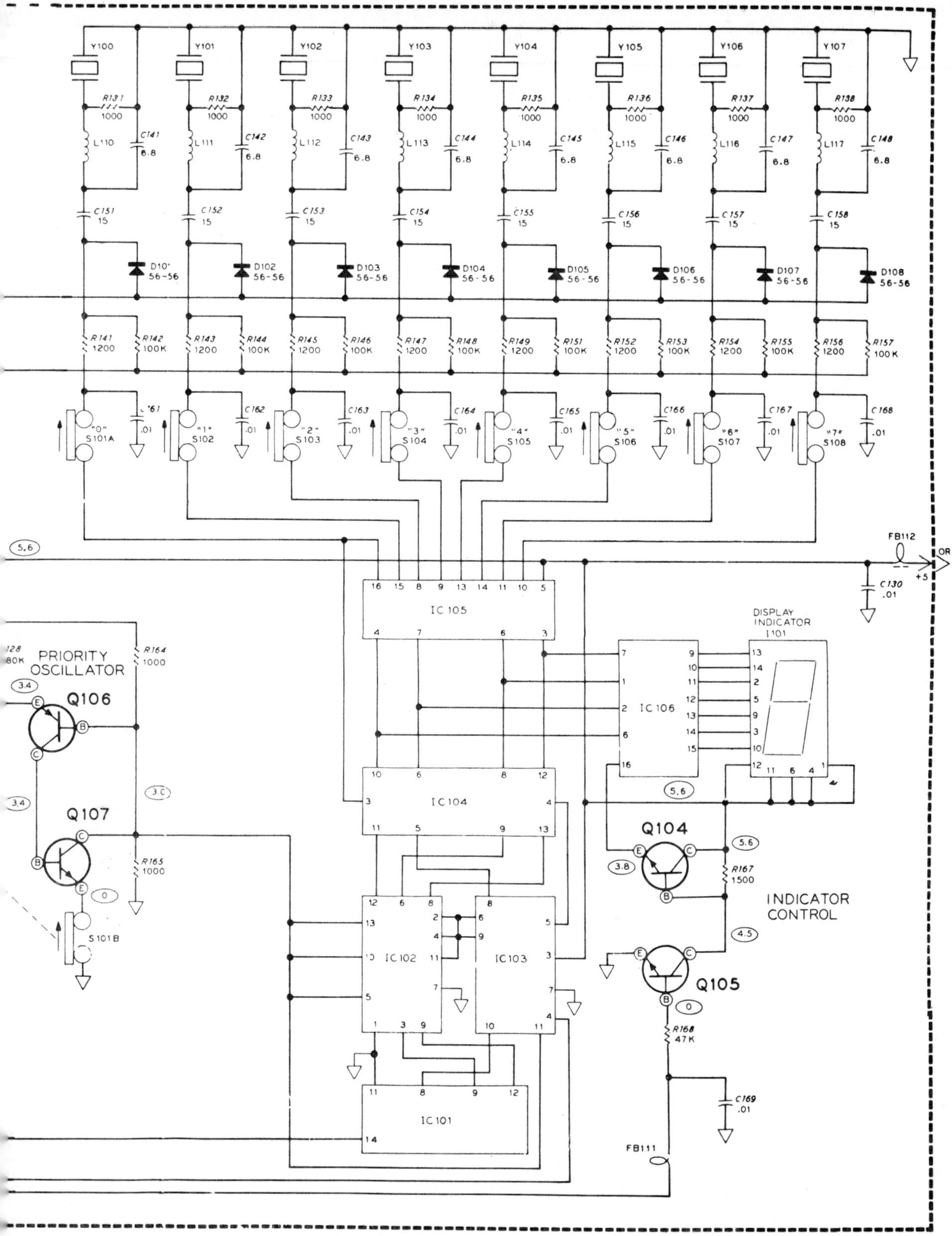

Y100
Y101
Y102
Y103
Y104
Y105
Y106
Y107
R131 1000
R132 1000
R133 1000
R134 1000
R135 1000
R136 1000
R137 1000
R138 1000
L110
L111
L112
L113
L114
L115
L116
L117
C141 6.8
C142 6.8
C143 6.8
C144 6.8
C145 6.8
C146 6.8
C147 6.8
C148 6.8
C151 15
C152 15
C153 15
C154 15
C155 15
C156 15
C157 15
C158 15
D102 56-56
D103 56-56
D104 56-56
D105 56-56
D106 56-56
D107 56-56
D108 56-56
R141 1200
R142 100K
R143 1200
R144 100K
R145 1200
R146 100K
R147 1200
R148 100K
R149 1200
R151 100K
R152 1200
R153 100K
R154 1200
R155 100K
R156 1200
R157 100K
"0" S101A
"1" S102
"2" S103
"3" S104
"4" S105
"5" S106
"6" S107
"7" S108
C162 .01
C163 .01
C164 .01
C165 .01
C166 .01
C167 .01
C168 .01
FB112
OR
+5
C130 .01
IC105
DISPLAY INDICATOR I101
IC106
IC104
IC102
IC103
IC101
PRIORITY OSCILLATOR
R164 1000
Q106
Q107
R165 1000
S101B
Q104
R167 1500
INDICATOR CONTROL
Q105
R168 47K
C169 .01
FB111

Fig. 7-1. Cont.

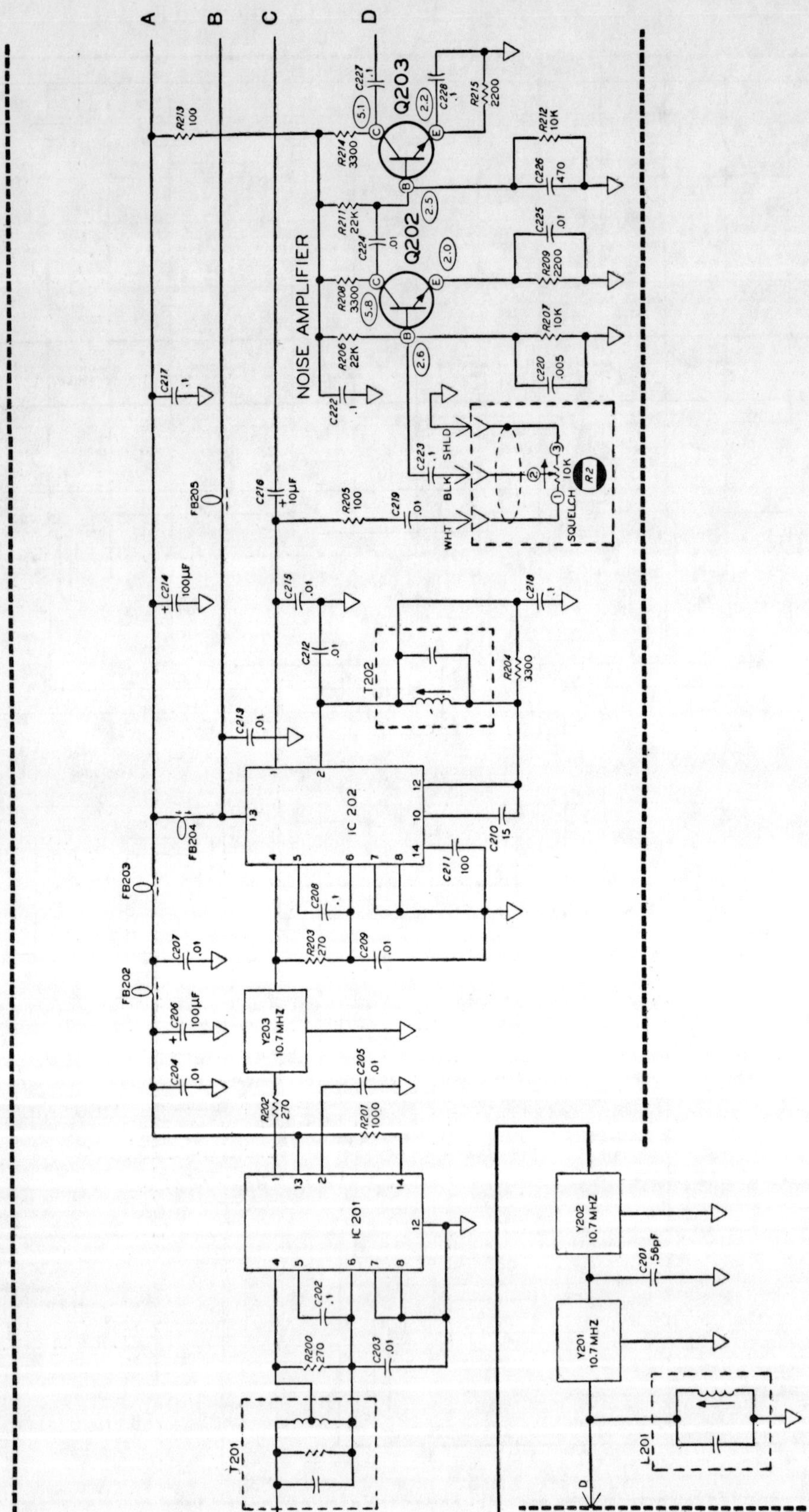
NOISE AMPLIFIER
Q202
Q203
IC 201
IC 202
T201
T202
Y201 10.7 MHZ
Y202 10.7 MHZ
Y203 10.7 MHZ
L201
SQUELCH
WHT
BLK
SHLD
A
B
C
D

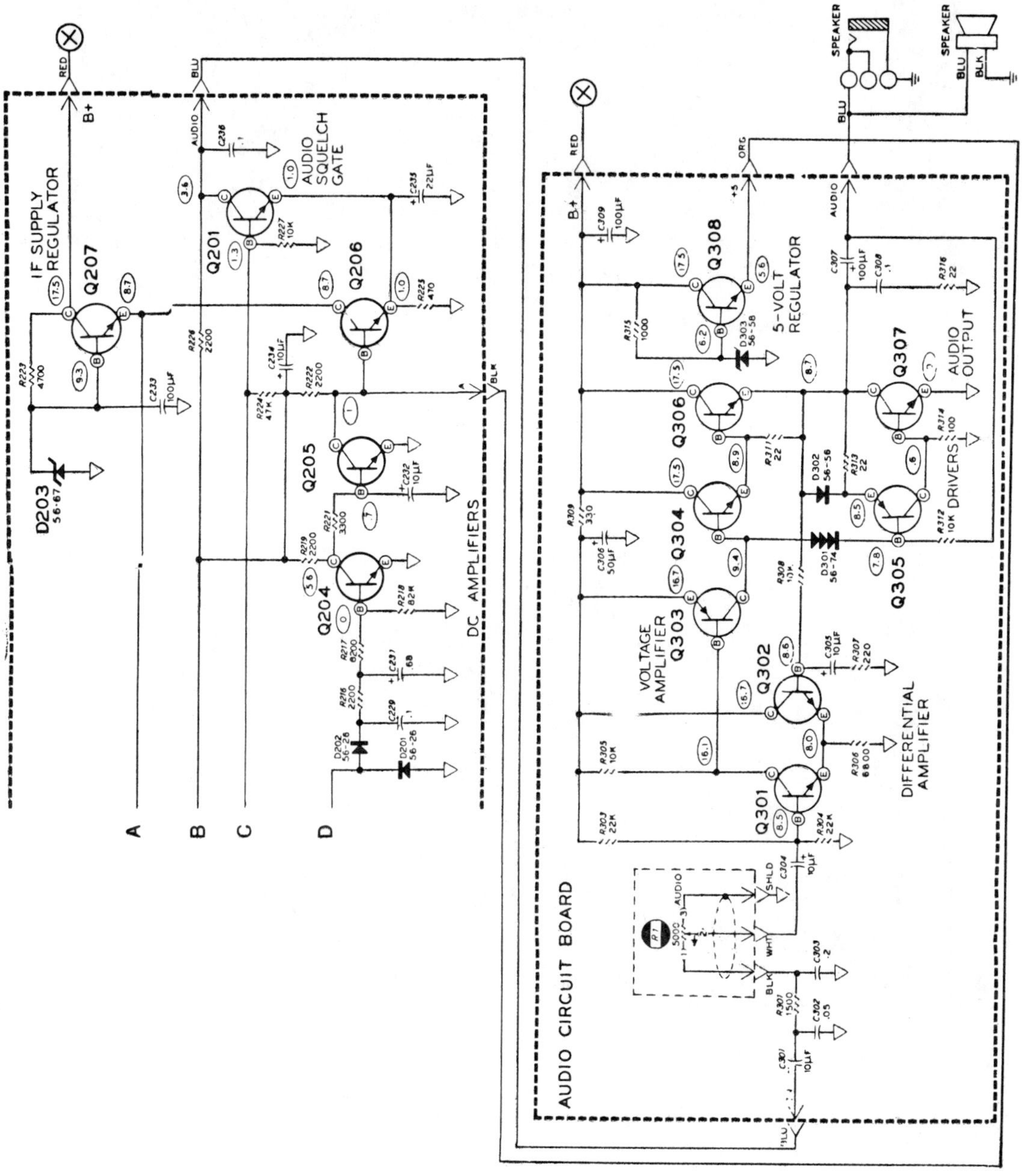
IF SUPPLY REGULATOR
Q207
D203
Q201
AUDIO SQUELCH GATE
Q206
Q205
Q204
DC AMPLIFIERS
A
B
C
D
RED
BLU
AUDIO
B+
AUDIO CIRCUIT BOARD
Q308
5-VOLT REGULATOR
Q306
Q304
Q303
VOLTAGE AMPLIFIER
Q302
Q301
DIFFERENTIAL AMPLIFIER
Q307
AUDIO OUTPUT
Q305
DRIVERS
SPEAKER
ORG
BLK
WHT
SHLD

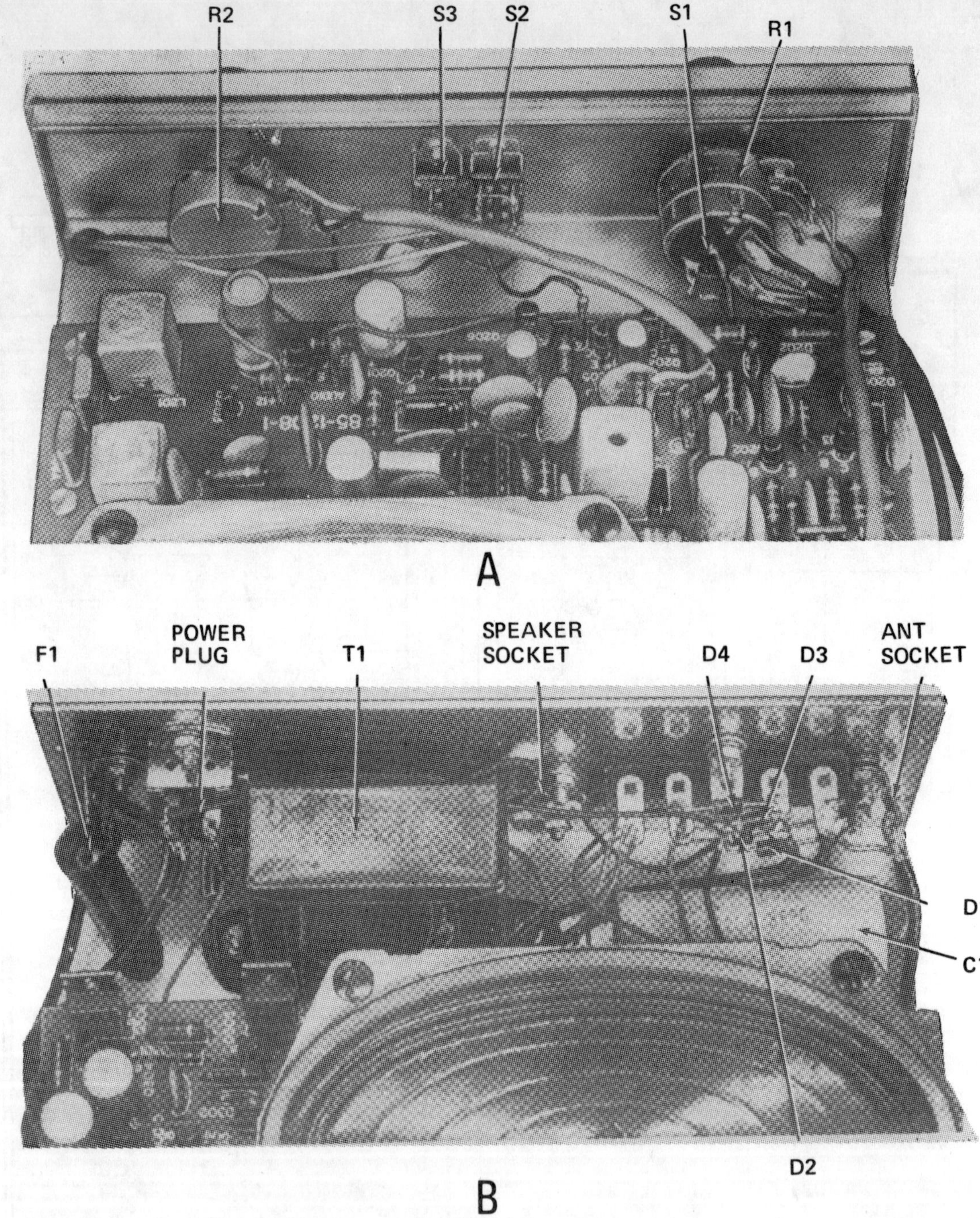

Fig. 7-3. Chassis photographs of Heath GR-110.

read half the indicated voltage on pin 13 of IC202.

7. Turn the AUDIO control counter-clockwise to its OFF position.

8. Remove the alligator test clip assembly from resistor R200.

9. Turn the monitor back on by turning the AUDIO control clockwise. This completes the IF alignment.

C—Alignment, Oscillator

1. Adjust the slug in coil L107 so it is level with the top of the coil form.

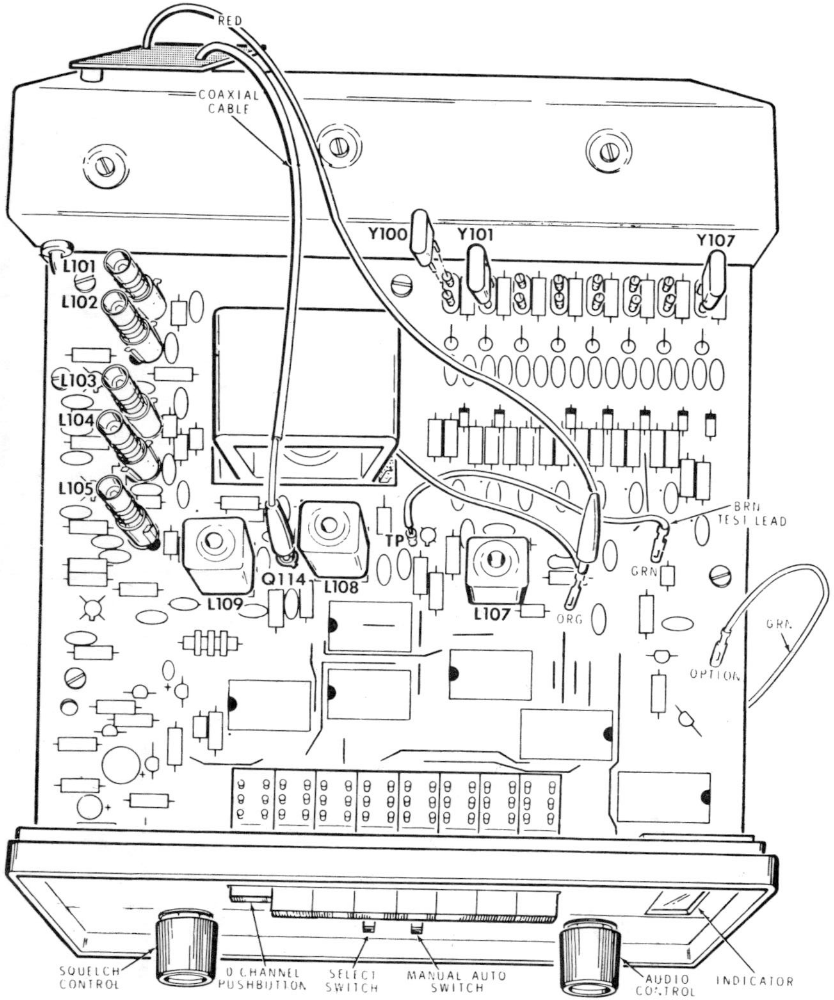

TOP VIEW

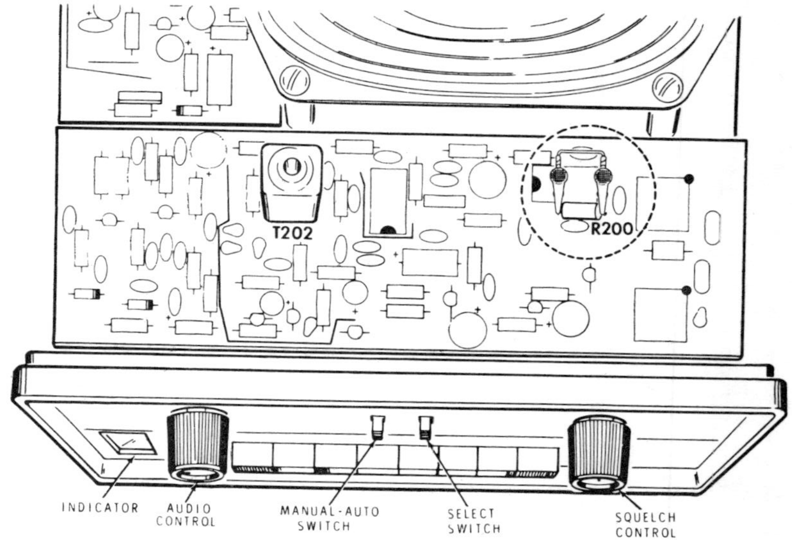

BOTTOM VIEW

Fig. 7-4. Internal layout of Heath GR-110.

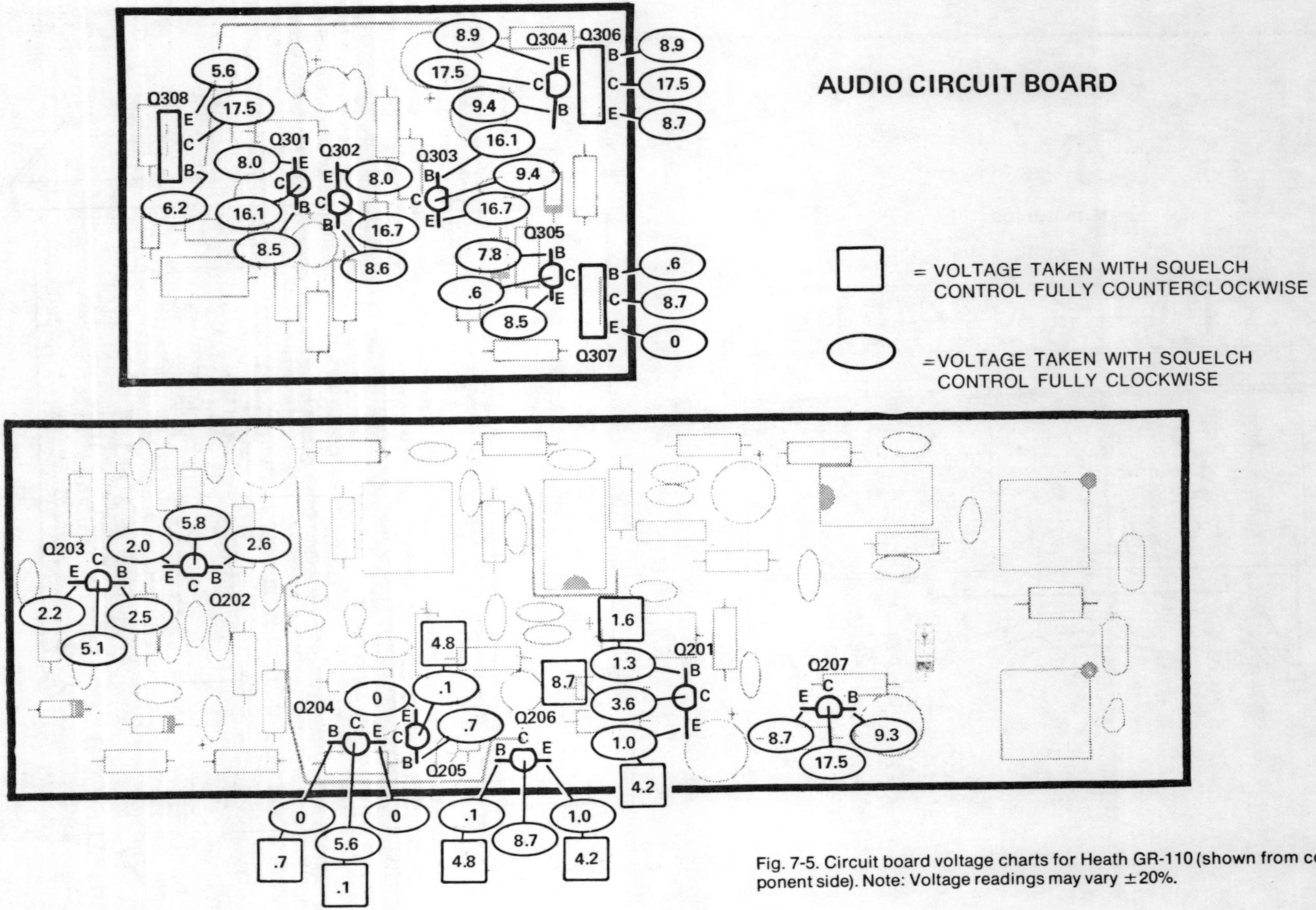

Fig. 7-5. Circuit board voltage charts for Heath GR-110 (shown from component side). Note: Voltage readings may vary ±20%.

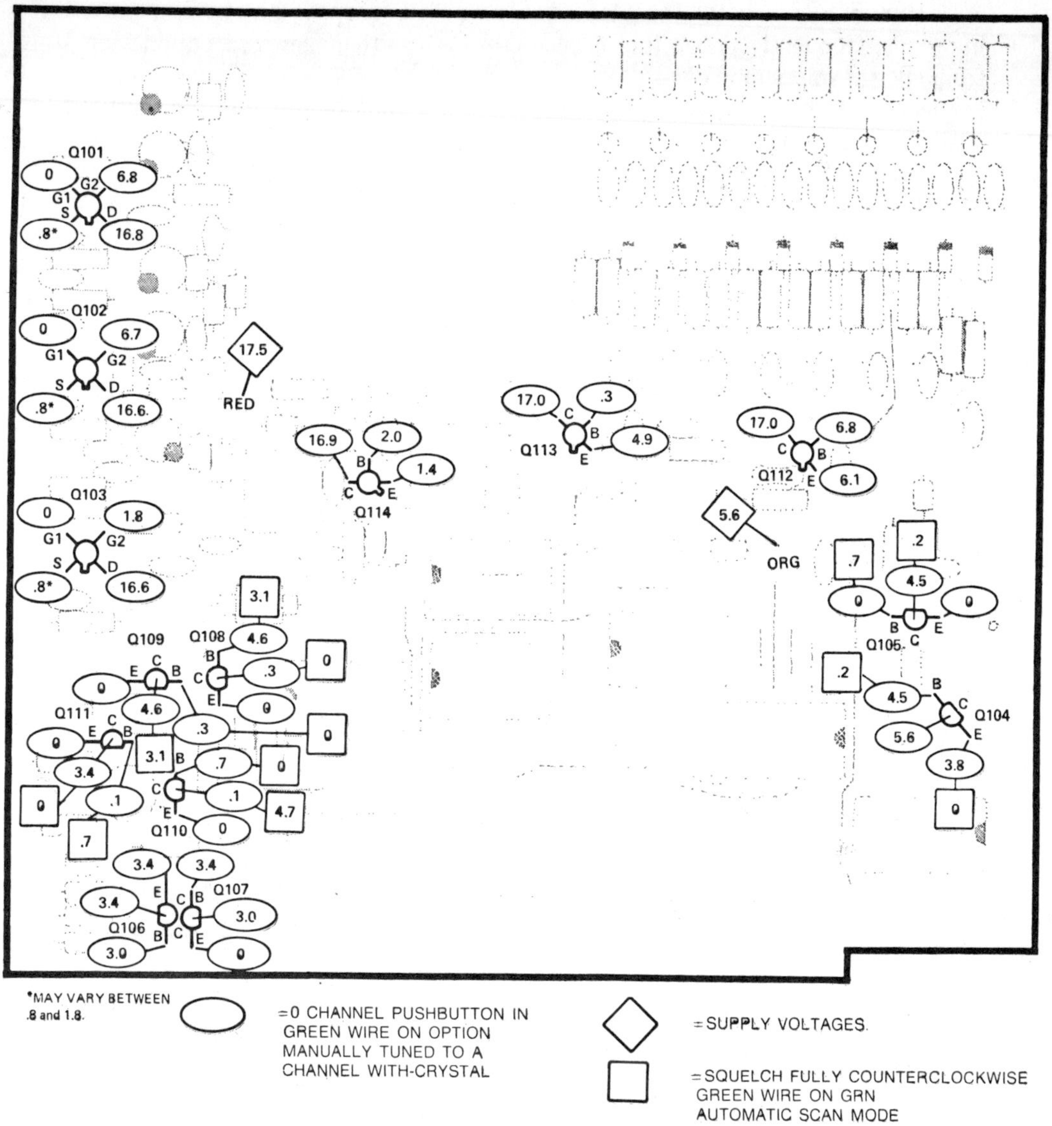

Fig. 7-6. Voltages on scan circuit board (shown from component side).

2. Connect the leads of an 11M input VTVM, or equivalent, to chassis ground and to connector pin TP near coil L108.
3. Place the MANUAL–AUTO switch to MANUAL; then sequence the monitor to the channel having the highest frequency crystal. Be sure the pushbutton for that channel is out.
4. Adjust coil L107 clockwise until a peak voltage is seen on the VTVM. Continue turning the slug until the voltage drops to zero. Now turn the slug counterclockwise until a voltage barely reappears on the VTVM. Stop at this point; it will not be necessary to make this adjustment again unless another crystal is installed in the monitor with a frequency higher than the highest channel crystal.

D—Alignment, RF

1. Connect the leads of an FM signal generator to the antenna of the monitor. Set the output frequency of the signal generator to the frequency of the highest crystal in the monitor. Turn the unmodulated output of the signal generator to zero.

As you proceed with the following steps, frequently check the output frequency of your signal generator.

2. Connect the leads of an AC VTVM to the lugs of the speaker.
3. Adjust the AUDIO control for 1.5V on the AC VTVM.
4. Turn the signal generator output up to produce minimum noise, or a minimum voltage level at the monitor output.
5. Adjust coils L108 and L109 for a minimum reading on the AC VTVM (minimum noise).
6. Adjust coils L101, L103, and L105 for a minimum reading on the AC VTVM. If the adjustment on coil L105 is very broad, first detune L103 slightly and tune coil L105 for a minimum reading. Then return coil L103 for a minimum reading.
7. Set the signal generator to the frequency of the lowest crystal installed in the monitor. The output level should remain at 0.5–1 μV.
8. Sequence the monitor to the channel with the lowest frequency crystal.
9. Adjust coils L102 and L104 for a minimum reading on the AC VTVM.
10. Adjust coil L101 (no more than one-fourth turn in either direction) for a minimum reading on the AC VTVM.
11. Once again sequence the monitor to the highest frequency channel. Tune the signal generator to that frequency.
12. Adjust coils L103 and L105 for a minimum reading on the AC VTVM.
13. Alternately repeat the low- and high-frequency adjustments until the AC VTVM readings are the same.

E—Priority Channel

If you have selected a crystal for a priority channel, one that will exclude all others when the station transmits, install the priority channel crystal in crystal pins Y100. This station will now be heard every time it transmits, unless the *0* pushbutton is depressed to exclude that channel.

Parts List

KEY No.	PART No.	PARTS Per Kit	DESCRIPTION
AUDIO CIRCUIT BOARD			
RESISTORS			
1/2-Watt, 10%			
A1	1-49	3	22 Ω (red-red-black)
A1	1-3	1	100 Ω (brown-black-brown)
A1	1-45	1	220 Ω (red-red-brown)
A1	1-4	1	330 Ω (orange-orange-brown)
A1	1-9	1	1000 Ω (brown-black-red)
A1	1-11	1	1500 Ω (brown-green-red)
A1	1-19	1	6800 Ω (blue-gray-red)
A1	1-20	3	10 kΩ (brown-black-orange)
A1	1-22	2	22 kΩ (red-red-orange)
CAPACITORS			
Disc			
B1	21-143	1	.05 μF
B1	21-95	1	.1 μF
B1	21-99	1	.2 μF

KEY No.	PART No.	PARTS Per Kit	DESCRIPTION
Electrolytic			
B2	25-98	1	50 μF
B3	25-257	3	10 μF vertical
B3	25-248	2	100 μF vertical
DIODES-TRANSISTORS			
C1	56-56	1	1N4149 diode
C1	56-58	1	1N709A zener diode
C1	56-74	1	MZ2362 diode

NOTE: Transistors are marked for identification in one of the following four ways.

1. Part number.
2. Type number.
3. Part number and type number.
4. Part number with a type number other than the one listed.

KEY No.	PART No.	PARTS Per Kit	DESCRIPTION
C2	417-91	3	2N5232A transistor
C2	417-201	2	X29A829 transistor
C3	417-175	3	2N5294 transistor
MISCELLANEOUS			
	85-1207-1	1	Audio circuit board
D1	205-1436	1	Heat sink

KEY No.	PART No.	PARTS Per Kit	DESCRIPTION
D2	250-285	1	4-40 x 1/4" screw
D3	252-15	1	4-40 nut
D4	254-9	1	#4 lockwasher
D5	432-121	9	Connector pin (2 extra included)
	344-51	1	Brown hookup wire
			Solder (Additional 3' rolls of solder, #331-6, can be ordered for 15 cents each.)

ITEMS FROM FINAL PACK

KEY No.	PART No.	PARTS Per Kit	DESCRIPTION
	597-260	1	Parts Order Form
	597-308	1	Kit Builders Guide
		1	Manual (See front cover for part number.)

IF CIRCUIT BOARD

RESISTORS

1/2-Watt, 10%

KEY No.	PART No.	PARTS Per Kit	DESCRIPTION
A1	1-3	2	100 Ω (brown-black-brown)
A1	1-42	3	270 Ω (red-violet-brown)
A1	1-6	1	470 Ω (yellow-violet-brown)
A1	1-9	1	1000 Ω (brown-black-red)
A1	1-44	6	2200 Ω (red-red-red)
A1	1-14	4	3300 Ω (orange-orange-red)
A1	1-16	1	4700 Ω (yellow-violet-red)
A1	1-73	1	8200 Ω (gray-red-red)
A1	1-20	3	10 kΩ (brown-black-orange)
A1	1-22	2	22 kΩ (red-red-orange)
A1	1-25	1	47 kΩ (yellow-violet-orange)
A1	1-102	1	82 kΩ (gray-red-orange)

CAPACITORS

Disc

KEY No.	PART No.	PARTS Per Kit	DESCRIPTION
B1	21-111	1	15 pF
B1	21-9	1	100 pF
B1	21-56	1	470 pF
B1	21-46	1	.005 μF
B1	21-47	12	.01 μF
B1	21-95	9	.1 μF

Electrolytic

KEY No.	PART No.	PARTS Per Kit	DESCRIPTION
B2	25-54	1	10 μF
B3	25-257	2	10 μF vertical
B3	25-233	1	22 μF vertical
B3	25-248	3	100 μF vertical

Other Capacitors

KEY No.	PART No.	PARTS Per Kit	DESCRIPTION
B4	25-200	1	.68 μF tantalum
B5	28-3	1	.56 pF phenolic (green-blue-gray)

DIODES-TRANSISTORS

KEY No.	PART No.	PARTS Per Kit	DESCRIPTION
C1	56-26	2	1N191 diode (brown-white-brown)
C1	56-67	1	VR-10A zener diode

NOTE: Transistors are marked for identification in one of the following four ways:

1. Part number.
2. Type number.
3. Part number and type number.
4. Part number with a type number other than the one listed.

KEY No.	PART No.	PARTS Per Kit	DESCRIPTION
C2	417-91	7	2N5232A transistor

TRANSFORMERS-FILTERS

KEY No.	PART No.	PARTS Per Kit	DESCRIPTION
D1	52-162	1	IF coil
D1	52-163	1	IF transformer
D1	52-154	1	IF transformer
D2	404-530	1	Ceramic filter
D3	404-535	2	Crystal filter

MISCELLANEOUS

KEY No.	PART No.	PARTS Per Kit	DESCRIPTION
	85-1208-3	1	IF circuit board
E1	432-121	7	Connector pin
E2	442-28	2	MC1357P integrated circuit
E3	475-10	4	Ferrite bead

SCAN CIRCUIT BOARD

RESISTORS

1/2-Watt, 10%

KEY No.	PART No.	PARTS Per Kit	DESCRIPTION
A1	1-3	2	100 Ω (brown-black-brown)
A1	1-4	1	330 Ω (orange-orange-brown)
A1	1-9	13	1000 Ω (brown-black-red)
A1	1-10	8	1200 Ω (brown-red-red)
A1	1-11	1	1500 Ω (brown-green-red)
A1	1-44	4	2200 Ω (red-red-red)
A1	1-14	1	3300 Ω (orange-orange-red)
A1	1-18	1	5600 Ω (green-blue-red)

KEY No.	PART No.	PARTS Per Kit	DESCRIPTION
A1	1-20	4	10 kΩ (brown-black-orange)
A1	1-22	2	22 kΩ (red-red-orange)
A1	1-24	2	33 kΩ (orange-orange-orange)
A1	1-25	3	47 kΩ (yellow-violet-orange)
A1	1-26	9	100 kΩ (brown-black-yellow)
A1	1-34	1	680 kΩ (blue-gray-yellow)

Other Resistors

A1	1-112	3	180 Ω, 5% (brown-gray-brown-gold)

CAPACITORS

Disc

B1	21-33	1	3.3 pF
B1	21-78	6	5 pF
B1	21-61	9	6.8 pF
B1	21-3	1	10 pF
B1	21-111	9	15 pF
B1	21-7	1	33 pF
B1	21-32	1	47 pF
B1	21-9	2	100 pF
B1	21-56	8	470 pF
B1	21-47	15	.01 μF

Other Capacitors

B2	20-110	1	75 pF mica
B3	25-149	1	5 μF electrolytic
B4	25-116	1	50 μF electrolytic
B5	25-200	1	.68 μF tantalum
B6	28-3	2	.56 pF phenolic (green-blue-gray)

DIODES-TRANSISTORS

C1	56-56	8	1N4149 diode

NOTE: Transistors are marked for identification in one of the following four ways:

1. Part number.
2. Type number.
3. Part number and type number.
4. Part number with a type number other than the one listed.

C2	417-91	6	2N5232A transistor
C2	417-201	2	X29A829 transistor
C3	417-154	3	2N2369 transistor
C4	417-274	3	40673 FET (field effect transistor) Do not remove the shorting spring until you are told to do so.

KEY No.	PART No.	PARTS Per Kit	DESCRIPTION

COILS-CHOKES

D1	40-1613	1	Antenna coil
D1	40-1625	2	RF coil
D2	40-1614	2	RF coil
D3	40-1615	1	Oscillator coil
D3	40-1616	1	Tripler coil
D3	40-1617	1	Output coil
D4	45-75	8	RF choke
D5	45-82	1	RF choke

INTEGRATED CIRCUITS

NOTE: If either the part number or the "Description" number is on a transistor or integrated circuit, you have the correct part. Do not be concerned about any other numbers. [On integrated circuits, the description number may have additional letters and numbers other than those given in the list. Example: SN(7490)N or MC(7490)P.]

E1	443-7	1	7490 integrated circuit
E1	443-12	2	7410 integrated circuit
E1	443-18	1	7404 integrated circuit
E2	443-35	1	7441 integrated circuit
E2	443-36	1	7447 integrated circuit

MISCELLANEOUS

F1	85-1210	1	Indicator circuit board
F2	64-607	1	Switch assembly
F3	411-283	1	Display indicator
F4	432-121	9	Connector pin
F5	432-134	20	Wire connector (3 extra included)
F6	434-225	4	14-pin IC socket
F7	434-226	2	16-pin IC socket
F8	475-10	11	Ferrite bead
	347-1	1	8-conductor cable

PART FROM FINAL PACK

	85-1209-2	1	Scan circuit board

CHASSIS

CONTROLS-CAPACITORS-TRANSFORMER-DIODE

A1	10-119	1	10 kΩ control
A2	19-141	1	Control-switch
A3	21-111	1	15 pF disc capacitor
A4	25-200	1	.68 μF tantalum capacitor
A5	25-154	1	2500 μF electrolytic capacitor
A6	54-115	1	Power transformer
A7	57-65	4	1N4002 diode

KEY No.	PART No.	PARTS Per Kit	DESCRIPTION

METAL PARTS

KEY No.	PART No.	PARTS Per Kit	DESCRIPTION
B1	90-577-1	1	Cabinet
B2	90-578-1	1	Bottom plate
B3	203-1444-1	1	Front panel
B4	204-1865-1	1	Gimbal bracket
B5	204-1866-1	1	Gimbal mount
B6	210-58	1	Bezel
B7	200-642-1	1	Chassis
B8	208-6	1	Capacitor clip

HARDWARE

#6 Hardware

KEY No.	PART No.	PARTS Per Kit	DESCRIPTION
C1	250-138	6	6-32 x 3/16" screw
C2	250-116	7	6-32 x 1/4" black screw
C3	250-170	6	#6 x 1/4" sheet metal screw
C4	250-357	2	6-32 x 3/8" black nylon screw
C4	250-89	10	6-32 x 3/8" screw
C5	252-3	11	6-32 nut
C6	254-1	22	#6 lockwasher
C7	255-60	4	6-32 x 1-1/8" spacer

Other Hardware

KEY No.	PART No.	PARTS Per Kit	DESCRIPTION
D1	250-285	19	4-40 x 1/4" screw
D2	250-16	1	8-32 x 3/16" setscrew
D3	250-1147	2	10-32 thumbscrew
D4	250-83	4	#10 x 1/2" sheet metal screw
D5	252-7	5	Control nut
D6	252-49	2	Thumbnut
D7	253-10	2	Control flat washer
D8	253-19	2	#10 flat washer
D9	253-136	2	Nylon washer
D10	254-5	2	Control lockwasher
D11	259-2	2	#8 solder lug

WIRE-CABLE-SLEEVING

KEY No.	PART No.	PARTS Per Kit	DESCRIPTION
	340-1	1	Heavy bare wire
	343-12	1	1-wire shielded cable
	344-2	1	Heavy black stranded wire
	344-90	1	Small stranded black wire
	344-92	1	Stranded red wire
	344-93	1	Stranded orange wire
	344-94	1	Stranded yellow wire
	344-95	1	Stranded green wire
	344-96	1	Stranded blue wire
	346-2	1	Clear sleeving
	346-3	1	Large black sleeving
	347-35	1	2-wire shielded cable
	346-19	1	Small sleeving
	89-1	1	Line cord

CONNECTORS-TERMINAL STRIPS-SOCKETS-PLUG

KEY No.	PART No.	PARTS Per Kit	DESCRIPTION
E1	432-120	24	Female connector (1 extra included)
E2	432-137	2	Push-on connector
E3	432-759	2	4-pin socket
E4	432-760	1	4-pin plug
E5	434-42	1	Phono socket
E6	434-231	1	Plastic phono socket
E7	438-4	2	Phono plug
E8	431-11	1	5-lug terminal strip
E9	431-43	1	3-lug terminal strip

SWITCHES-KNOBS-FUSES

KEY No.	PART No.	PARTS Per Kit	DESCRIPTION
F1	60-86	1	Spring return switch
F2	60-78	1	Slide switch
F3	462-159	1	Knob with setscrew opening
F3	462-302	1	Knob
F4	455-50	1	Knob insert
F5	421-17	1	3/4 ampere (3/4 A) pigtail fuse
F6	421-25	1	1-1/2 ampere slow-blow fuse
F7	423-10	1	Fuseholder assembly

MISCELLANEOUS

KEY No.	PART No.	PARTS Per Kit	DESCRIPTION
	73-39	1	Foam gasket material
G1	73-43	1	Plastic grommet
G2	73-59	2	Rubber grommet
G3	260-16	2	Alligator clip
G4	261-29	5	Mounting foot
G5	401-139	1	Speaker
G6	390-362	1	Fuse label
G7	390-926	1	Caution label
	390-1112	1	Warning label
G8	391-34	1	Blue and white label
G9	446-606	1	Readout window
G10	490-1	1	Alignment tool
G11	490-5	1	Nut starter
	171-7386	1	Alignment generator

NOTES:

1. Set aside the alignment generator parts. They will not be used until after the kit is assembled.

Chapter 8

E.F. Johnson Monitors

E. F. Johnson is currently producing three different models, with several variations in center frequencies carrying different suffixes in the model number (see below). When modification other than tuning is necessary to change center frequency, such data is provided.

One advantage of equipment manufactured by E. F. Johnson is parts availability. In fact, if you do warranty service on Johnson CB equipment, you probably already have many replacement parts in stock. There are over 700 service centers scattered over the country, and many of them carry a good stock of parts. Johnson warranty is for one year on parts and labor and is worth considering if you are not already with it.

The schematics in this chapter are annotated with voltage readings. If voltage readings are more than 10% off suggested values, you probably can consider them significant.

UNITS COVERED

CRYSTAL DATA FOR ALL MODELS

30–50 MHz. Third-overtone crystal with injection 10.7 MHz above signal frequency (crystal frequency = channel frequency + 10.7 MHz). Maximum impedance, 35Ω. HC-18/U holder with plug-in pin leads; ±0.002% tolerance.

150–170 MHz. Third harmonic of third-overtone crystal with injection 10.7 MHz below signal frequency.

Crystal frequency =

$$\frac{\text{channel frequency} - 10.7\text{ MHz}}{3}$$

Maximum impedance, 35Ω. HC-18/U holder with plug-in pin leads; ±0.0015% tolerance.

450–470 MHz. Ninth harmonic of third-overtone crystal with injection 10.7 MHz below the signal frequency.

Crystal frequency =

$$\frac{\text{channel frequency} - 10.7\text{ MHz}}{9}$$

Maximum impedance, 35Ω. HC-18/U holder with plug-in pin leads; ±0.0015% tolerance.

To insure best performance, use only Johnson-approved crystals which meet rigid Johnson specifications as well as all FCC requirements.

CRYSTAL INSTALLATION AND PROGRAMMING

To add or change crystals:

1. Remove the cabinet locking brackets, which secure the rear panel to the cabinet. Remove the whip antenna.
2. Carefully slide the chassis out of the cabinet.
3. The crystal sockets are marked with numbers corresponding to the numbered front-panel pushbuttons. Insert the crystals to be added in the positions you prefer.
4. Program the 2-band receiver by inserting the programming wire for each channel into the appropriate jack.

For 30–50 and 150–170 MHz *Duo-Scans*, use the jacks toward the front of the receiver for 30–50 MHz crystals; use the jacks toward the rear of the receiver for 150–170 MHz crystals.

For 150–170 and 450–470 MHz *Duo-Scans*, use the jacks toward the front of the receiver for 150–170 MHz crystals; use the jacks toward the rear of the receiver for 450–470 MHz crystals.

Single-band monitor receivers have no programming provisions.

5. Carefully reinstall the receiver in its cabinet. Press the speaker towards the bottom of the printed-circuit board so that it will enter the cabinet. Replace the locking brackets and tighten the two screws at the rear of the cabinet. Connect the antenna and power lead. Install the receiver in its mounting bracket.
6. Check unit operation to be sure you have installed the crystals and programming wires properly.

JOHNSON UHF–VHF DUO-SCAN

Specifications

Frequency	153–160 MHz and 456–464 MHz
Channels	Eight (in any combination of one or both bands)
Sensitivity (EIA 12 dB SINAD)	
High-band VHF	0.4 μV
UHF	0.6 μV
Selectivity (20 dB quieting)	
High-band VHF	−60 dB at ±30 kHz
UHF	−60 dB at ±50 kHz
Spurious and image rejection	
High-band VHF	−50 dB
UHF	−50 dB
Intermodulation rejection	
High-band VHF	−60 dB
UHF	−55 dB
Audio output power (at less than 10% distortion)	3W
Built-in power supply requirements	11V to 16V DC. or 117V AC
Dimensions	2 × 6¼ × 9 in.
Weight	2 lb 4 oz

Service Data

Crystal layout	Fig. 8-2A
Schematic	Schematics 8-1 (early production). 8-5 (late production)
Alignment data, steps	A, B, C, D, E, F

The standard UHF–VHF *Duo-Scan* is factory aligned on 460 MHz and 157 MHz. To obtain peak performance, selected crystal frequencies should be within ±3.5 MHz of 157 MHz and ±4 MHz of 260 MHz. For reception outside these limits, retune front end for new center frequency.

JOHNSON UHF MONO-SCAN

Specifications

Frequency range	456–464 MHz
Channels	Eight
Sensitivity (EIA 12 dB SINAD)	0.6 μV

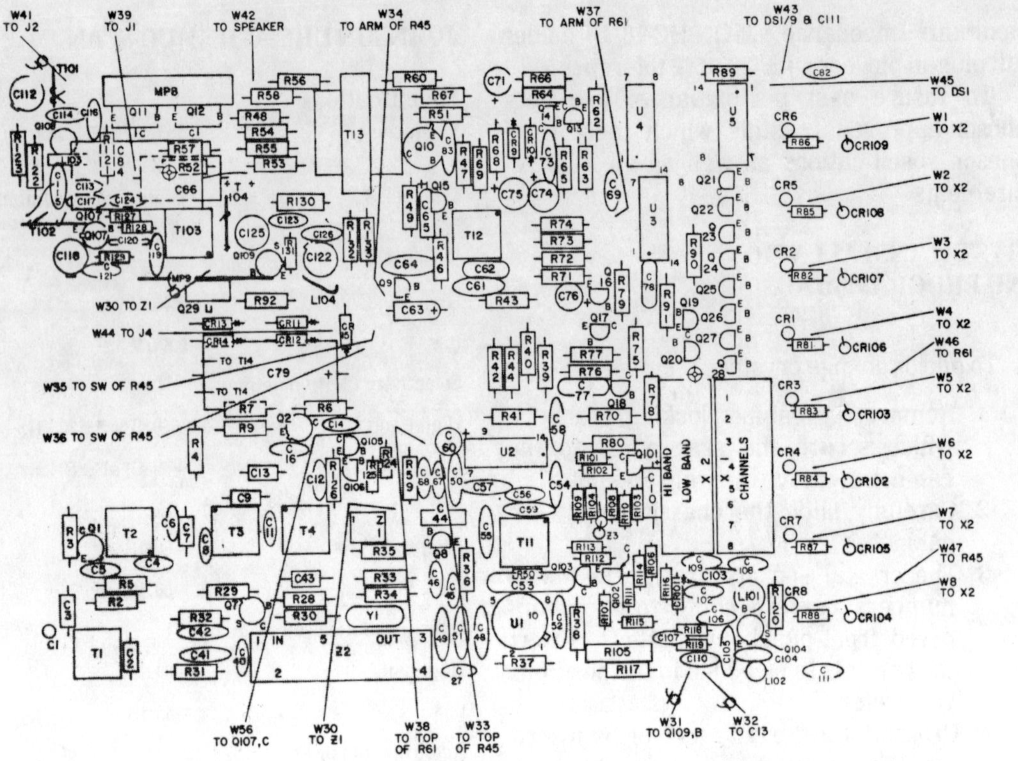

SOLDER SIDE
(BOTTOM VIEW)

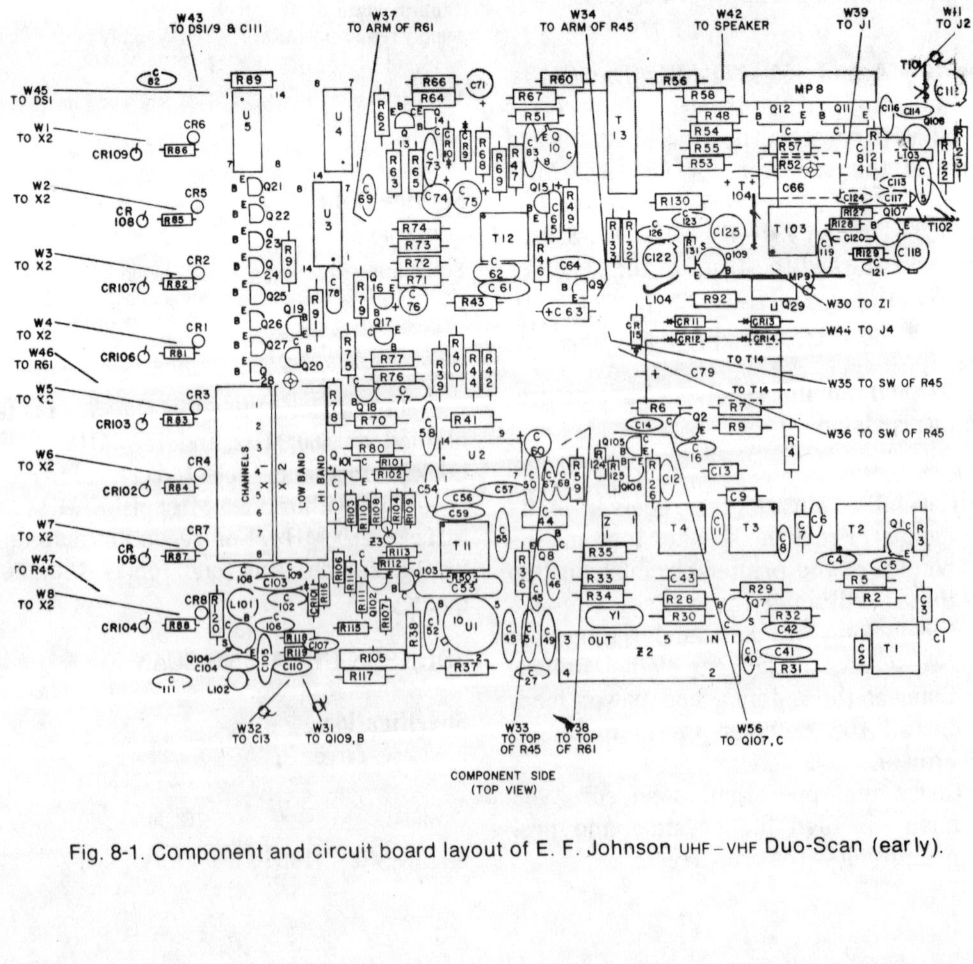

COMPONENT SIDE
(TOP VIEW)

Fig. 8-1. Component and circuit board layout of E. F. Johnson UHF–VHF Duo-Scan (early).

A

B

Fig. 8-2.(A) Crystal and programming-wire location in Johnson Mono-Scan and Duo-Scan. (B) Crystal location in Mini-Scan.

Selectivity (20 dB quieting)	−60 dB at ±50 kHz
Spurious and image rejection	−50 dB
Intermodulation rejection	−55 dB
Audio output power (at less than 10% distortion)	3W
Built-in power supply requirements	11V to 16V DC, or 117V AC
Dimensions	2 × 6¼ × 9 in.
Weight	2 lb 4 oz

Service Data

Crystal layout	Fig. 8-2A
Schematic	Schematic 8-2
Internal layout	Fig. 8-3
Alignment data, steps	A, G, H, E, F

The standard UHF *Mono-Scan* is factory aligned on 460 MHz, and selected crystals should fall within ±4 MHz of 460 MHz. If peak reception is to be obtained outside these limits, retune the receiver's front end for the new center frequency.

JOHNSON VHF MONO-SCAN

Specifications

Frequency range	153–160 MHz
Channels	Eight
Sensitivity (EIA 12 dB SINAD)	
High-band VHF	0.4 μV
Selectivity (30 dB quieting)	
High-band VHF	−60 dB at ±30 kHz
Spurious and image rejection	
High-band VHF	−50 dB
Intermodulation rejection	
High-band VHF	−60 dB
Audio output power (at less than 10% distortion)	3W
Built-in power supply requeirments	11V to 16V DC, or 117V AC
Dimensions	2 × 6¼ × 9 in.
Weight	2 lb 4 oz

Service Data

Crystal layout	Fig. 8-2A
Schematic	Schematic 8-3
Internal layout	Fig. 8-4
Alignment data, steps	B, C, E, I, F

Standard VHF *Mono-Scan* is aligned for a pass band of within 3.5 MHz of 157.00 MHz. Retuning of front end is necessary for peak performance outside these limits.

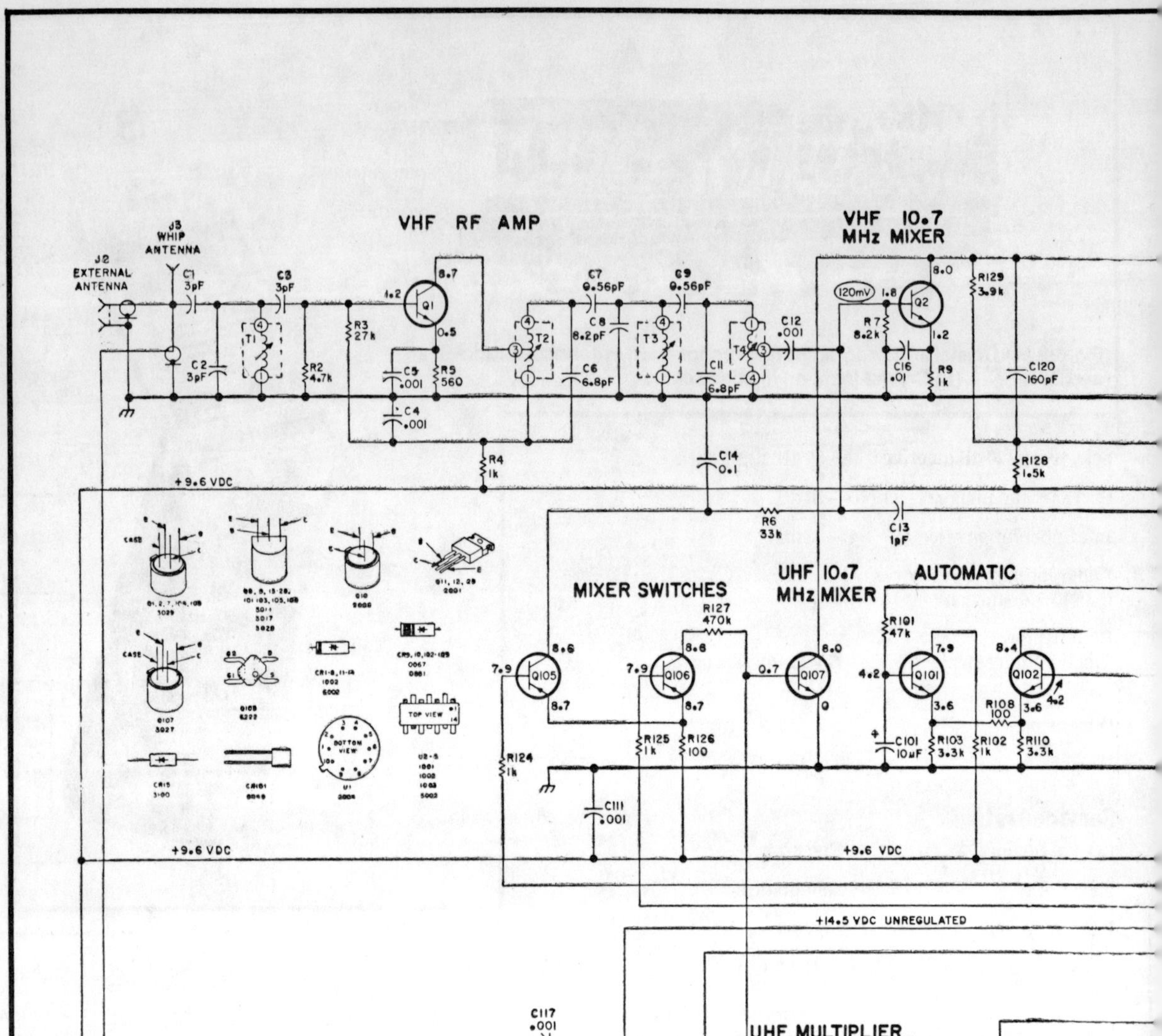
VHF RF AMP
VHF 10.7 MHz MIXER
J2 EXTERNAL ANTENNA
J3 WHIP ANTENNA
C1 3pF
C3 3pF
C2 3pF
T1
R3 27k
R2 4.7k
Q1
C5 .001
R5 560
C4 .001
R4 1k
C7 0.56pF
C9 0.56pF
C8 8.2pF
C6 6.8pF
T21
T3
T4
C11 6.8pF
C12 .001
C14 0.1
120mV
Q2
R7 8.2k
C16 .01
R9 1k
R129 3.9k
C120 160pF
R128 1.5k
R6 33k
C13 1pF
+9.6 VDC
MIXER SWITCHES
UHF 10.7 MHz MIXER
AUTOMATIC
R127 470k
Q105
Q106
Q107
Q101
Q102
R101 47k
R108 100
R124 1k
R125 1k
R126 100
C111 .001
C101 10µF
R103 3.3k
R102 1k
R110 3.3k
TOP VIEW
BOTTOM VIEW
+14.5 VDC UNREGULATED
UHF RF AMP
UHF MULTIPLIER
T101
C112
Q108
C115 47pF
C117 .001
L103 20µH
T102
C118
R123 47k
C114 47pF
R122 68k
C116 .001
C113 .001
R121 22
L104
R131 15k
58mV
Q109
C124 .01
T104
C125
C121
R132 68k
C126 .001
R133 560
R130 100
C123 .001
C82 .01
AC POWER SUPPLY
10 VDC REGULATOR
J5 DC INPUT
J4
T14
BLK
BRN
CR11
CR12
CR13
CR14
S10 ON/OFF
C79 1000
C80 1000
R92 150
Q29
CR15

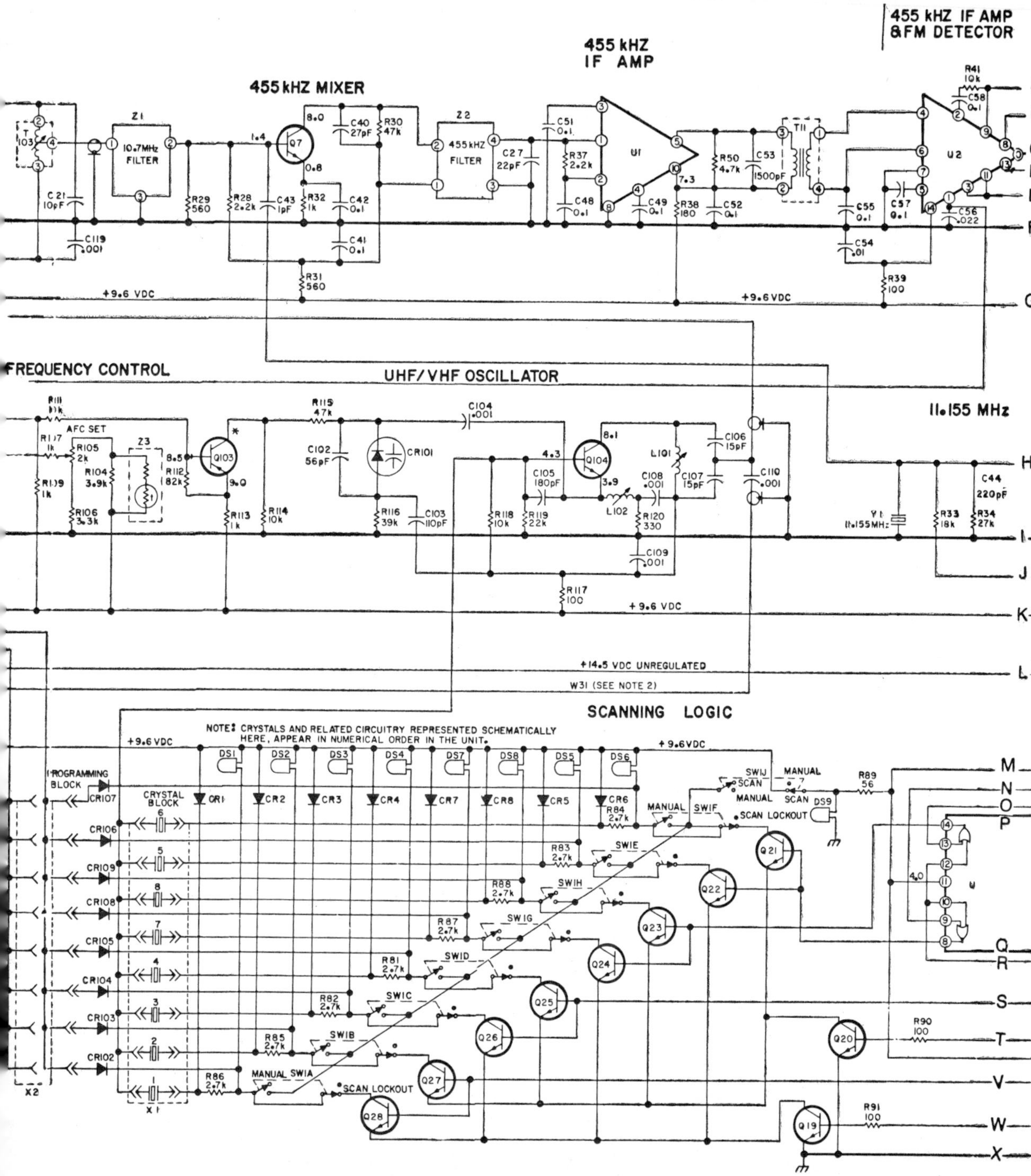

Schematic 8-1. E. F. Johnson UHF – VHF Duo-Scan (early model).

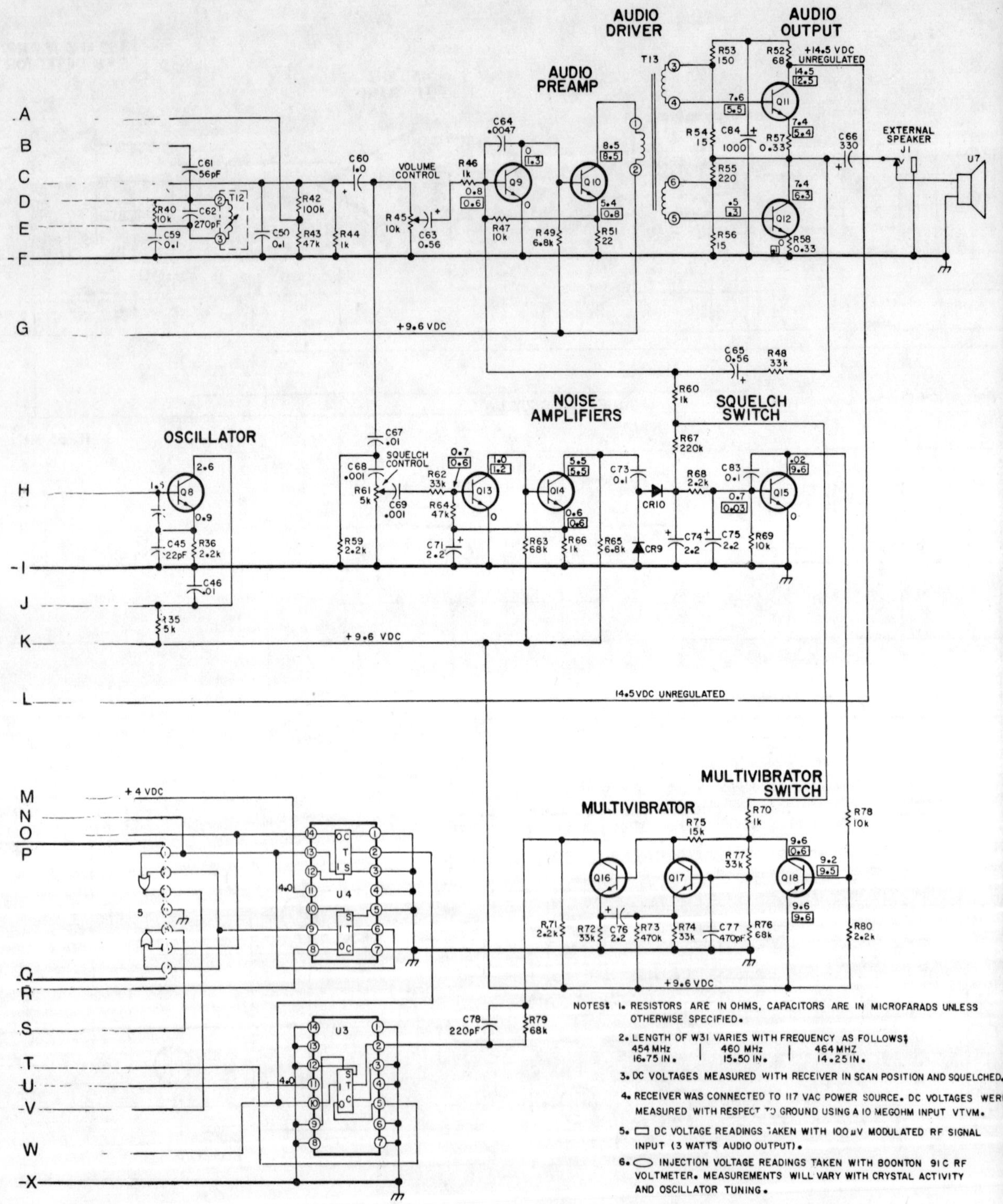

NOTES: 1. RESISTORS ARE IN OHMS, CAPACITORS ARE IN MICROFARADS UNLESS OTHERWISE SPECIFIED.

2. LENGTH OF W31 VARIES WITH FREQUENCY AS FOLLOWS:

454 MHz	460 MHz	464 MHz
16.75 IN.	15.50 IN.	14.25 IN.

3. DC VOLTAGES MEASURED WITH RECEIVER IN SCAN POSITION AND SQUELCHED.

4. RECEIVER WAS CONNECTED TO 117 VAC POWER SOURCE. DC VOLTAGES WERE MEASURED WITH RESPECT TO GROUND USING A 10 MEGOHM INPUT VTVM.

5. ▭ DC VOLTAGE READINGS TAKEN WITH 100 µV MODULATED RF SIGNAL INPUT (3 WATTS AUDIO OUTPUT).

6. ⬭ INJECTION VOLTAGE READINGS TAKEN WITH BOONTON 91C RF VOLTMETER. MEASUREMENTS WILL VARY WITH CRYSTAL ACTIVITY AND OSCILLATOR TUNING.

7. "*" SET FOR 3.5 VDC NO SIGNAL INPUT.

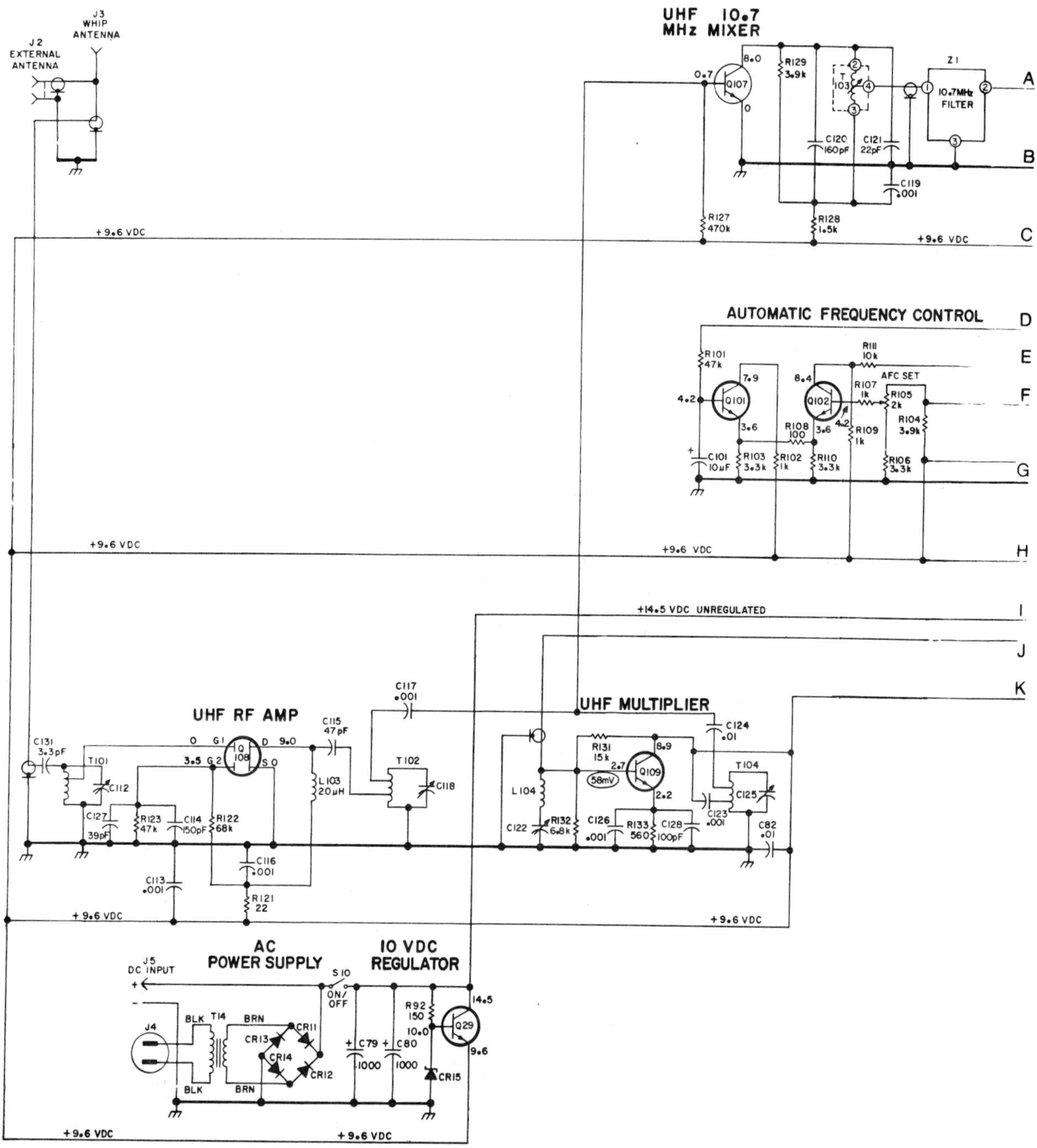

Schematic 8-2. Johnson UHF Mono-Scan (for units with A, or later, model designator on serial number sticker).

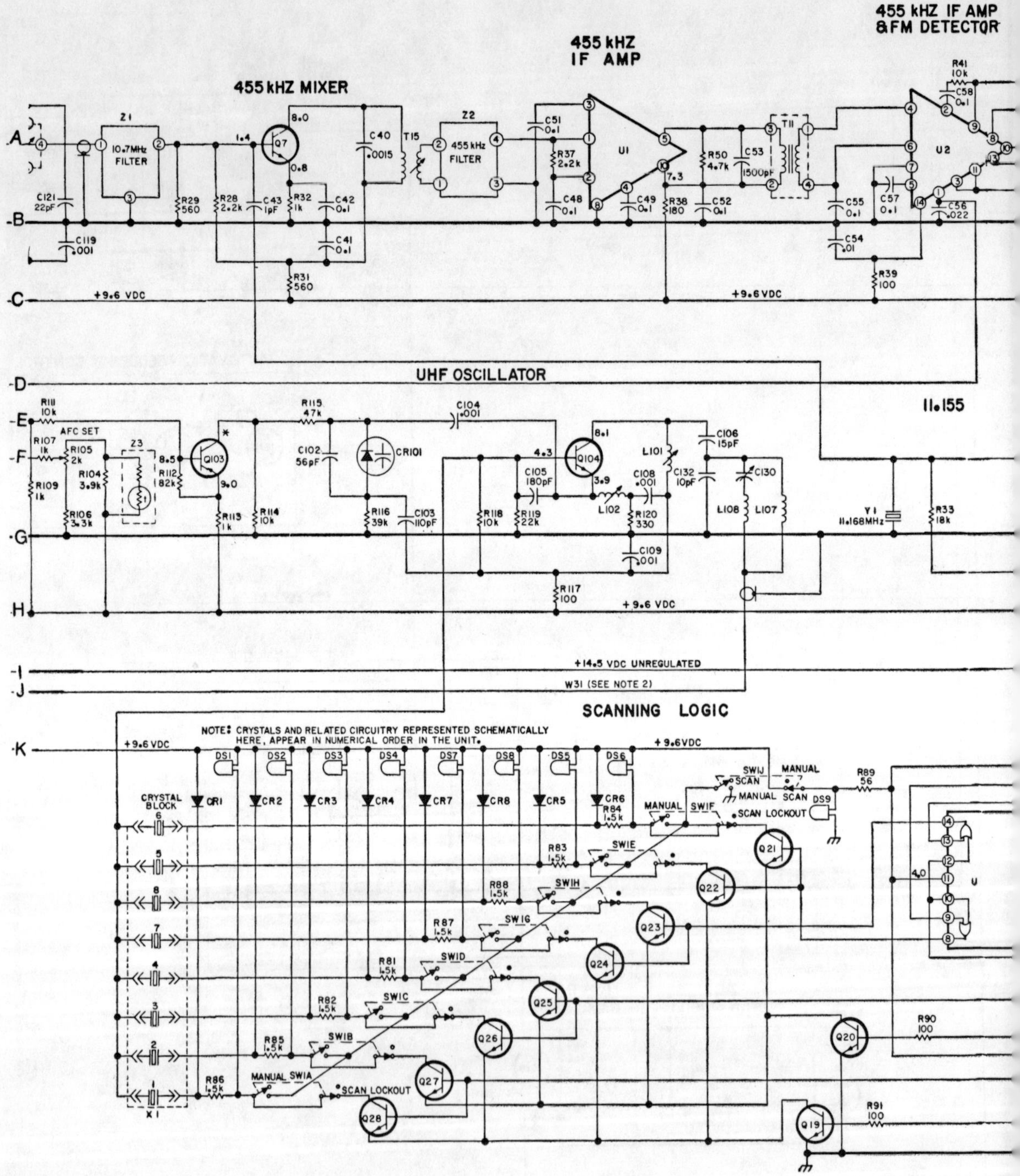
455 kHZ MIXER
455 kHZ IF AMP
455 kHZ IF AMP &FM DETECTOR
UHF OSCILLATOR
11.155
SCANNING LOGIC
NOTE: CRYSTALS AND RELATED CIRCUITRY REPRESENTED SCHEMATICALLY HERE, APPEAR IN NUMERICAL ORDER IN THE UNIT.
+9.6 VDC
+14.5 VDC UNREGULATED
W31 (SEE NOTE 2)
AFC SET
CRYSTAL BLOCK
MANUAL
SCAN
SCAN LOCKOUT
MANUAL SW1A

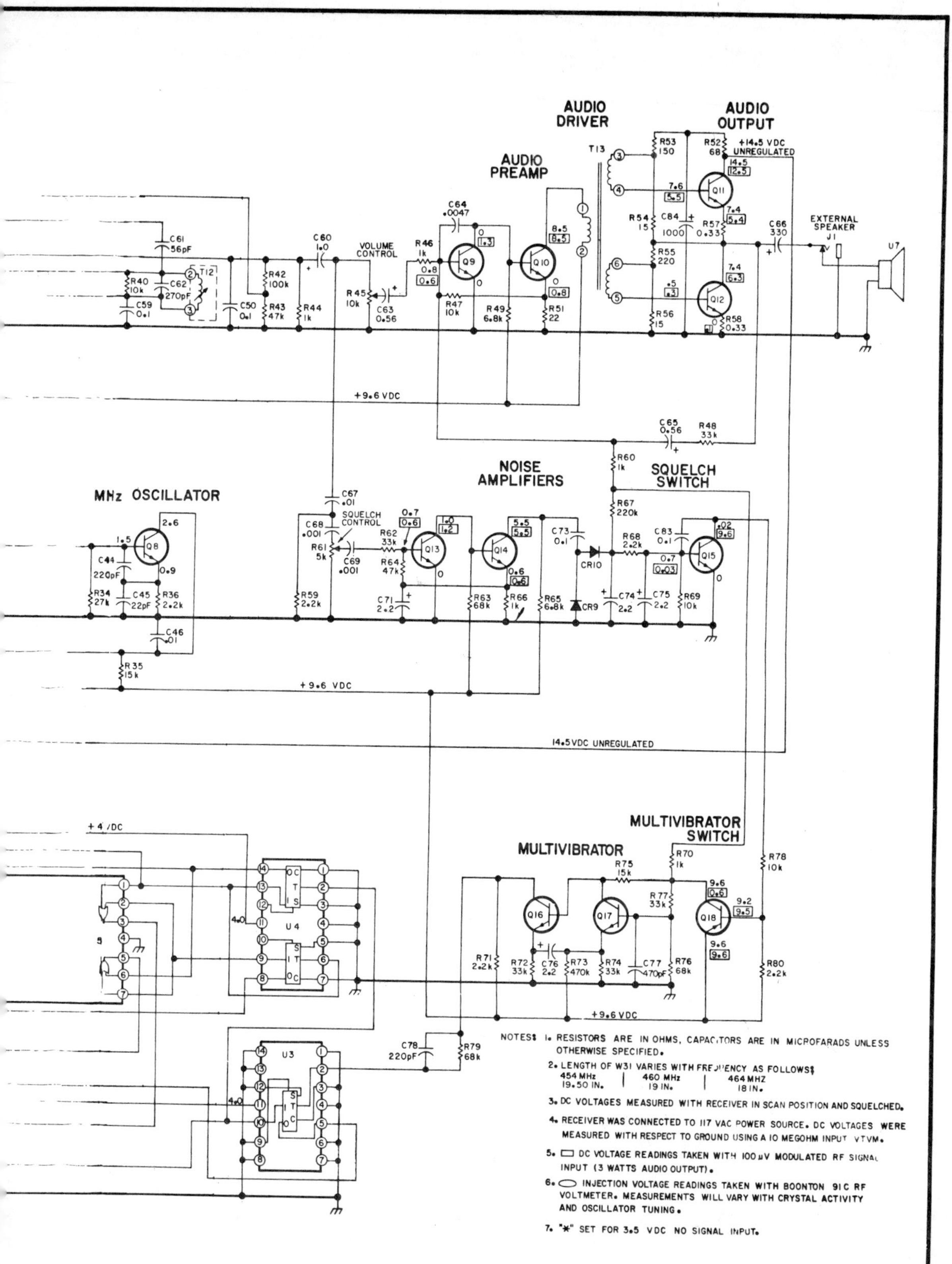

NOTES: 1. RESISTORS ARE IN OHMS, CAPACITORS ARE IN MICROFARADS UNLESS OTHERWISE SPECIFIED.

2. LENGTH OF W31 VARIES WITH FREQUENCY AS FOLLOWS:

454 MHz	460 MHz	464 MHZ
19.50 IN.	19 IN.	18 IN.

3. DC VOLTAGES MEASURED WITH RECEIVER IN SCAN POSITION AND SQUELCHED.

4. RECEIVER WAS CONNECTED TO 117 VAC POWER SOURCE. DC VOLTAGES WERE MEASURED WITH RESPECT TO GROUND USING A 10 MEGOHM INPUT VTVM.

5. ☐ DC VOLTAGE READINGS TAKEN WITH 100 µV MODULATED RF SIGNAL INPUT (3 WATTS AUDIO OUTPUT).

6. ⬭ INJECTION VOLTAGE READINGS TAKEN WITH BOONTON 91C RF VOLTMETER. MEASUREMENTS WILL VARY WITH CRYSTAL ACTIVITY AND OSCILLATOR TUNING.

7. "*" SET FOR 3.5 VDC NO SIGNAL INPUT.

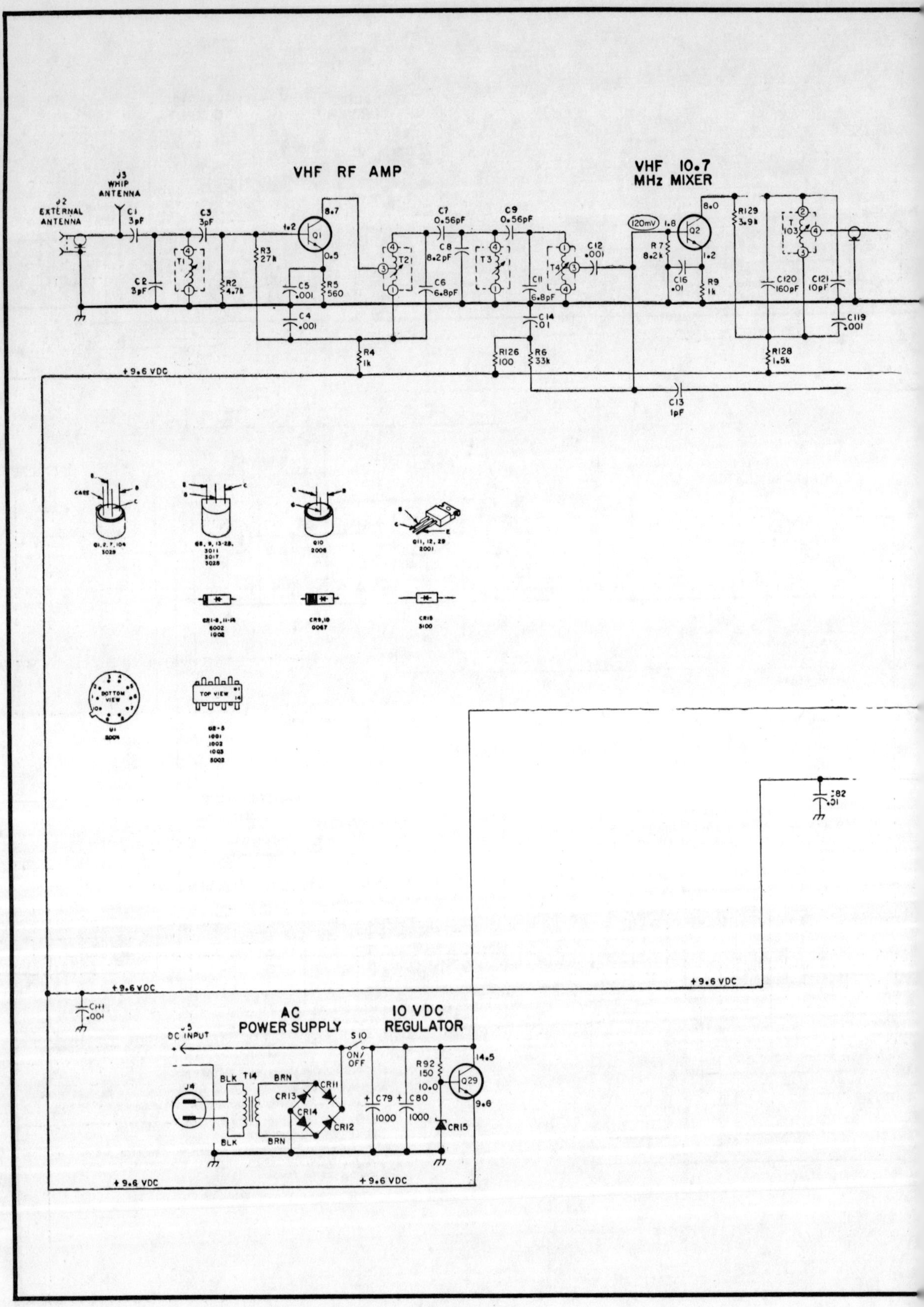
VHF RF AMP
VHF 10.7 MHz MIXER
J2 EXTERNAL ANTENNA
J3 WHIP ANTENNA
+9.6 VDC
AC POWER SUPPLY
10 VDC REGULATOR
J5 DC INPUT
S10 ON/OFF
BOTTOM VIEW
TOP VIEW

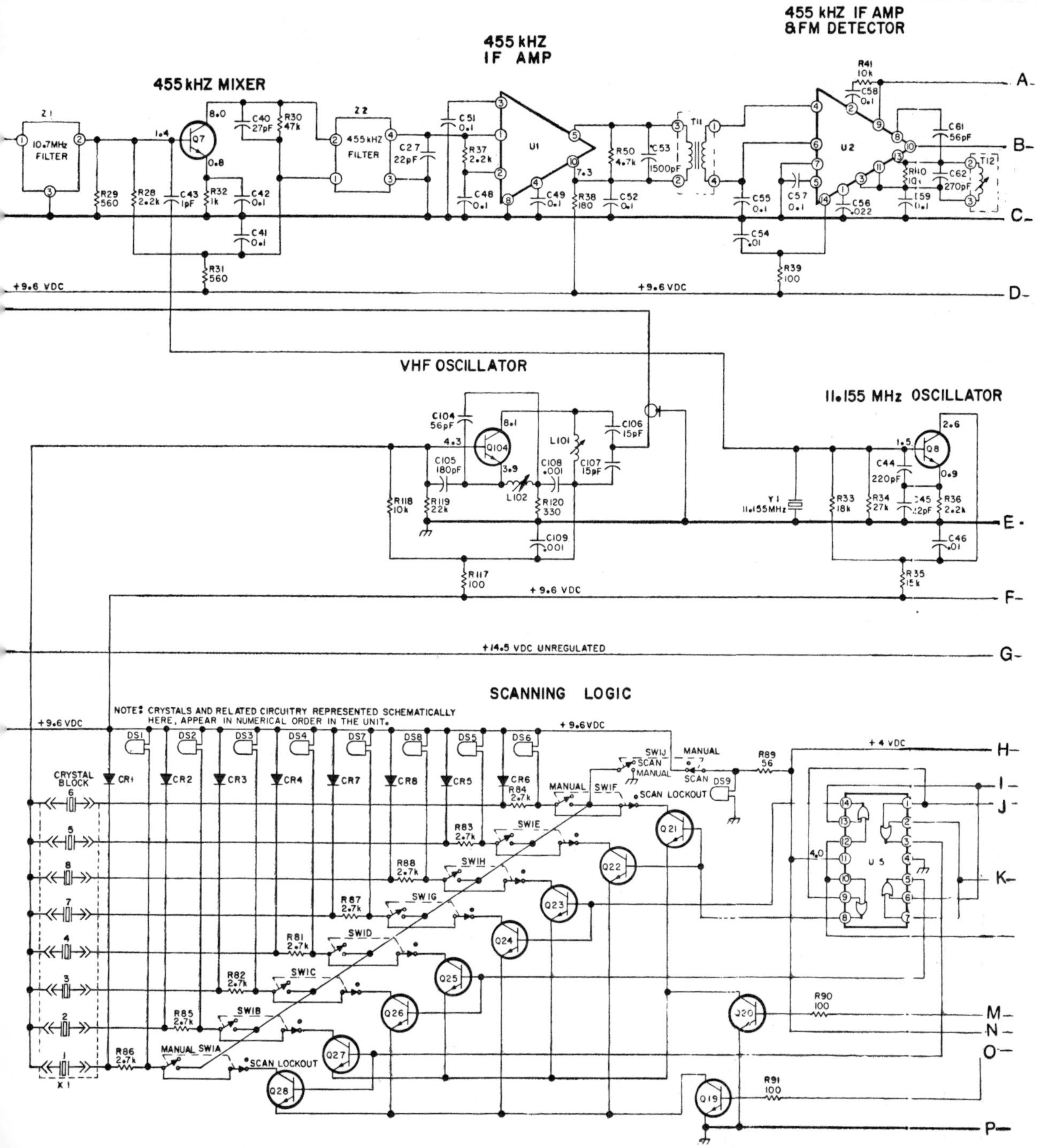

Schematic 8-3. Johnson VHF Mono-Scan (for units with A, or later, model designator on serial number sticker).

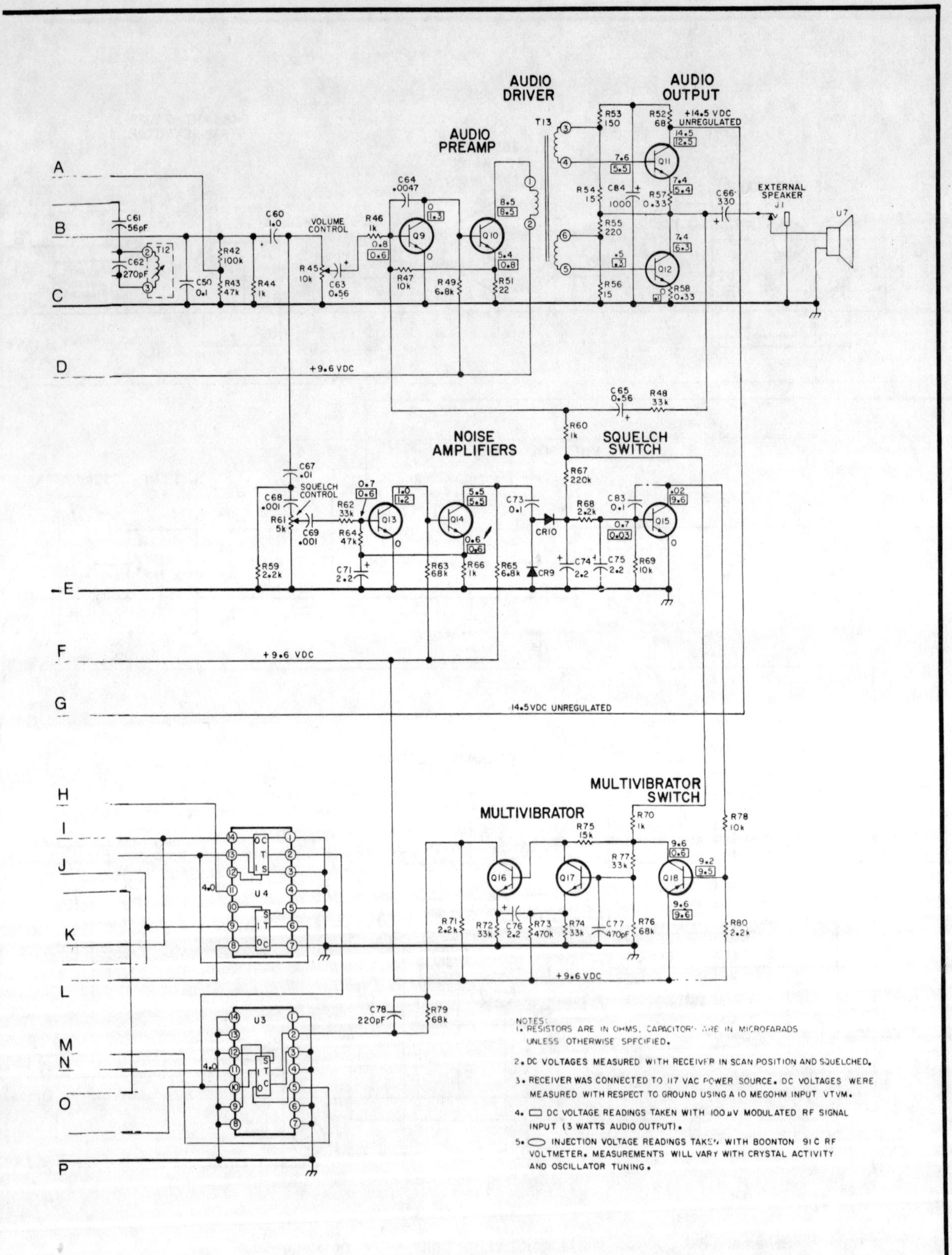

NOTES:

1. RESISTORS ARE IN OHMS, CAPACITORS ARE IN MICROFARADS UNLESS OTHERWISE SPECIFIED.
2. DC VOLTAGES MEASURED WITH RECEIVER IN SCAN POSITION AND SQUELCHED.
3. RECEIVER WAS CONNECTED TO 117 VAC POWER SOURCE. DC VOLTAGES WERE MEASURED WITH RESPECT TO GROUND USING A 10 MEGOHM INPUT VTVM.
4. ▭ DC VOLTAGE READINGS TAKEN WITH 100 µV MODULATED RF SIGNAL INPUT (3 WATTS AUDIO OUTPUT).
5. ⬭ INJECTION VOLTAGE READINGS TAKEN WITH BOONTON 91C RF VOLTMETER. MEASUREMENTS WILL VARY WITH CRYSTAL ACTIVITY AND OSCILLATOR TUNING.

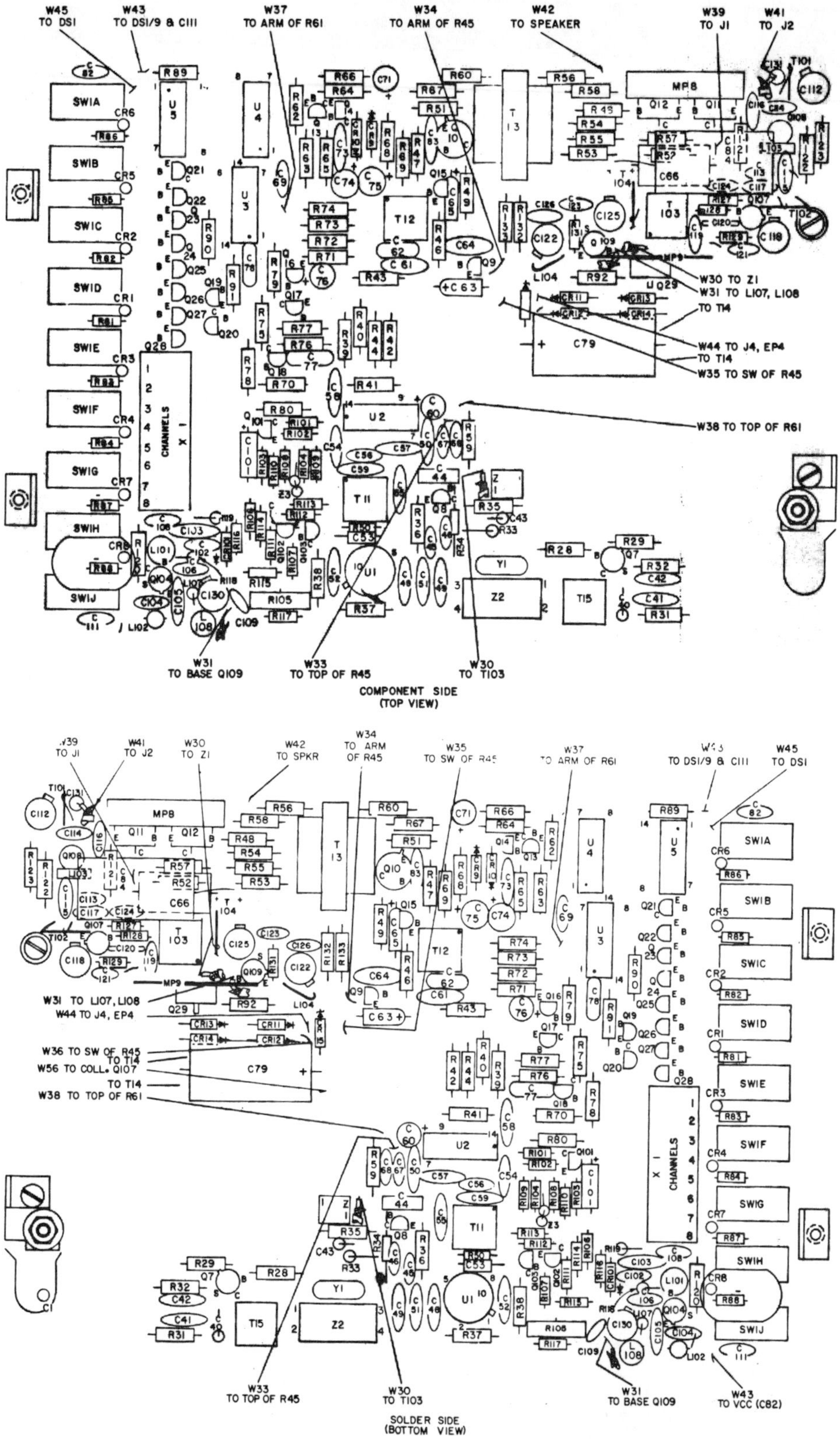

Fig. 8-3. Component and circuit board layout for E. F. Johnson UHF Mono-Scan.

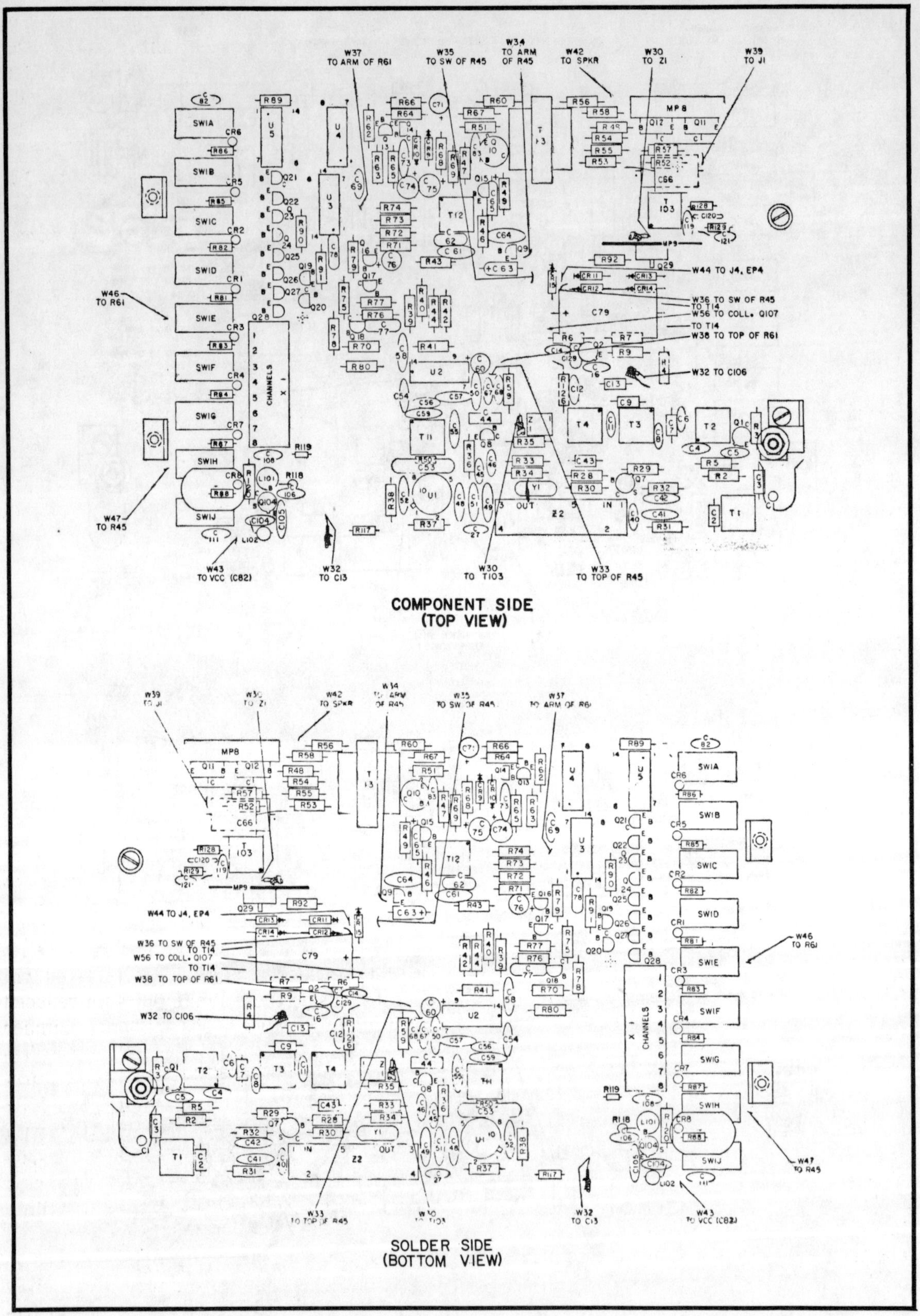

Fig. 8-4. Component and circuit board layout of E. F. Johnson VHF Mono-Scan.

JOHNSON HIGH–LOW DUO-SCAN

Specifications

Frequency ranges and corresponding catalog numbers	39.5–44.5 and 153–160 MHz, 241-0390-030; 33–39 and 153–160 MHz, 241-0390-036; 37–43 and 153–160 MHz, 241-0390-040; 41–47 and 153–160 MHz, 241-0390-044
Channels	Eight (in any combination of one or both bands)
Sensitivity (EIA 12 dB SINAD)	
Low-band VHF	0.4 μV
High-band VHF	0.4 μV
Selectivity (20 dB quieting)	
Low-band VHF	−60 dB at ±20 kHz
High-band VHF	−60 dB at ±30 kHz
Spurious and image rejection	
Low-band VHF	−50 dB
High-band VHF	−50 dB
Intermodulation rejection	
Low-band VHF	−55 dB
High-band VHF	−60 dB
Audio output power (at less than 10% distortion)	3W
Dimensions	2 × 6¼ × 9 in.
Weight	2 lb 4 oz

Service Data

Crystal Layout	Fig. 8-2A
Schematic	Schematic 8-4
Internal Layout	Fig. 8-5
Alignment Data Steps	A-2, J, K, E, F

High-Low VHF The standard Johnson high–low *Duo-Scan* monitor receiver is factory aligned on a low-band frequency of 42.000 MHz and a high-band frequency of 156.5 MHz. To obtain peak receiver performance, selected crystal frequencies should be within ±2.5 MHz of 42 MHz and ±3.5 MHz of 156 MHz. The receiver will operate outside these limits but with reduced sensitivity.

To maintain peak receiver performance on frequencies outside these limits, the receiver should be realigned on the required center frequency. Align the monitor receiver on the available frequency which is close to the center of the frequencies actually installed. If the receiver is required to operate on a center frequency below 35 MHz, change capacitors C23 and C26 to 22 pF and 15 pF, respectively, before realignment is started.

JOHNSON VHF MINI-SCAN

Specifications

Frequency range	146–174 MHz
Channel capability	Four
Sensitivity (EIA SINAD)	0.4 μV
Selectivity	−55 dB at ±30 kHz (EIA)
Spurious rejection	−50 dB (−30 dB image rejection)
Intermodulation rejection	−55 dB
Maximum frequency spread	10 MHz (6 dB or less degradation)
Audio output (10% or less distortion)	100 mV
Power required	Four AA cells
Dimensions	5.9 × 2.6 × 1.4 in.
Weight	11.5 oz (with batteries)

The standard Johnson *Mini-Scan* monitor receiver, part number 241-0352-102, is factory aligned on a center frequency of 156.5 MHz. To obtain peak receiver performance, selected crystal frequencies should be within +5.5 MHz and 4.5 MHz of 156.5 MHz. The receiver will operate outside these limits but with reduced sensitivity. Receivers bearing part number 241-0352-101 are factory aligned on a center frequency of 150.25 MHz. Selected monitor frequency crystals should be within +4.75 MHz and 5.25 MHz of this frequency. Receivers with part number 241-0352-103 are factory aligned on a center frequency of 167.25 MHz. Selected monitor frequency crystals should be within +4.75 MHz and −5.25 MHz of this frequency.

To maintain peak receiver performance on frequencies outside these limits, the receiver should be realigned on the required center frequency. Align the receiver on the available fre-

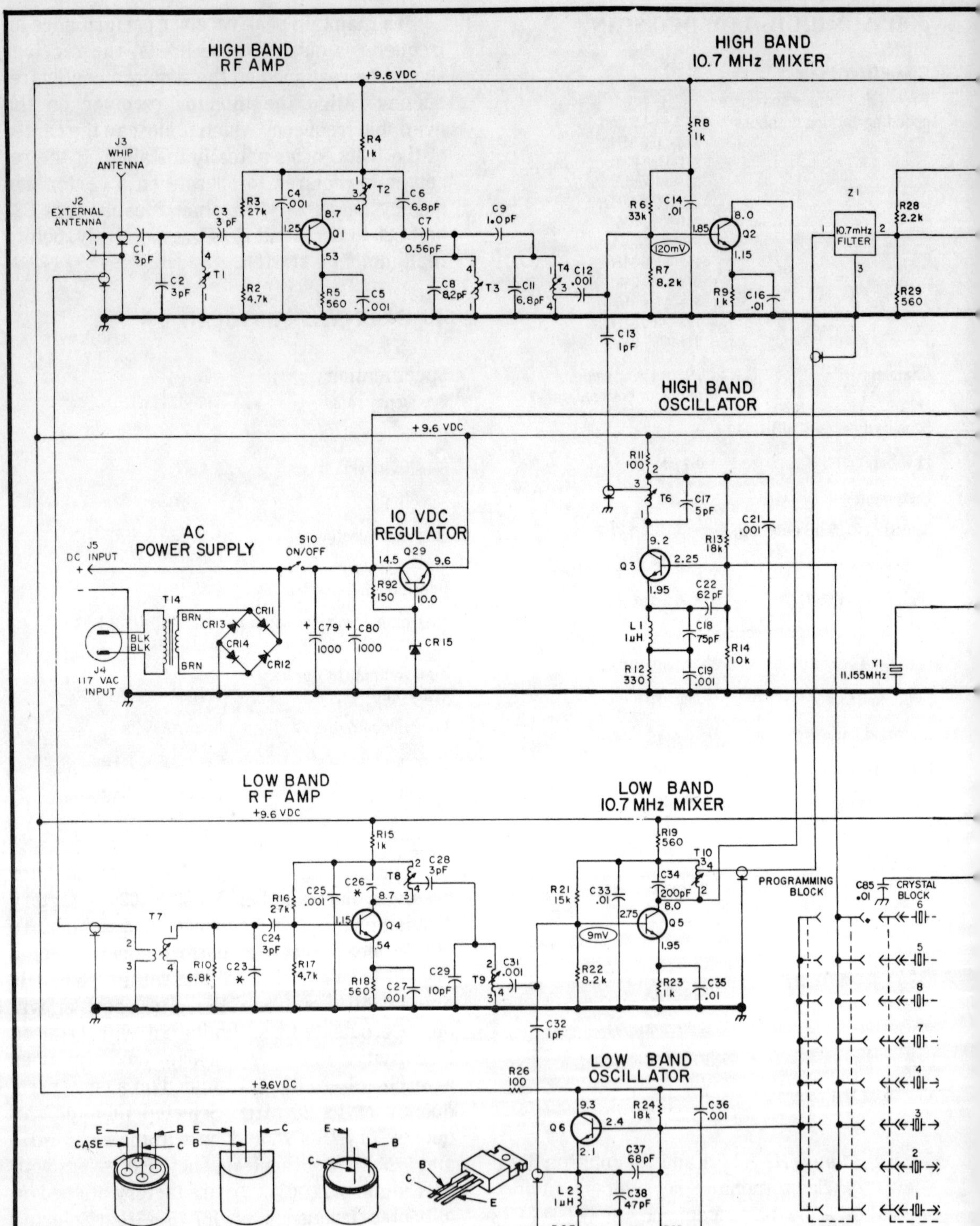

HIGH BAND
RF AMP
HIGH BAND
10.7 MHz MIXER
HIGH BAND
OSCILLATOR
AC
POWER SUPPLY
10 VDC
REGULATOR
LOW BAND
RF AMP
LOW BAND
10.7 MHz MIXER
LOW BAND
OSCILLATOR
PROGRAMMING
BLOCK
CRYSTAL
BLOCK
Q1-6
3004
3005
3006
3007
Q7-9,13-28
3011
3017
3028
Q10
2006
Q11,12,29
2001

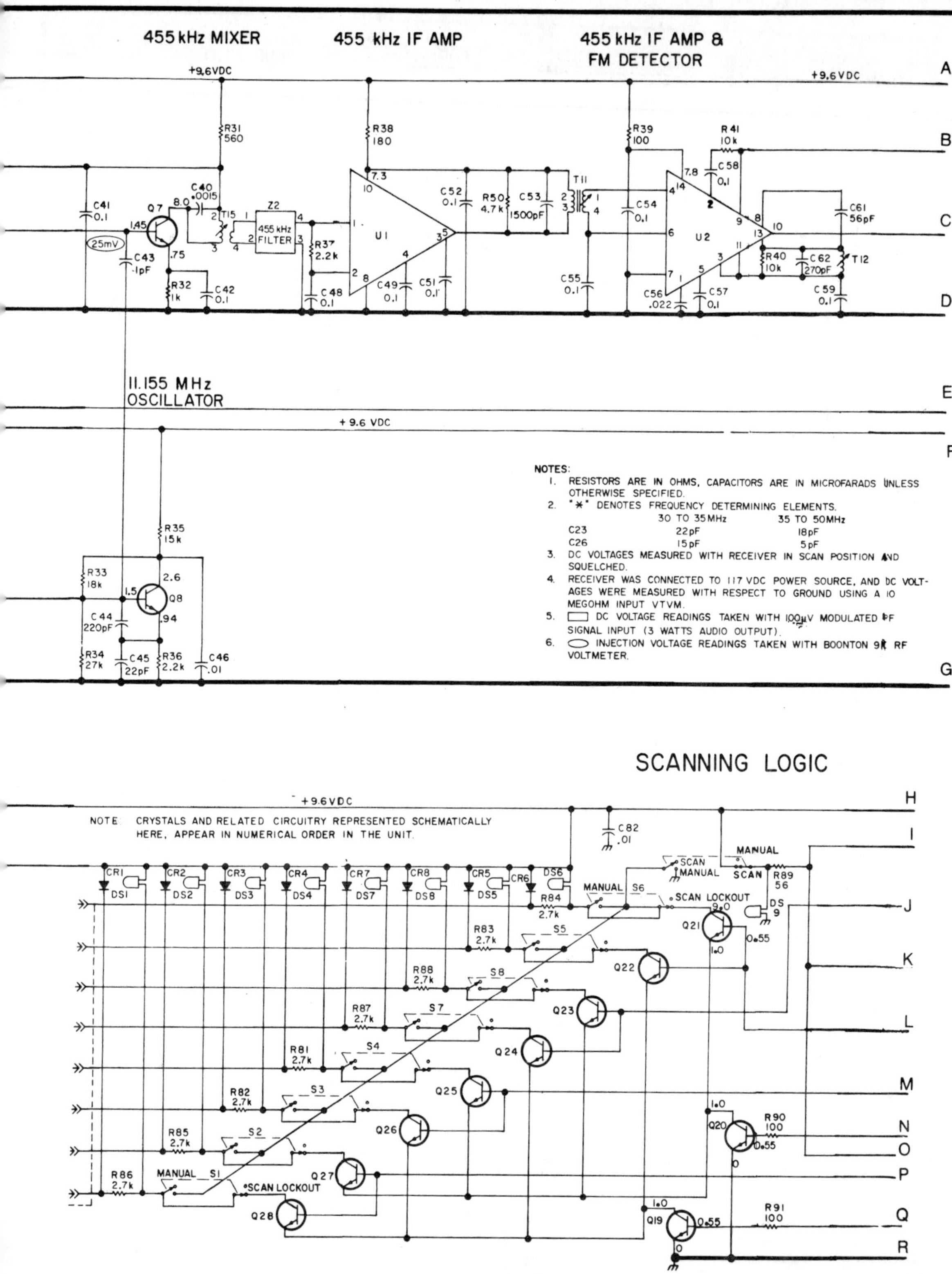

Schematic 8-4. Johnson High–Low VHF Duo-Scan (for units with C, or later, model designator on serial number sticker).

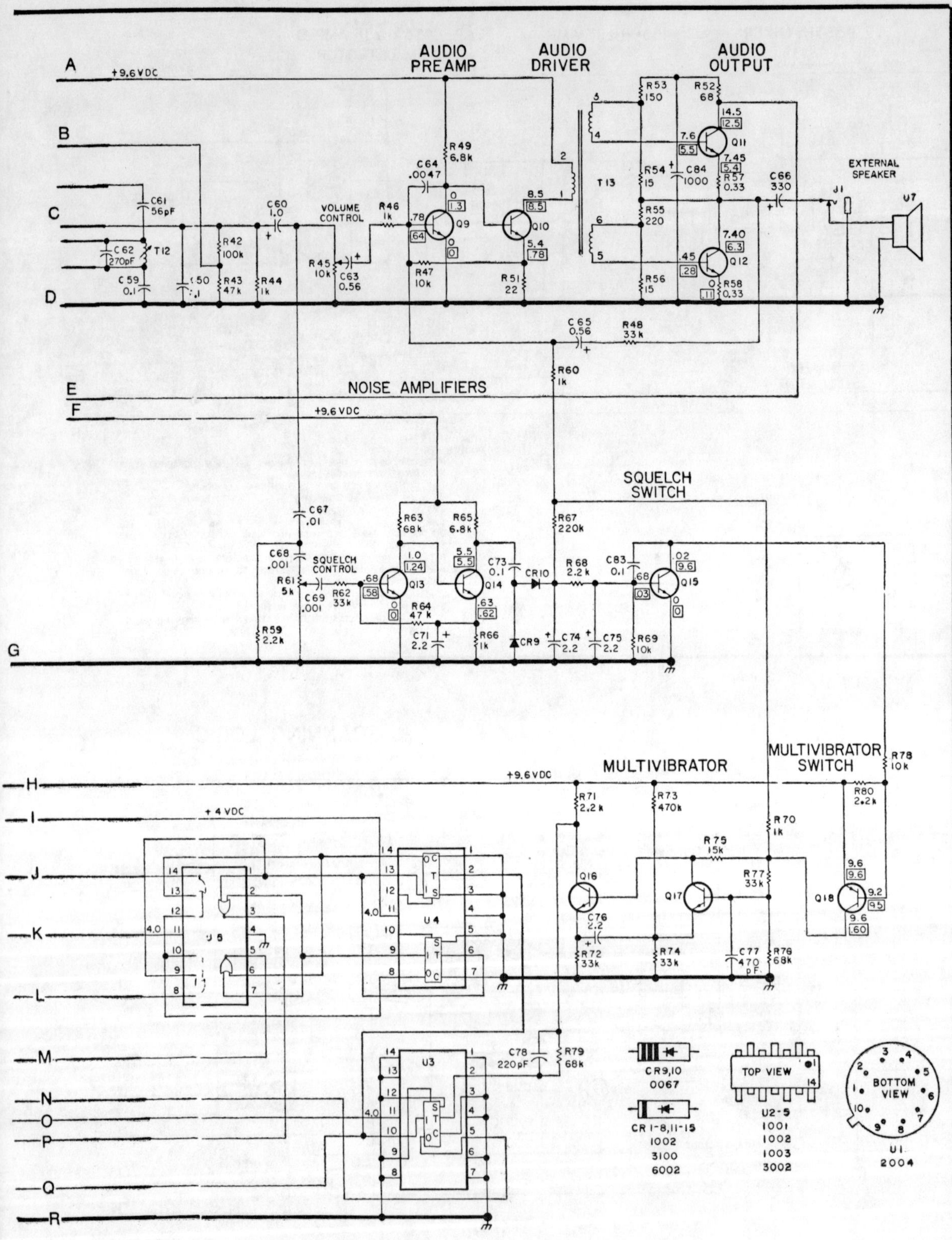
AUDIO PREAMP
AUDIO DRIVER
AUDIO OUTPUT
EXTERNAL SPEAKER
VOLUME CONTROL
NOISE AMPLIFIERS
SQUELCH SWITCH
SQUELCH CONTROL
MULTIVIBRATOR
MULTIVIBRATOR SWITCH
+9.6 VDC
+4 VDC
TOP VIEW
BOTTOM VIEW
CR9,10 0067
CR1-8,11-15 1002 3100 6002
U2-5 1001 1002 1003 3002
U1 2004

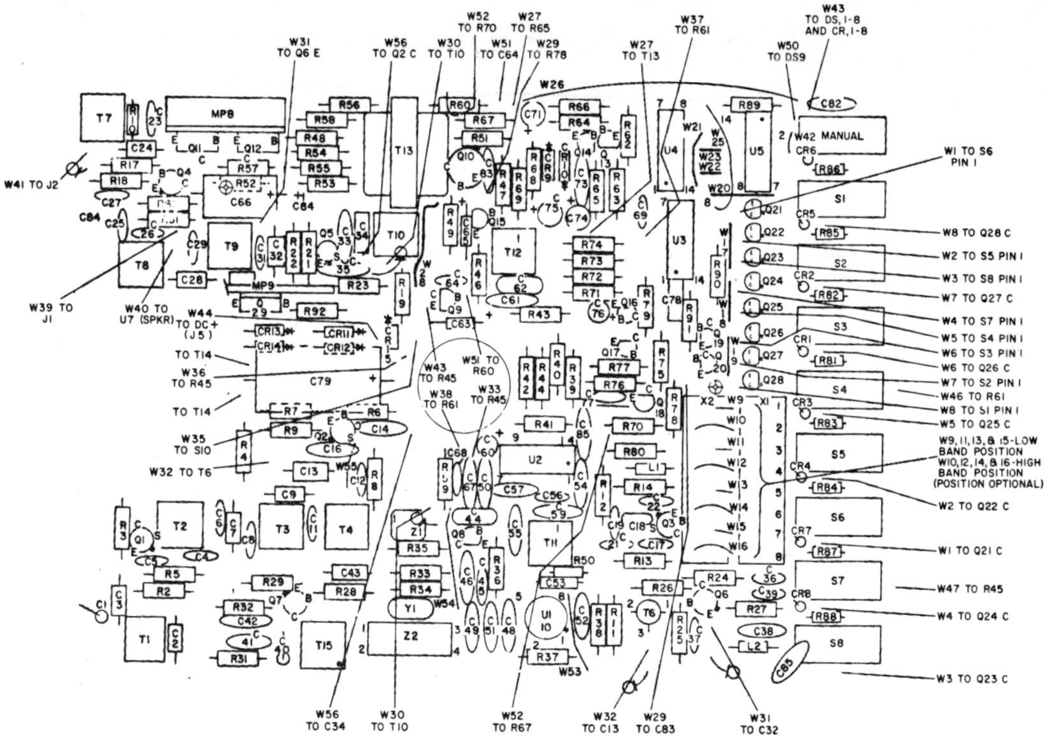

SOLDER SIDE
(Bottom View)

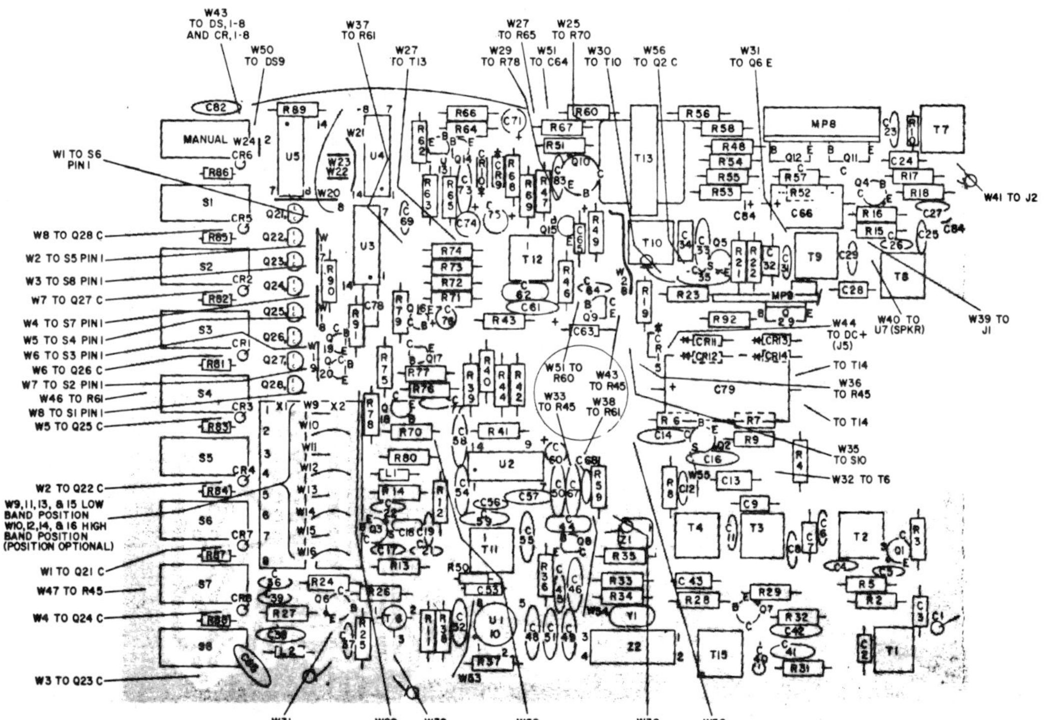

COMPONENT SIDE
(Top View)

Fig. 8-5. Component and circuit board layout of E. F. Johnson High–Low VHF Duo-Scan.

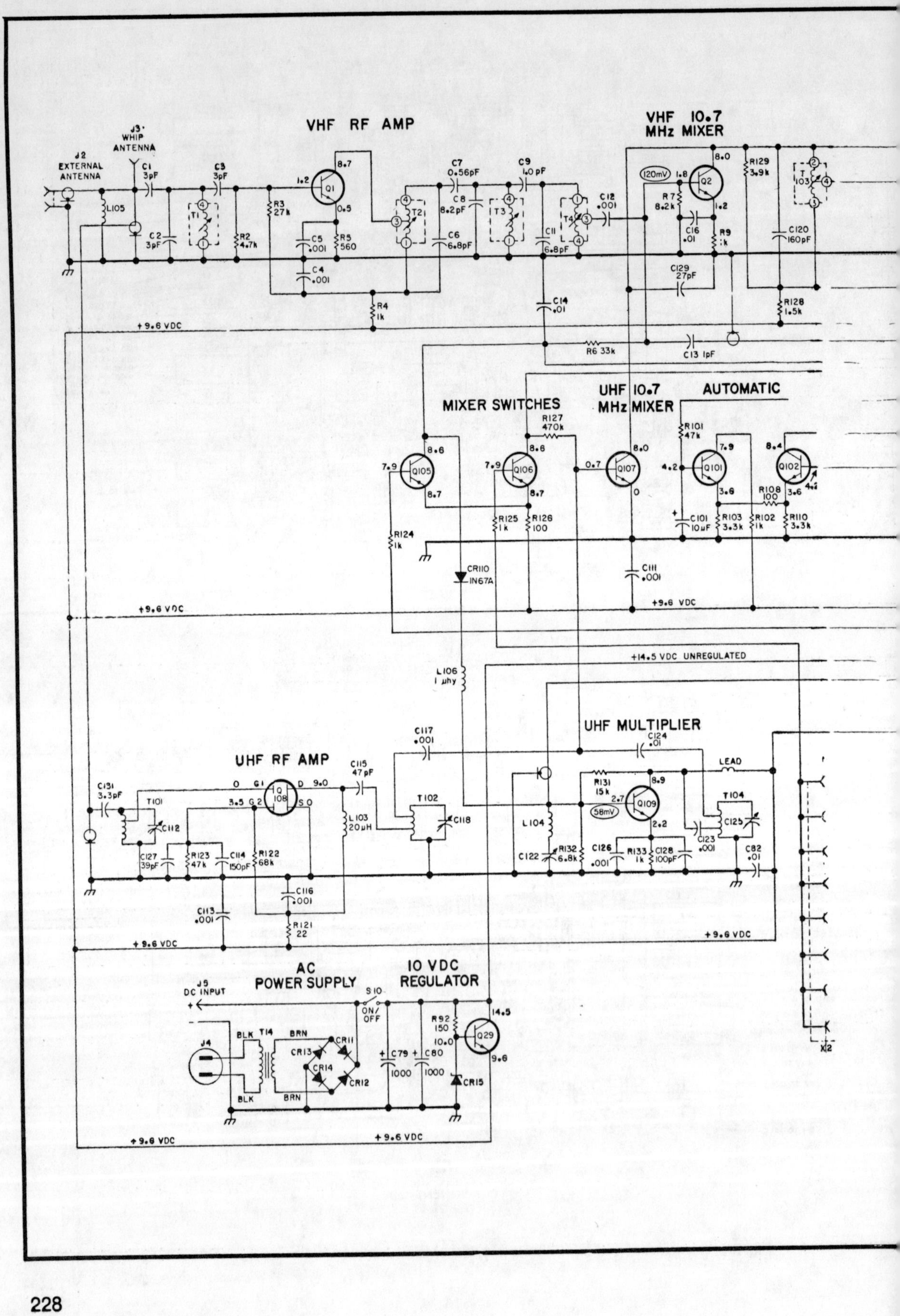

VHF RF AMP
VHF 10.7 MHz MIXER
J2 EXTERNAL ANTENNA
J3 WHIP ANTENNA
C1 3pF
C3 3pF
C2 3pF
L105
T1
R2 4.7k
R3 27k
Q1
C5 .001
R5 560
C4 .001
R4 1k
C7 0.56pF
C9 1.0pF
C8 8.2pF
C6 6.8pF
T2
T3
T4
C11 6.8pF
C12 .001
C14 .01
120mV
Q2
R7 8.2k
C16 .01
R9 1k
C129 27pF
R129 3.9k
C120 160pF
T103
R128 1.5k
R6 33k
C13 1pF
+9.6 VDC
MIXER SWITCHES
UHF 10.7 MHz MIXER
AUTOMATIC
R127 470k
Q105
Q106
Q107
Q101
Q102
R101 47k
R108 100
R124 1k
R125 1k
R126 100
C101 10µF
R103 3.3k
R102 1k
R110 3.3k
C111 .001
CR110 IN67A
+14.5 VDC UNREGULATED
L106 1 µhy
UHF MULTIPLIER
C117 .001
C124 .01
LEAD
UHF RF AMP
C131 3.3pF
T101
C112
C127 39pF
R123 47k
C114 150pF
Q108
C115 47pF
L103 20µH
R122 68k
C116 .001
C113 .001
R121 22
T102
C118
R131 15k
58mV
Q109
L104
C122
R132 6.8k
C126 .001
R133 1k
C128 100pF
C123 .001
T104
C125
C82 .01
X2
AC POWER SUPPLY
10 VDC REGULATOR
J5 DC INPUT
J4
BLK
T14
BRN
CR13
CR11
CR14
CR12
S10 ON/OFF
C79 1000
C80 1000
R92 150
Q29
CR15

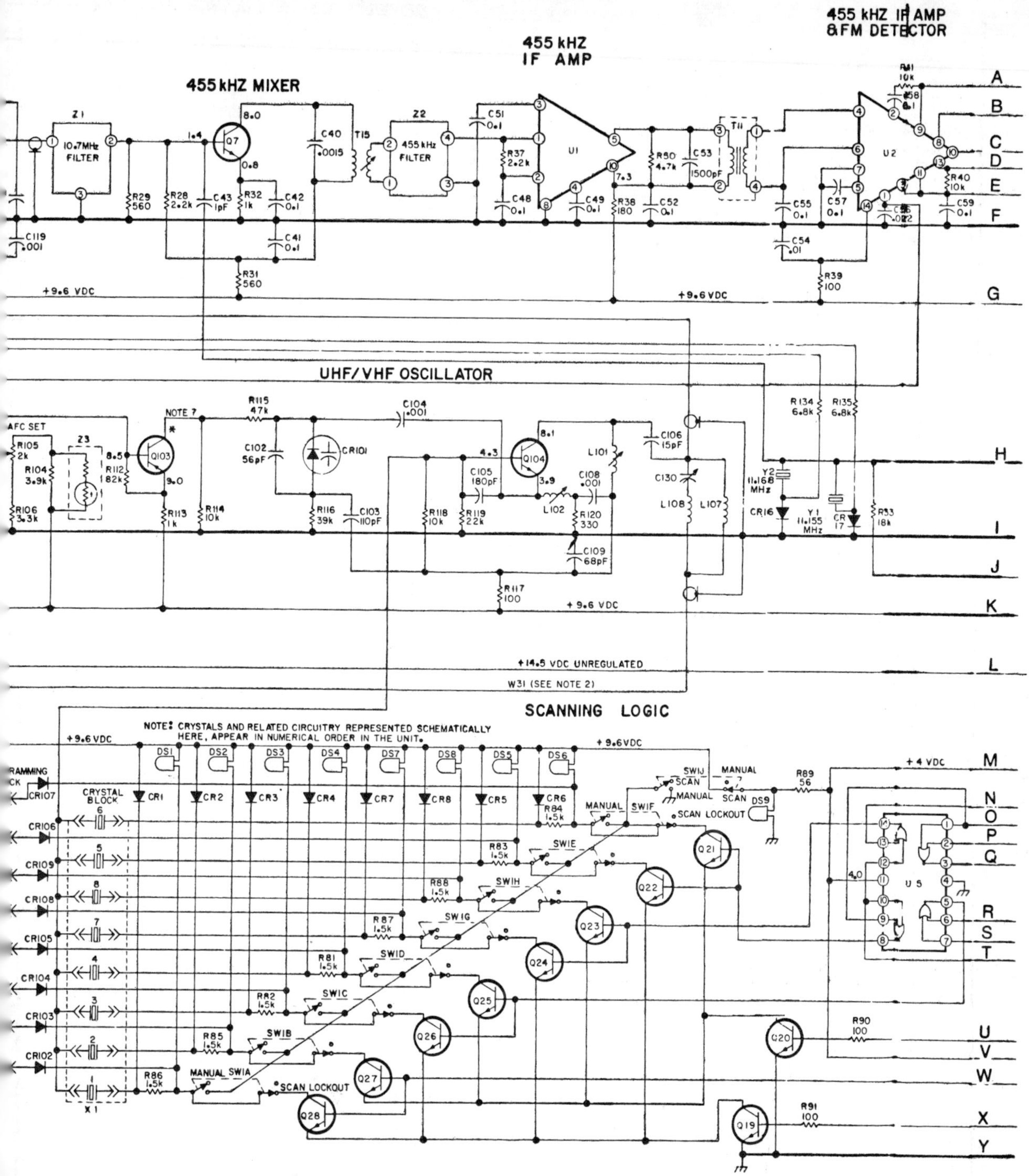

Schematic 8-5. Johnson UHF–VHF Duo-Scan (for units with C, or later, model designator on serial number sticker).

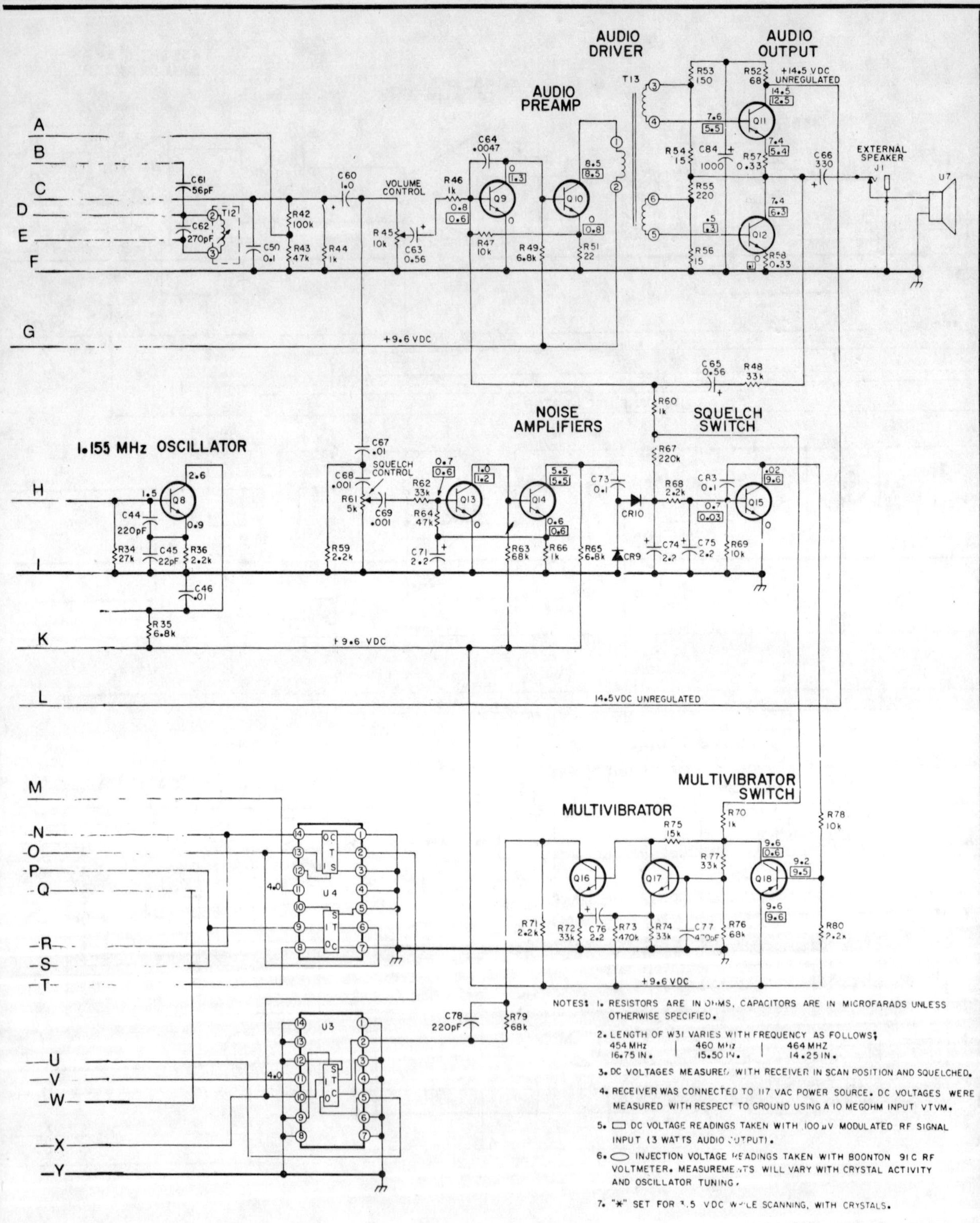

NOTES: 1. RESISTORS ARE IN OHMS, CAPACITORS ARE IN MICROFARADS UNLESS OTHERWISE SPECIFIED.

2. LENGTH OF W31 VARIES WITH FREQUENCY AS FOLLOWS:

454 MHz	460 MHz	464 MHz
16.75 IN.	15.50 IN.	14.25 IN.

3. DC VOLTAGES MEASURED WITH RECEIVER IN SCAN POSITION AND SQUELCHED.

4. RECEIVER WAS CONNECTED TO 117 VAC POWER SOURCE. DC VOLTAGES WERE MEASURED WITH RESPECT TO GROUND USING A 10 MEGOHM INPUT VTVM.

5. ▭ DC VOLTAGE READINGS TAKEN WITH 100 µV MODULATED RF SIGNAL INPUT (3 WATTS AUDIO OUTPUT).

6. ⬭ INJECTION VOLTAGE READINGS TAKEN WITH BOONTON 91C RF VOLTMETER. MEASUREMENTS WILL VARY WITH CRYSTAL ACTIVITY AND OSCILLATOR TUNING.

7. "✱" SET FOR 3.5 VDC WHILE SCANNING, WITH CRYSTALS.

quency which is close to the center of the frequencies actually installed.

When referred to alignment data steps, follow the appropriate steps in the order in which they are presented. Doublecheck to be certain that you use the proper data and alignment procedure.

A-1—AFC Adjustment

1. Connect a DC VTVM to the collector of Q103.
2. With no signal input (*scan* mode), adjust AFC control R105 for a 3.5V DC reading.

A-2—Oscillator Adjustment (High Band)

1. Connect an RF voltmeter probe to the base of Q2.
2. Adjust T6 for maximum RF voltmeter reading. A typical reading is 100 to 200 mV.

B—Oscillator Adjustment

1. Manually select the UHF or VHF center-frequency crystal.
2. Connect an RF voltmeter probe to the base of Q2.
3. Adjust L101 for a maximum RF voltmeter reading. A typical reading is between 85 and 500 mV.

C—Alignment, VHF RF and IF Amplifiers

1. Connect an RF signal generator to the antenna connector. Set the generator output to the VHF band center frequency.
2. Connect an AC voltmeter across the external-speaker jack. Adjust T1, T2, T3, T4, T15, and T103 for a minimum AC voltmeter reading (maximum quieting sensitivity). With an RF voltmeter connected to pin 1 of integrated circuit U1, a typical reading of approximately 90 mV should be indicated with a 0.5 μV signal input applied to the antenna connector.
3. Connect the RF voltmeter probe to pin 4 of IC U2. Adjust T11 for a maximum RF voltmeter reading. A typical reading is approximately 75 mV with a 0.5 μV signal input at the antenna connector.

D—Alignment, UHF RF Amplifier and Mixer

1. Connect an RF signal generator to the antenna connector. Set the generator output to the UHF center frequency.
2. Connect an AC voltmeter across the external-speaker jack. Adjust C125, C118, and C112 for minimum AC voltmeter reading (maximum quieting sensitivity). With an RF voltmeter connected to pin 1 of IC U1, a typical reading is approximately 40 mV with 0.5 μV at the antenna connector.

E—Alignment, Discriminator

1. Externally modulate the RF signal generator with a 1 kHz audio signal at a deviation level of 3.5 kHz.
2. Set the RF signal generator to UHF or VHF center frequency and connect it to the antenna connector.
3. Adjust T12 for maximum undistorted audio output.

F—Quieting-Sensitivity Check

1. Adjust the receiver VOLUME control for a 0 dB audio output reference level.
2. Set the RF signal generator to the center frequency, with 0.7 μV at the antenna connector. The 0 dB audio output level should drop a minimum of 20 dB for both the UHF and VHF center frequencies.
3. Repeat steps 1 and 2 on the remaining crystals with the signal generator set to each specific crystal frequency and the output level set to 1.0 μV.

G—Oscillator Adjustment

1. Manually select the center-frequency crystal.
2. Connect an RF voltmeter probe to the base of Q107.
3. Adjust L101 for a maximum RF voltmeter reading.

H—Alignment, RF Amplifier, Mixer, and IF Amplifier

1. Connect the RF signal generator to the antenna connector and set the generator output to the crystal closest to the center frequency.

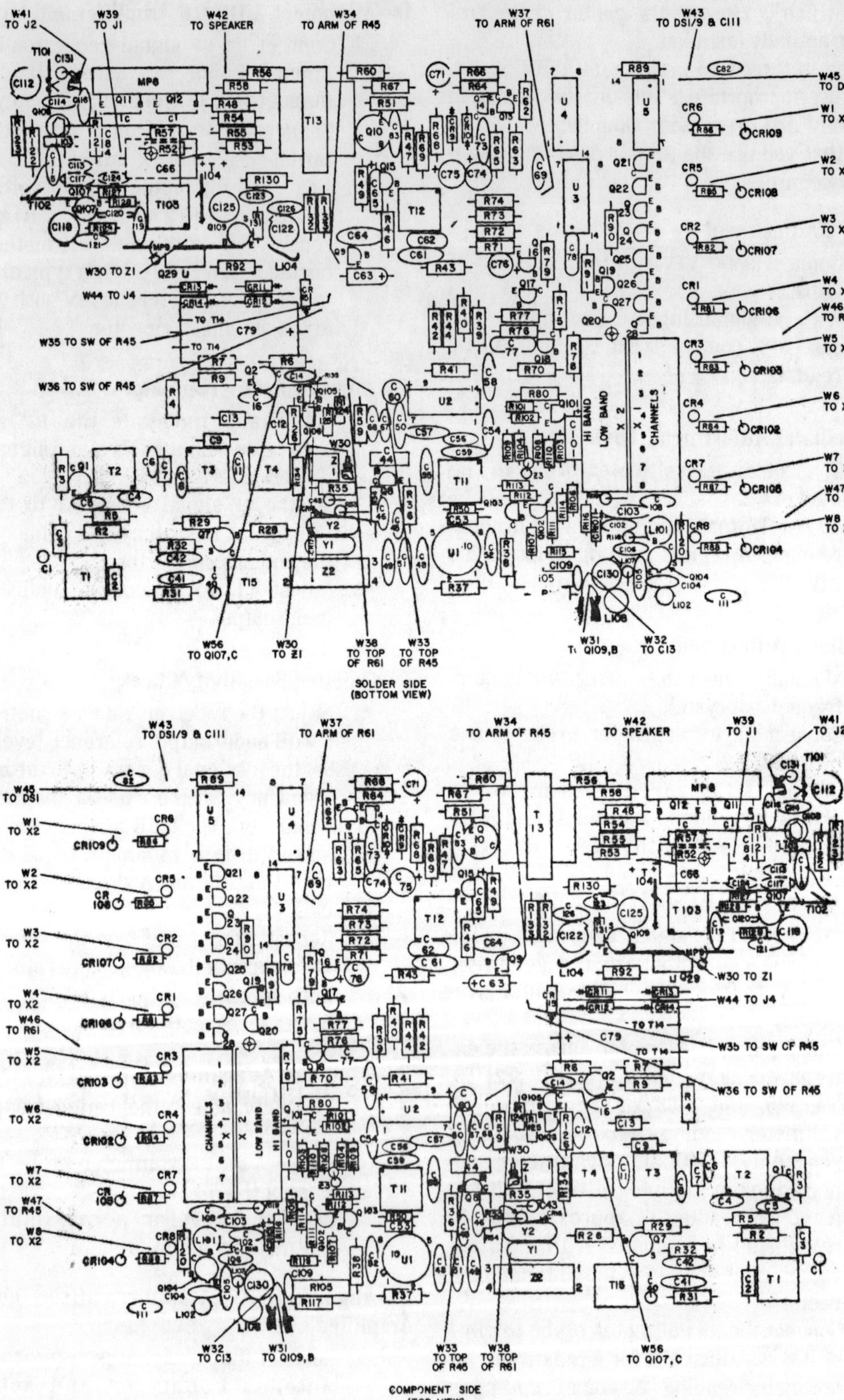

Fig. 8-6. Component and circuit board layout of E. F. Johnson UHF-VHF Duo-Scan (late).

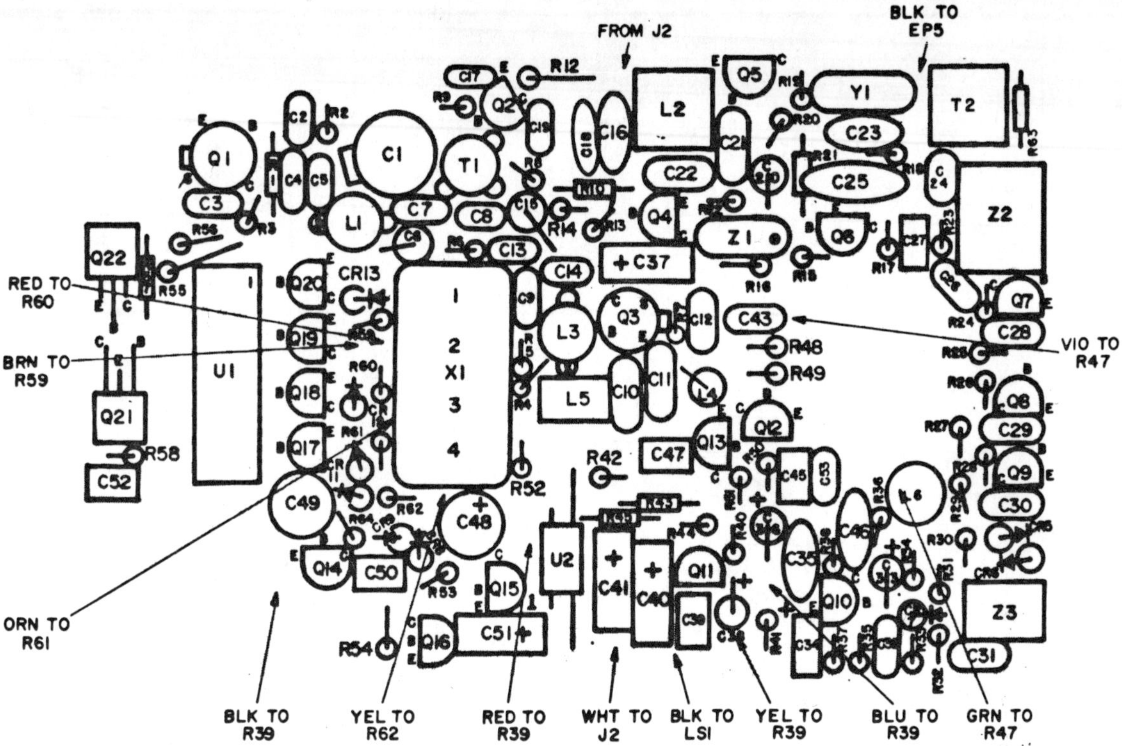

(SOLDER SIDE VIEW)

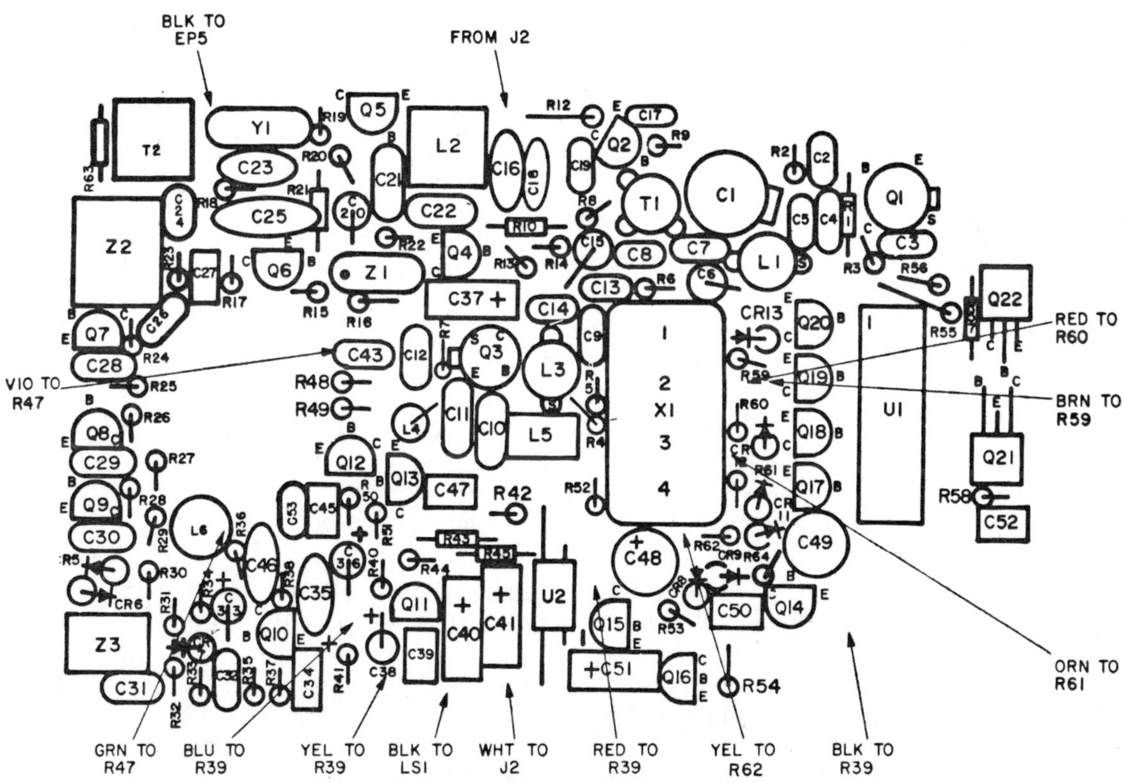

(COMPONENT SIDE VIEW)

Fig. 8-7. Component and circuit board layout of E. F. Johnson VHF Mini-Scan.

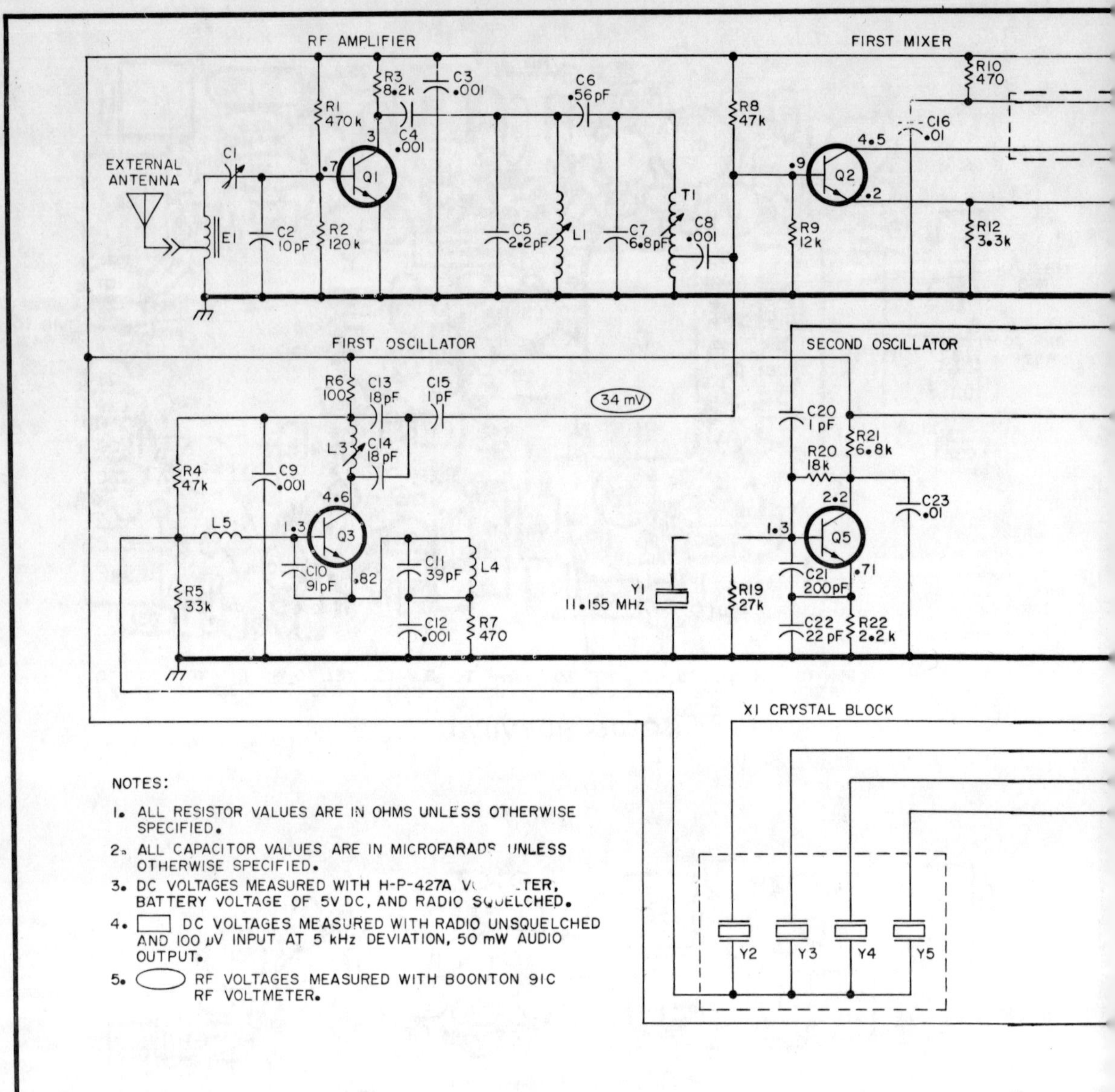

2. Connect the RF voltmeter to pin 1 of IC U1. Adjust C125, C118, C121, T15, and T103 for a maximum RF voltmeter indication. A typical reading of approximately 3 mV with 0.5 μV at the antenna connector should be indicated.
3. Connect the RF voltmeter to pin 4 of IC U2. Adjust T11 for a maximum RF voltmeter reading. A typical reading of approximately 45 mV with 0.5 μV at the antenna connector should be indicated.

I—Alignment, Ceramic Filter

The 455 kHz ceramic filter is factory adjusted and will not normally require realignment. However, if the ceramic filter or associated components are replaced, proceed as follows:

1. Connect an oscilloscope across external-speaker jack J1.
2. Connect the RF signal generator to the antenna connector. Set the generator

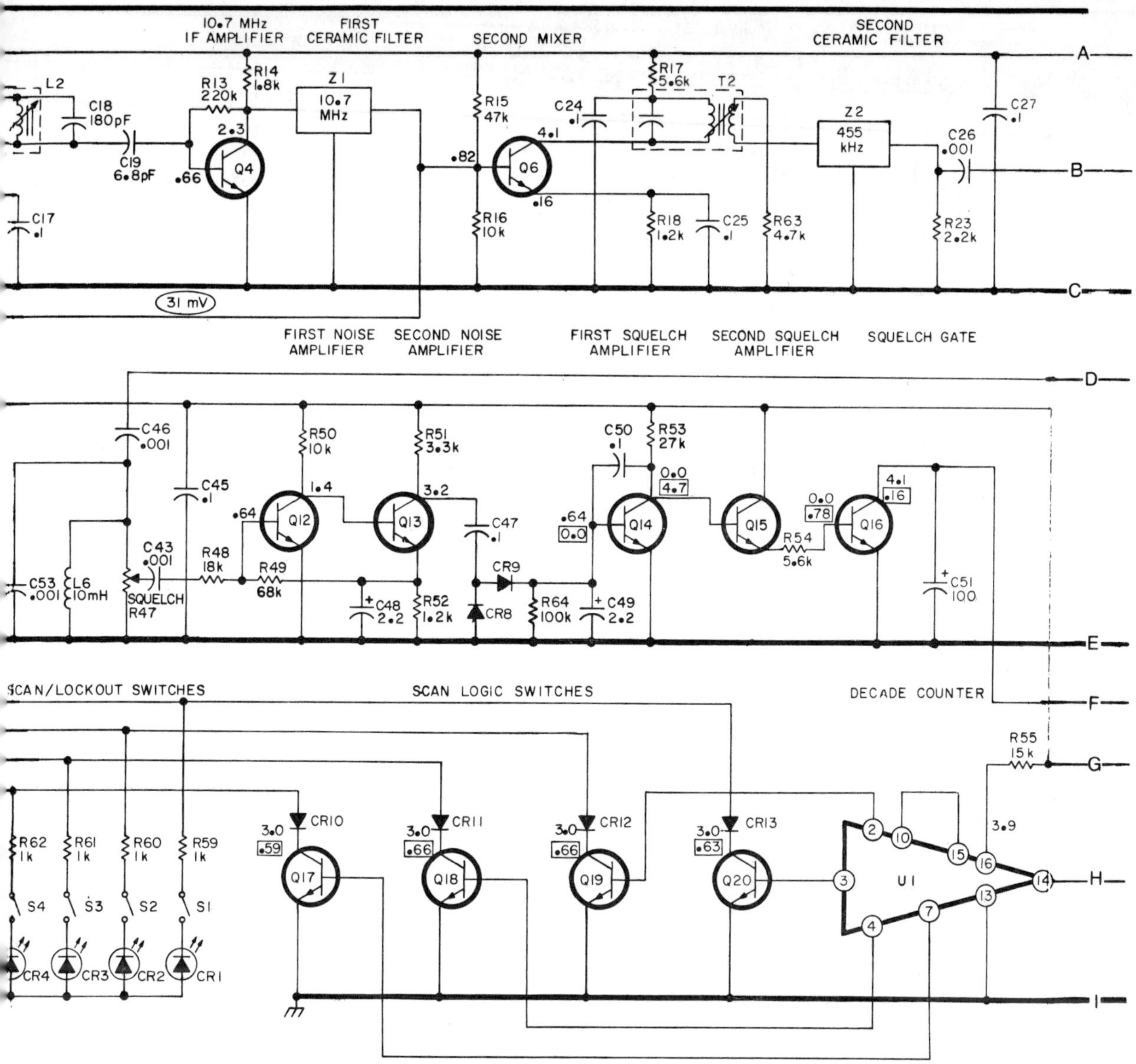

Schematic 8-6. Johnson Mini-Scan monitor receiver (for units with A, or later, model designator on serial number sticker).

output level for 5 μV and externally modulate with a 1 kHz audio signal at a 7–8 kHz deviation level.

3. Adjust the ceramic filter for clean audio waveform.

J—Alignment, RF and IF Amplifiers (High Band)

1. Connect an RF signal generator to the receiver antenna connector. Set the generator output to the high-band center frequency.
2. Connect an AC voltmeter across the external-speaker jack. Adjust T1, T2, T3, T4, T10, and T15 for a minimum AC voltmeter reading (maximum quieting sensitivity). With an RF voltmeter connected to pin 1 of integrated circuit U1, a typical reading of approximately 90 mV should be indicated with a 0.5 μV

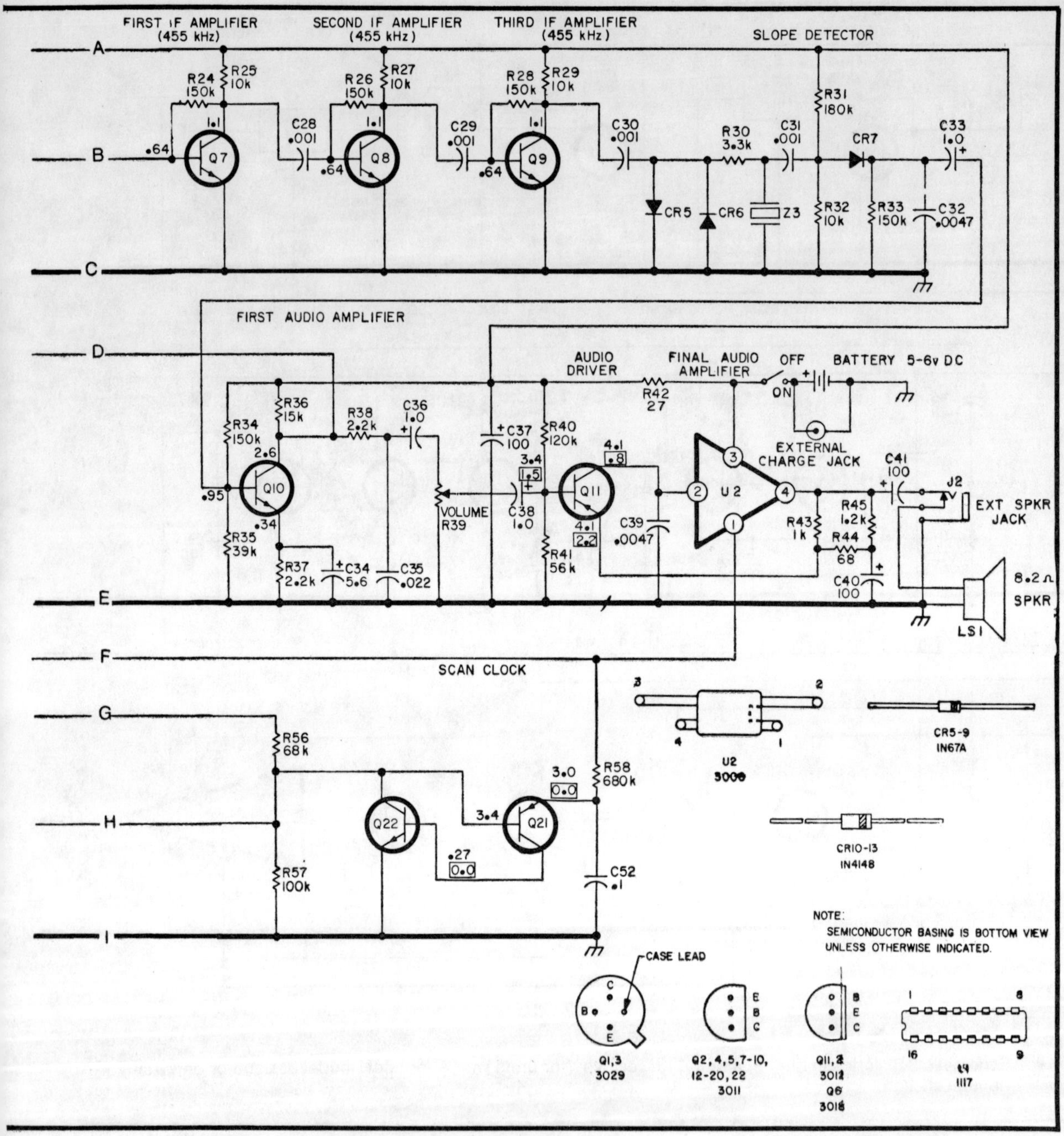

signal input applied to the antenna connector.

3. Connect the RF voltmeter probe to pin 4 of IC U2. Adjust T11 for a maximum RF voltmeter reading. A typical reading is approximately 75 mV with a 0.5 μV signal input.

K—Alignment, RF Amplifier (Low Band)

1. Connect an RF signal generator to the receiver antenna connector.
2. Connect an AC voltmeter across the external-speaker jack. Adjust T7, T8, and T9 for minimum AC voltmeter reading (maximum quieting sensitivity). With

an RF voltmeter connected to pin 1 of IC U1, a typical reading is approximately 40 mV with 0.5 μV at the antenna connector.

L—Alignment, First Oscillator

1. Connect an RF signal generator to the antenna connector and an AC VTVM across the speaker.
2. Select the center-frequency channel and apply 5V DC to the receiver.
3. Connect an RF voltmeter probe to the base of Q2.
4. Adjust L3 for a maximum RF voltmeter reading (a typical reading is 25 to 40 mV).

M—Alignment, RF Amplifier, IF Amplifier, and Discriminator

1. With no RF signal input, adjust L2 for maximum audio noise.
2. With the RF signal generator output connected to the antenna connector, adjust (in order) C1, L1, L3, T1, and L2 for maximum quieting sensitivity.
3. Remove the RF signal and adjust T2 for maximum audio noise.
4. Repeat step 2; then adjust L2 and C1 last for best quieting sensitivity.

Parts List

SYMBOL NO.	DESCRIPTION	PART NO.
HIGH—LOW VHF DUO-SCAN		
	ACCESSORY PACKAGE	023-3066-010
E1	Antenna, collapsible	501-0010-001
EP1	6 ft. power cable	597-1002-011
EP3	Connector package	023-2209-001
ML4	Reduced schematic	564-3001-392
ML8	Hardware envelope	023-2615-001
MP4	Rubber bumper	574-1006-003
W9	Battery cable	023-1652-001
	ASSEMBLIES	
A1	Cabinet Assembly	023-2920-002
A2	Rear-Panel Assembly	023-3171-001
	Includes:	
J1	CC tini-jax NTT312	515-2001-001
J3	Cord/connector AC	597-1002-012
T14	AC power transformer	592-3001-008
U9	Rear panel	023-2965-003
A3	Front-Panel Assembly	023-3074-011
	Includes:	
HW3	Nut, spanner	013-1313-001
MP1	Knob	013-1086-008
MP2	Front panel	032-0307-001
MP3	Front panel overlay	559-2058-111
R45	Potentiometer 10K volume	562-0018-008
R61	Potentiometer 5K squelch	562-0018-007
U7	Speaker Assembly	023-2927-002
	Includes:	
LS1	Speaker, 4 ohm	589-1003-005
MP26	Foam ring	018-0960-001
	CAPACITORS	
C27	22 pF J 200V N750 ceramic	510-3020-220
C40	27 pF J 200V N750 ceramic	510-3020-270
C41	0.10 MF Z 10V Y5U disc	510-3008-104
C42	Same as C41	
C43	1 pF J 500V composition	510-9002-109
C44	220 pF J 100V 1DM15	510-0001-221
C45	22 pF J 200V N750 ceramic	510-3020-220
C46	0.010 MF M 16V Y5S disc	510-3010-103
C48	0.10 MF Z 10V Y5U disc	510-3008-104
C49	Same as C48	
C50	Same as C48	
C51	Same as C48	
C52	Same as C48	

SYMBOL NO.	DESCRIPTION	PART NO.
C53	1500 pF J 500V 4DM20	510-0008-152
C54	0.10 MF Z 10V Y5U disc	510-3008-104
C55	Same as C54	
C56	0.022 MF M 16V Y5S disc	510-3010-223
C57	0.10 MF Z 10V Y5U disc	510-3008-104
C58	Same as C57	
C59	Same as C57	
C60	1 MF M 35V dipped	510-2045-109
C61	56 pF J 200V NPO ceramic	510-3013-560
C62	270 pF J 100V 1DM15	510-0001-271
C63	0.56 MF M 35V tubular	510-2005-568
C64	4700 pF M 500V Y5U disc	510-3004-472
C65	0.56 MF M 35V tubular	510-2005-568
C66	330 MF 10V aluminum	510-4003-007
C67	0.010 MF M 16V Y5S disc	510-3010-103
C68	1000 pF M 1KV Y5S disc	510-3061-102
C69	Same as C68	
C71	2.2 MF M 35V dipped	510-2045-229
C73	0.10 MF Z 10V Y5U disc	510-3008-104
C74	2.2 MF M 35V dipped	510-2045-229
C75	Same as C74	
C76	Same as C74	
C77	470 pF J 500V 4DM15	510-0004-471
C78	220 pF J 100V 1DM15	510-0001-221
C79	1000 MF 16V aluminum	510-4006-005
C80	Same as C79	
C82	0.010 MF M 16V Y5S disc	510-3010-103
C83	0.10 MF Z 10V Y5U disc	510-3008-104
C84	1000 MF 16V aluminum	510-4006-005
C101	100 MF 25V aluminum	510-4006-002
C102	56 pF J 200V N750 ceramic	510-3020-560
C103	110 pF J 200V N750 ceramic	510-3020-111
C104	1000 pF M 1KV Y5S disc	510-3061-102
C105	180 pF J 200V N750 ceramic	510-3020-181
C106	15 pF J 200V N750 ceramic	510-3020-150
C108	1000 pF M 1KV disc	510-3061-102
C109	10 pF J 200V N750 ceramic	510-3020-100
C111	1000 pF M 1KV Y5S diss	510-3061-102
C112	1.7-11.0 pF air variable	187-0106-005
C113	1000 pF M 1KV Y5S disc	510-3061-102
C114	150 pF J 500V 4DM15	510-0004-151
C115	47 pF J 200V N750 ceramic	510-3020-470
C116	1000 pF M 1KV Y5S disc	510-3061-102
C117	Same as C116	
C118	1.7-11.0 pF air variable	187-0106-005
C119	1000 pF M 1KV Y5S disc	510-3061-102
C120	160 pF J 100V 1DM10	510-0002-161
C121	22 pF J 200V N750 ceramic	510-3020-220
C122	1.7-11.0 pF air variable	187-0106-005
C123	1000 pF M 1KV Y5S disc	510-3061-102

SYMBOL NO.	DESCRIPTION	PART NO.
C124	0.010 MF M 16V Y5S disc	510-3010-103
C125	1.7-11.0 pF air variable	187-0106-005
C126	1000 pF M 1KV Y5S disc	510-3061-102
C127	39 pF J 200V NPO ceramic	510-3013-390
C128	100 pF M 50V Y5U ceramic	510-3002-101
C130	1.7-11.0 pF air variable	187-0106-005
C131	3.3 pF J 500V composition	510-9002-339
	DIODES	
CR1	1N4149 75V 75MA SI SW	523-0006-002
CR2	Same as CR1	
CR3	Same as CR1	
CR4	Same as CR1	
CR5	Same as CR1	
CR6	Same as CR1	
CR7	Same as CR1	
CR8	Same as CR1	
CR9	1N67A 80V 30 MA germanium	523-1000-067
CR10	Same as CR9	
CR11	1N4003 200V 1A rectifier	523-0001-002
CR12	Same as CR11	
CR13	Same as CR11	
CR14	Same as CR11	
CR15	10V J 1W zener	523-2003-100
CR101	MV2209 VVC	523-0009-049
	INDUCTORS	
L101	0.14-0.21 UH variable inductor	542-1006-001
L102	RF inductor	542-1006-020
L103	20 UHY choke	542-3002-002
L104	RF inductor	023-3165-004
L107	1.0 UH choke	542-3002-001
L108	0.22-0.37 UH variable inductor	542-1006-017
	TRANSISTORS	
Q7	SI NPN 175 MHz amp TO72	576-0003-029
Q8	SI NPN SW	576-0003-028
Q9	SI NPN Gen. Purp. TO92	576-0003-011
Q10	SI NPN 60V 1.5A 1W TO5	576-0002-006
Q11	SI NPN 60V 4A 36W X75	576-0002-001
Q12	Same as Q11	
Q13	SI NPN Gen. Purp. TO92	576-0003-011
Q14	Same as Q13	
Q15	Same as Q13	
Q16	SI NPN SW	576-0003-028
Q17	Same as Q16	
Q18	SI PNP 50 MHz amp TO92	576-0003-017
Q19	SI NPN SW	576-0003-028
Q20	Same as Q19	
Q21	Same as Q19	
Q22	Same as Q19	
Q23	Same as Q19	
Q24	Same as Q19	
Q25	Same as Q19	
Q26	Same as Q19	
Q27	Same as Q19	
Q28	Same as Q19	
Q29	SI NPN 60V 4A 36W X75	576-0002-001
Q101	SI NPN Gen. Purp. TO92	576-0003-011
Q102	Same as Q101	
Q103	SI PNP 50 MHz amp TO92	576-0003-017
Q104	SI NPN 175 MHz amp TO72	576-0003-029
Q107	SI NPN 470 MHz mix TO72	576-0003-027
Q108	SI mosfet 470 MHz	576-0006-222
Q109	SI NPN 175 MHz amp TO72	576-0003-029
	RESISTORS	
R28	2.2K ohm K 1/2 W	569-1004-222
R29	560 ohm K 1/2 W	569-1004-561
R30	47K ohm K 1/2 W	569-1004-473
R31	560 ohm K 1/2 W	569-1004-561
R32	1.0K ohm K 1/2 W	569-1004-102
R33	18K ohm K 1/2 W	569-1004-183
R34	27K ohm K 1/2 W	569-1004-273
R35	15K ohm K 1/2 W	569-1004-153
R36	2.2K ohm K 1/2 W	569-1004-222
R37	Same as R36	

SYMBOL NO.	DESCRIPTION	PART NO.
R38	180 ohm K 1/2 W	569-1004-181
R39	100 ohm K 1/2 W	569-1004-101
R40	10K ohm K 1/2 W	569-1004-103
R41	Same as R40	
R42	100K ohm K 1/2 W	569-1004-104
R43	47K ohm K 1/2 W	569-1004-473
R44	1.0K ohm K 1/2 W	569-1004-102
R46	Same as R44	
R47	10K ohm K 1/2 W	569-1004-103
R48	33K ohm K 1/2 W	569-1004-333
R49	6.8K ohm K 1/2 W	569-1004-682
R50	4.7K ohm K 1/4 W	569-1002-472
R51	22 ohm K 1/2 W	569-1004-220
R52	68 ohm K 1/2 W	569-1004-680
R53	150 ohm K 1/2 W	569-1004-151
R54	15 ohm K 1/2 W	569-1004-150
R55	220 ohm K 1/2 W	569-1004-221
R56	15 ohm K 1/2 W	569-1004-150
R57	0.33 ohm K 1/2 W	569-2003-338
R58	Same as R57	
R59	2.2K ohm K 1/2 W	569-1004-222
R60	1.0K ohm K 1/2 W	569-1004-102
R62	33K ohm K 1/2 W	569-1004-333
R63	68K ohm K 1/2 W	569-1004-683
R64	47K ohm K 1/2 W	569-1004-473
R65	6.8K ohm K 1/2 W	569-1004-682
R66	1.0K ohm K 1/2 W	569-1004-102
R67	220K ohm K 1/2 W	569-1004-224
R68	2.2K ohm K 1/2 W	569-1004-222
R69	10K ohm K 1/2 W	569-1004-103
R70	1.0K ohm K 1/2 W	569-1004-102
R71	2.2K ohm K 1/2 W	569-1004-222
R72	33K ohm K 1/2 W	569-1004-333
R73	470K ohm K 1/2 W	569-1004-474
R74	33K ohm K 1/2 W	569-1004-333
R75	15K ohm K 1/2 W	569-1004-153
R76	68K ohm K 1/2 W	569-1004-683
R77	33K ohm K 1/2 W	569-1004-333
R78	10K ohm K 1/2 W	569-1004-103
R79	68K ohm K 1/2 W	569-1004-683
R81	1.5K ohm K 1/4 W	569-1002-152
R82	Same as R81	
R83	Same as R81	
R84	Same as R81	
R85	Same as R81	
R86	Same as R81	
R87	Same as R81	
R88	Same as R81	
R89	56 ohm K 1/2 W	569-1004-560
R90	100 ohm K 1/2 W	569-1004-101
R91	Same as R90	
R92	150 ohm K 1/2 W	569-1004-151
R101	47K ohm K 1/4 W	569-1002-473
R102	1.0K ohm K 1/4 W	569-1002-102
R103	3.3K ohm K 1/4 W	569-1002-332
R104	3.9K ohm K 1/4 W	569-1002-392
R105	2K 1/8 W PC trim pot	562-0004-202
R106	3.3K ohm K 1/4 W	569-1002-332
R107	1.0K ohm K 1/4 W	569-1002-102
R108	100 ohm K 1/4 W	569-1002-101
R109	1.0K ohm K 1/4 W	569-1002-102
R110	3.3K ohm K 1/4 W	569-1002-332
R111	10K ohm K 1/4 W	569-1002-103
R112	82K ohm K 1/4 W	569-1002-823
R113	1.0K ohm K 1/4 W	569-1002-102
R114	10K ohm K 1/4 W	569-1002-103
R115	47K ohm K 1/4 W	569-1002-473
R116	39K ohm K 1/4 W	569-1002-393
R117	100 ohm K 1/4 W	569-1002-101
R118	10K ohm K 1/4 W	569-1002-103
R119	22K ohm K 1/4 W	569-1002-223
R120	330 ohm K 1/2 W	569-1004-331
R121	22 ohm K 1/2 W	569-1004-220
R122	68K ohm K 1/2 W	569-1004-683
R123	47K ohm K 1/2 W	569-1004-473
R127	470K ohm K 1/4 W	569-1002-474
R128	1.5K ohm K 1/4 W	569-1002-152
R129	3.9K ohm K 1/4 W	569-1002-392
R131	15K ohm K 1/4 W	569-1002-153
R132	6.8K ohm K 1/2 W	569-1004-682
R133	1.0K ohm K 1/2 W	569-1004-102

SYMBOL NO.	DESCRIPTION	PART NO.
	SWITCH	
SW1	Push button switch (S1-S9)	583-4008-004
	SWITCH BUTTONS	
	Switch button, white (9 required)	032-0306-010
	TRANSFORMERS	
T11	10 MM 455 kHz IF	592-5006-006
T12	10 MM 455 kHz IF	592-5006-008
T13	Input/driver	592-1015-001
T15	455 kHz	592-5006-006
T101	RF transformer	023-3165-001
T102	RF transformer	023-3165-002
T103	RF transformer	023-3165-003
T103	10.7 MHz	592-5009-011
	INTEGRATED CIRCUITS (IC'S)	
U1	IC CA3011 linear amplifier	544-2002-004
U2	IC MC1351PQ FM	544-2003-002
U3	IC MC723P	544-3001-002
U4	IC MC790P 2X J-K FF	544-3001-003
U5	IC MC724P 4X gate	544-3001-001
	PC BOARD	
U6	PC board	035-0180-002
	COAX CABLE	
W30	RG 188A/U coax cable	071-0321-930
W31	Same as W30	
W41	Same as W30	
	CRYSTAL SOCKET	
X1	Crystal socket	126-0110-108
	CRYSTAL (SECOND OSCILLATOR)	
Y1	11.155 MHz, 32 pF, HC-18/C	519-0009-001
	FILTERS	
Z1	10.7 MHz filter	532-2001-011
Z2	455 kHz filter	532-2002-011

COMPONENT CODES

Capacitors	Resistors
1DM10, 15 - 1 coat dipped mica, size 10, 15	CC - carbon composition
4DM10, 15 - 4 coat dipped mica, size 10, 15	WW - wire wound
D - ±.5 pF, J - ±5%, M - ±20%	K - ±10%
Z - +80 -20%	MF - ±20% at 25°C, ±7.5% over temp. range

Crystal Ordering Information

We recommend crystals manufactured by M-Tron Industries, Inc. Yankton, South Dakota 57078

Use the following information when ordering M-Tron crystals:

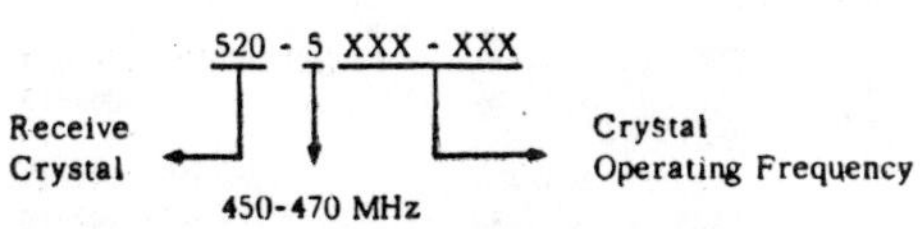

For Example: A crystal with an operating frequency of 451.175 MHz will have a part number of 520-5451-175.

UHF MONO-SCAN

SYMBOL NO.	DESCRIPTION	PART NO.
	ASSEMBLIES	
A1	Cabinet Assembly	023-2920-002
A2	Rear Panel Assembly Includes:	023-3171-001
J1	CC tini-jax NTT312	515-2001-001
J3	Cord/connector AC	597-1002-012
T14	AC power transformer	592-3001-008
U9	Rear panel	023-2965-003
A3	Front-Panel Assembly Includes:	023-3074-010
HW3	Nut spanner	013-1313-001
MP1	Knob	013-1086-008
MP2	Lens red	032-0183-008
MP2	Front panel	032-0307-001
MP3	Front panel overlay	559-2058-012
NP3	Front panel overlay	559-2058-003
R45	Potentiometer 10K vol	562-0018-008
R61	Potentiometer 5K squ	562-0018-007
	BUTTONS	
MP16	Push button	032-0306-001
MP17	Push button	032-0306-002
MP18	Push button	032-0306-003
MP19	Push button	032-0306-004
MP20	Push button	032-0306-005
MP21	Push button	032-0306-006
MP22	Push button	032-0306-007
MP23	Push button	032-0306-008
MP24	Push button	032-0306-009
	CAPACITORS	
C1	3 pF J 1500V composition	510-9002-309
C2	Same as C1	
C3	Same as C1	
C4	1000 pF M 1KV Y5S disc	510-3061-102
C5	Same as C4	
C6	6.8 pF J 200V N750 ceramic	510-3020-689
C7	0.56 pF K 500V composition	510-9003-568
C8	8.2 pF J 200V N750 ceramic	510-3020-829
C9	0.56 pF K 500V composition	510-9003-568
C11	6.8 pF J 200V N750 ceramic	510-3020-689
C12	1000 pF M 1kV Y5S disc	510-3061-102
C13	1 pF J 500V composition	510-9002-109
C14	0.010 MF M 16V Y5S disc	510-3010-103
C16	Same as C14	
C27	22 pF J 200V N750 ceramic	510-3020-220
C40	27 pF J 200V N750 ceramic	510-3020-270
C41	0.10 MF Z 10V Y5U disc	510-3008-104
C42	Same as C41	
C43	1 pF J 500V composition	510-9002-109
C44	220 pF J 100V 1DM15	510-0001-221
C45	22 pF J 200V N750 ceramic	510-3020-220
C46	0.010 MF M 16V Y5S disc	510-3010-103
C48	0.10 MF Z 10V Y5U disc	510-3008-104
C49	Same as C48	
C50	Same as C48	
C51	Same as C48	
C52	Same as C48	
C53	150 pF J 500V 4DM20	510-0008-152
C54	0.10 MF Z 10V Y5U disc	510-3008-104
C55	Same as C54	
C56	0.022 MF M 16V Y5S disc	510-3010-223
C57	0.10 MF Z 10V Y5U disc	510-3008-104
C58	Same as C57	
C59	Same as C57	
C60	1 MF M 35V dipped	510-2045-109
C61	56 pF J 200V NPO ceramic	510-3013-560
C62	270 pF J 100V 1DM15	510-0001-271

SYMBOL NO.	DESCRIPTION	PART NO.
C63	0.56 MF M 35V tubular	510-2005-568
C64	4700 pF M 500V Y5U disc	510-3004-472
C65	0.56 MF M 35V tubular	510-2005-568
C66	330 MF 10V aluminum	510-4003-007
C67	0.01 MF M 16V Y5S disc	510-3010-103
C68	1000 pF M 1KV Y5S disc	510-3061-102
C69	Same as C68	
C71	2.2 MF M 35V dipped	510-2045-229
C73	0.10 MF Z 10V Y5U disc	510-3008-104
C74	2.2 MF M 35V dipped	510-2045-229
C75	Same as C74	
C76	Same as C74	
C77	470 pF J 500V 4DM15	510-0004-471
C78	220 pF J 100V 1DM15	510-0001-221
C79	1000 MF 16V aluminum	510-4006-005
C80	Same as C79	
C82	0.010 MF M 16V Y5S disc	510-3010-103
C83	0.10 MF Z 10V Y5U disc	510-3008-104
C84	1000 MF 16V aluminum	510-4006-005
C104	56 pF J 200V N750 ceramic	510-3020-560
C105	180 pF J 200V N750 ceramic	510-3020-181
C106	15 pF J 200V N750 ceramic	510-3020-150
C108	1000 pF M 1KV Y5S disc	510-3061-102
C111	1000 pF M 1KV Y5S disc	510-3061-102
C119	Same as C111	
C120	160 pF J 100V 1DM10	510-0002-161
C121	10 pF J 200V N750 ceramic	510-3020-100

DIODES

SYMBOL NO.	DESCRIPTION	PART NO.
CR1	1N4149 75V 75MA SI SW	523-0006-002
CR2	Same as CR1	
CR3	Same as CR1	
CR4	Same as CR1	
CR5	Same as CR1	
CR6	Same as CR1	
CR7	Same as CR1	
CR8	Same as CR1	
CR9	1N67A 80V 30MA germanium	523-1000-067
CR10	Same as CR9	
CR11	1N4003 200V 1A rectifier	523-0001-002
CR12	Same as CR11	
CR13	Same as CR11	
CR14	Same as CR11	
CR15	10V J 1W zener	523-2003-100

INDUCTORS

SYMBOL NO.	DESCRIPTION	PART NO.
L101	0.14-0.21 UH variable inductor	542-1006-001
L102	RF inductor	542-1006-020
L107	1.00H choke	542-3002-001

TRANSISTORS

SYMBOL NO.	DESCRIPTION	PART NO.
Q1	SI NPN 175 MHz amp TO72	576-0003-029
Q2	Same as Q1	
Q7	Same as Q1	
Q8	SI NPN SW	576-0003-028
Q9	SI NPN Gen. Purp. TO92	576-0003-011
Q10	SI NPN 60V 1.5A 1W TO5	576-0002-006
Q11	SI NPN 60V 4A 36 W	576-0002-001
Q12	Same as Q11	
Q13	SI NPN Gen. Purp. TO92	576-0003-011
Q14	Same as Q13	
Q15	Same as Q13	
Q16	SI NPN SW	576-0003-028
Q17	Same as Q16	
Q18	SI PNP 50 MHz amp TO92	576-0003-017
Q19	SI NPN SW	576-0003-028
Q20	Same as Q19	
Q21	Same as Q19	
Q22	Same as Q19	
Q23	Same as Q19	
Q24	Same as Q19	
Q25	Same as Q19	
Q26	Same as Q19	
Q27	Same as Q19	

SYMBOL NO.	DESCRIPTION	PART NO.
Q28	Same as Q19	
Q29	SI NPN 60V 4A 36 W X75	576-0002-001
Q104	SI NPN 175 MHz amp TO72	576-0003-029

RESISTORS

SYMBOL NO.	DESCRIPTION	PART NO.
R2	4.7K ohm K 1/2 W	569-1004-472
R3	27K ohm K 1/2 W	569-1004-273
R4	1.0K ohm K 1/2 W	569-1004-102
R5	560 ohm K 1/2 W	569-1004-561
R6	33K ohm K 1/2 W	569-1004-333
R7	8.2K ohm K 1/2 W	569-1004-822
R9	1.0K ohm K 1/2 W	569-1004-102
R28	2.2K ohm K 1/2 W	569-1004-222
R29	560 ohm K 1/2 W	569-1004-561
R30	47K ohm K 1/2 W	569-1004-473
R31	560 ohm K 1/2 W	569-1004-561
R32	1.0K ohm K 1/2 W	569-1004-102
R33	18K ohm K 1/2 W	569-1004-183
R34	27K ohm 1/2 W	569-1004-272
R35	15K ohm K 1/2 W	569-1004-152
R36	2.2K ohm K 1/2 W	569-1004-222
R37	Same as R36	
R38	180 ohm K 1/2 W	569-1004-181
R39	100 ohm K 1/2 W	569-1004-101
R40	10K ohm K 1/2 W	569-1004-103
R41	Same as R40	
R42	100K ohm K 1/2 W	569-1004-104
R43	47K ohm K 1/2 W	569-1004-473
R44	1.0K ohm K 1/2 W	569-1004-102
R46	Same as R44	
R47	10K ohm K 1/2 W	569-1004-103
R48	33K ohm K 1/2 W	569-1004-333
R49	6.8K ohm K 1/2 W	569-1004-682
R50	4.7K ohm K 1/4 W	569-1002-472
R51	22 ohm K 1/2 W	569-1004-220
R52	68 ohm K 1/2 W	569-1004-680
R53	150 ohm K 1/2 W	569-1004-151
R54	15 ohm K 1/2 W	569-1004-150
R55	220 ohm K 1/2 W	569-1004-221
R56	15 ohm K 1/2 W	569-1004-150
R57	0.33 ohm K 1/2 W	569-2003-338
R58	Same as R57	
R59	2.2K ohm K 1/2 W	569-1004-222
R60	1.0K ohm K 1/2 W	569-1004-102
R62	33K ohm K 1/2 W	569-1004-333
R63	68K ohm K 1/2 W	569-1004-683
R64	47K ohm K 1/2 W	569-1004-473
R65	6.8K ohm K 1/2 W	569-1004-682
R66	1.0K ohm K 1/2 W	569-1004-102
R67	220K ohm K 1/2 W	569-1004-224
R68	2.2K ohm K 1/2 W	569-1004-222
R69	10K ohm K 1/2 W	569-1004-103
R70	1.0K ohm K 1/2 W	569-1004-102
R71	2.2K ohm K 1/2 W	569-1004-222
R72	33K ohm K 1/2 W	569-1004-333
R73	470K ohm K 1/2 W	569-1004-474
R74	33K ohm K 1/2 W	569-1004-333
R75	15K ohm K 1/2 W	569-1004-152
R76	68K ohm K 1/2 W	569-1004-683
R77	33K ohm K 1/2 W	569-1004-333
R78	10K ohm K 1/2 W	569-1004-103
R79	68K ohm K 1/2 W	569-1004-683
R80	2.2K ohm K 1/2 W	569-1004-222
R81	1.5K ohm K 1/4 W	569-1002-152
R82	Same as R81	
R83	Same as R81	
R84	Same as R81	
R85	Same as R81	
R86	Same as R81	
R87	Same as R81	
R88	Same as R81	
R89	56 ohm K 1/2 W	569-1004-560
R90	100 ohm K 1/2 W	569-1004-101
R91	Same as R90	
R92	150 ohm K 1/2 W	569-1004-151
R117	100 ohm K 1/4 W	569-1002-101
R118	10K ohm K 1/4 W	569-1002-103
R119	22K ohm K 1/4 W	569-1002-223

SYMBOL NO.	DESCRIPTION	PART NO.
R120	330 ohm K 1/2 W	569-1004-331
R126	100 ohm K 1/2 W	569-1004-101
R128	1.5K ohm K 1/4 W	569-1002-152
R129	3.9K ohm K 1/4 W	569-1002-392

SWITCH

SYMBOL NO.	DESCRIPTION	PART NO.
SW1	Push button switch	583-4008-004

TRANSFORMERS

SYMBOL NO.	DESCRIPTION	PART NO.
T1	10MM 150 MHz	592-5009-016
T2	Same as T1	
T3	Same as T1	
T4	Same as T1	
T11	10MM 455 kHz IF	592-5006-006
T12	10MM 455 kHz IF	592-5006-008
T13	Input/driver	592-1015-001
T103	10.7 MHz	592-5009-011

INTEGRATED CIRCUITS (IC'S)

SYMBOL NO.	DESCRIPTION	PART NO.
U1	IC CA3011 linear amplifier	544-2002-004
U2	IC MC1351PQ FM	544-2003-002
U3	IC MC723P	544-3001-002
U4	IC MC790P 2X J-K PF	544-3001-003
U5	IC MC724P 4X gate	544-3001-001

PC BOARD

SYMBOL NO.	DESCRIPTION	PART NO.
U6	PC board	035-0180-002
U7	Speaker	023-2927-002
	Includes:	
LS1	Speaker 4 ohm	589-1003-005
MP26	Foam ring	018-0960-001
U10	Accessory Package	023-3066-010
	Includes:	
E1	Whip antenna	501-0010-001
EP1	6 ft. power cable	597-1002-011
EP3	Connector package	023-2209-001
ML4	Reduced schematic	564-3001-392
ML8	Hardware envelope	023-2615-001
MP4	Rubber bumper	574-1006-003
W9	Battery cable	023-1652-001

CRYSTAL
(SECOND OSCILLATOR)

SYMBOL NO.	DESCRIPTION	PART NO.
Y1	11.155 MHz 32 pF HC-18/U	519-0009-001

FILTERS

SYMBOL NO.	DESCRIPTION	PART NO.
Z1	10.7 MHz filter	532-2001-011
Z2	455 kHz filter	532-2002-011

HIGH–LOW VHF DUO-SCAN

SYMBOL NO.	DESCRIPTION	PART NO.
	ACCESSORY PACKAGE	023-3066-010
	Includes:	
E1	Antenna, collapsible	501-0010-001
EP1	6 ft. power cable	597-1002-001
EP3	Connector package	023-2209-001
	Schematic	
ML8	Hardware envelope	023-2615-001
MP4	Rubber bumper	574-1006-003
W9	Battery cable	023-1652-001
	ASSEMBLIES	
A1	Cabinet Assembly	023-2920-002

SYMBOL NO.	DESCRIPTION	PART NO.
A2	Rear-Panel Assembly	023-2965-003
	Includes:	
	Bracket, circuit bd. mtg.	016-1888-001
J2	Recepticál, coaxial J2	142-0101-002
	Bracket, circuit bd. mtg.	016-1888-002
	Rear panel	017-1654-003
A3	Front Panel Assembly	023-3074-001
	Includes:	
	Front panel	032-0307-001
	Overlay	559-2058-112
	Lens (red translucent)	032-0183-008
	Knobs	013-1086-008
	Spanner nuts	013-1313-001
	Set screw, #6-32	011-0019-006
R45	Potentiometer, 10kΩ, volume	562-0018-008
R61	Potentiometer, 5kΩ, squelch	562-0018-007
U7	Speaker Assembly	023-2927-002
	Includes:	
LS1	Speaker, 4 ohm	589-1003-005
MP26	Foam ring	018-0960-001

BRACKETS

SYMBOL NO.	DESCRIPTION	PART NO.
BKT1	Bracket, cabinet mounting	016-1884-004
BKT2	Bracket, circuit bd. mtg. (front)	016-1889-001
BKT3	Bracket, antenna mounting	016-1918-001

CAPACITORS

SYMBOL NO.	DESCRIPTION	PART NO.
C1	3 pF ± 5%, 500V, composition	510-9002-309
C2	Same as C1	
C3	Same as C1	
C4	0.001 ±20%, 1 kV, Y5S, seramic	510-3061-102
C5	Same as C4	
C6	6.8 pF ±5% 200V, N750, cer.	510-3020-689
C7	56 pF, ±5%, 500V, composition	510-9002-568
C8	8.2 pF, ±5%, 200V, N750, cer.	510-3020-829
C9	1.0 pF ±5%, 500V, composition	510-9002-109
C11	6.8 pF ±5%, 200V, N750, cer.	510-3020-689
C12	0.001 ±20%, 1kV, Y5S, ceramic	510-3061-102
C13	1.0 pF ±5%, 500V, composition	510-9002-109
C14	0.01 μF ±20%, 16V, Y5S, cer.	510-3010-103
C16	Came as C14	
C17	5 pF ±5%, 200V, N750, ceramic	510-3020-509
C18	75 pF ±5%, 100V, mica	510-0002-750
C19	0.001 ±20%, 1kV, Y5S, ceramic	510-3061-102
C21	Same as C19	
C22	62 pF ±5%, 200V, ceramic	510-3016-620
C24	5.6 pF ±5%, 500V, composition	510-9002-569
C25	0.001 ±20%, 1kV, Y5S, ceramic	510-3061-102
C27	Same as C25	
C28	3 pF ±5%, 500V, composition	510-9002-309
C29	10 pF ±5%, 200V, N750, cer.	510-3020-100
C31	0.001 ±20%, 1kV, Y5S, ceramic	510-3061-102
C32	1.0 pF ±5%, 500V, composition	510-9002-109
C33	0.01 μF ±20%, 16V, Y5S, cer.	510-3010-103
C34	200 pF ±5%, 100V, mica	510-0001-201
C35	0.01 μF ±20%, 16V, Y5S, cer.	510-3010-103
C36	0.001 ±20%, 1kV, Y5S, ceramic	510-3016-102
C37	68 pF ±5%, 200V, N750, ceramic	510-3020-680
C38	47 pF ±5%, 200V, N750, ceramic	510-3020-470
C39	0.001 ±20%, 1kV, Y5S, ceramic	510-3061-102
C40	1500 pF	510-1101-152
C41	0.1 μF +80/-20%, 10V, Y5U, ceramic	510-3008-104
C42	Same as C41	
C43	1.0 pF ±5%, 500V, composition	510-9002-109
C44	220 pF ±5%, 100V, mica	510-0001-221
C45	22 pF ±5%, 200V, N750, ceramic	510-3020-220
C46	0.01 μF ±20%, 16V, Y5S, cer.	510-3010-103
C48	0.1 μF +80/-20%, 10V, Y5U, cer.	510-3008-104
C49	Same as C48	
C50	Same as C48	
C51	Same as C48	
C52	Same as C48	
C53	1500 pF	510-1101-152
C54	0.1 μF +80/-20%, 10V, Y5U, cer.	510-3008-104
C55	Same as C54	
C56	0.022 μF ±20%, 16V, Y5S, cer.	510-3010-223

SYMBOL NO.	DESCRIPTION	PART NO.
C57	0.1 μF +80/-20%, 10V, Y5U, cer.	5'0-3008-104
C58	Same as C57	
C59	Same as C57	
C60	1 μF ±20%, 35V, dipped tant.	510-2045-109
C61	56 pF ±5%, 200V, NPO	510-3013-560
C62	270 pF ±5%, 100V, mica	510-0001-271
C63	0.56 μF ±20%, 35V, tantalum	510-2005-568
C64	0.0047 ±20%, 500V, Y5U, cer.	510-3004-472
C65	0.56 μF ±20%, 35V, tantalum	510-2005-568
C66	330 μF +100/-10%, 10V, electrolytic	510-4003-007
C67	0.01 μF ±20%, 16V, Y5S, cer.	510-3010-103
C68	0.001 ±20%, 1kV, Y5S, ceramic	510-3061-102
C69	Same as C68	
C71	2.2 μF ±20%, 35V, dipped tant.	510-2045-229
C73	0.1 μF +80/-20%, 10V, Y5U, cer.	510-3008-104
C74	2.2 μF ±20%, 35V, dipped tant.	510-2045-229
C75	Same as C74	
C76	Same as C74	
C77	470 pF ±20%, 1000V, Y5F	510-3060-471
C78	220 pF ±5%, 100V, mica	510-0001-221
C79	1000 μF +100/-10%, 16V, electrolytic	510-4006-005
C80	Same as C79	
C82	0.01 μF ±20%, 16V, Y5S, cer.	510-3010-103
C83	0.1 pF, ±20%, 16V, Y5S, cer.	510-3010-104
C84	1000 μF +100/-10%, 16V, electrolytic	510-4006-005
C85	0.01 μF ±20%, 16V, Y5S, cer.	510-3010-103

DIODES

SYMBOL NO.	DESCRIPTION	PART NO.
CR1	Silicon switching, 75V, 75 mA, FA111	523-0006-002
CR2	Same as CR1	
CR3	Same as CR1	
CR4	Same as CR1	
CR5	Same as CR1	
CR6	Same as CR1	
CR7	Same as CR1	
CR8	Same as CR1	
CR9	1N67, 100V, 30 mA	523-1000-067
CR10	Same as CR9	
CR11	Silicon, 200V, 1A power supply	523-0001-002
CR12	Same as CR11	
CR13	Same as CR11	
CR14	Same as CR11	
CR15	Zener, 1 watt, 10V ±5%, 25 mA	523-2003-100

LAMPS

SYMBOL NO.	DESCRIPTION	PART NO.
DS1	Pilot lamp, 14.4 V, 0.12A, clear	549-3001-003
DS2	Same as DS1	
DS3	Same as DS1	
DS4	Same as DS1	
DS5	Same as DS1	
DS6	Same as DS1	
DS7	Same as DS1	
DS8	Same as DS1	
DS9	Same as DS1	

JACKS

SYMBOL NO.	DESCRIPTION	PART NO.
J1	Jack, external speaker	515-2001-001
J4	Connector, male power (AC) assembly	597-1002-012
J5	Connector (DC)	515-4100-001

INDUCTORS

SYMBOL NO.	DESCRIPTION	PART NO.
L1	Inductor, 1 μH, 5%	542-3002-001
L2	Same as L1	

SWITCH BUTTONS

SYMBOL NO.	DESCRIPTION	PART NO.
MP17	Switch button, white (9 required)	032-0306-010

TRANSISTORS

SYMBOL NO.	DESCRIPTION	PART NO.
Q1	SI NPN 175 MHz amp TO72	576-0003-029
Q2	Same as Q1	
Q3	Same as Q1	
Q4	Same as Q1	
Q5	Same as Q1	
Q6	Same as Q1	
Q7	Same as Q1	
Q8	Switch, low osc.	576-0003-028
Q9	SI NPN Gen. Pur. TO92	576-0003-011
Q10	SI NPN 60V, 1.5A, 1W TO5	576-0002-006
Q11	SI NPN 60V, 4A, 36W X75	576-0002-001
Q12	Same as Q11	
Q13	SI NPN Gen. Pur. TO92	576-0003-011
Q14	Same as Q13	
Q15	Same as Q13	
Q16	Switch, low osc.	576-0003-028
Q17	Same as Q16	
Q18	SI PNP 50 MHz amp TO92	576-0003-017
Q19	Switch, low osc.	576-0003-028
Q20	Same as Q19	
Q21	Same as Q19	
Q22	Same as Q19	
Q23	Same as Q19	
Q24	Same as Q19	
Q25	Same as Q19	
Q26	Same as Q19	
Q27	Same as Q19	
Q28	Same as Q19	
Q29	SI NPN 60V, 4A 36W X75	576-0002-001

RESISTORS

SYMBOL NO.	DESCRIPTION	PART NO.
R2	4.7 KΩ, ±10%, 1 2 watt	569-1004-472
R3	27kΩ, ±10%, 1 2 watt	569-1004-273
R4	1 kΩ, 1 2 watt	569-1004-102
R5	560Ω ±10%, 1 2 watt	569-1004-561
R6	33 kΩ ±10%, 1 2 watt	569-1004-333
R7	8.2k, ±10%, 1 2 watt	569-1004-822
R8	1 kΩ, 1 2 watt	569-1004-102
R9	Same as R8	
R10	6.8 kΩ, ±10%, 1 4 watt	569-1002-682
R11	100Ω ±10%, 1 2 watt	569-1004-101
R12	330Ω, ±10%, 1 2 watt	569-1004-331
R13	18 kΩ, ±10%, 1 2 watt	569-1004-183
R14	10 kΩ, ±10%, 1 2 watt	569-1004-103
R15	27 kΩ, ±10%, 1 2 watt	569-1004-102
R16	27 kΩ, ±10%, 1 2 watt	569-1004-273
R17	4.7 kΩ, ±10%, 1 2 watt	569-1004-472
R18	560Ω ±10%, 1 2 watt	569-1004-561
R19	Same as R18	
R21	15 kΩ, ±10%, 1 2 watt	569-1004-153
R22	8.2k, ±10%, 1 2 watt	569-1004-822
R23	1 kΩ, 1 2 watt	569-1004-102
R24	18 kΩ, ±10%, 1 2 watt	569-1004-183
R25	10 kΩ, ±10%, 1 2 watt	569-1004-103
R26	100Ω ±10%, 1 2 watt	569-1004-101
R27	330Ω, ±10%, 1 2 watt	569-1004-331
R28	2.2 kΩ, ±10%, 1 2 watt	569-1004-222
R29	560Ω, ±10%, 1 2 watt	569-1004-561
R31	Same as R29	
R32	1 kΩ, 1 2 watt	569-1004-102
R33	18 kΩ, ±10%, 1 2 watt	569-1004-183
R34	27 kΩ, ±10%, 1 2 watt	569-1004-273
R35	15 kΩ, ±10%, 1 2 watt	569-1004-153
R36	2.2 kΩ, ±10%, 1 2 watt	569-1004-103
R37	Same as R36	
R38	180Ω, ±10%, 1 2 watt	569-1004-181
R39	100Ω, ±10%, 1 2 watt	569-1004-101
R40	10 kΩ, ±10%, 1 2 watt	569-1004-103
R41	Same as R40	
R42	100 kΩ, ±10%, 1 2 watt	569-1004-104
R43	47 kΩ ±10%, 1 2 watt	569-1004-473
R44	1 kΩ, 1 2 watt	569-1004-102
R46	Same as R44	
R47	10 kΩ, ±10%, 1 2 watt	569-1004-103
R48	33 kΩ, ±10%, 1 2 watt	569-1004-333
R49	6.8 kΩ, ±10%, 1 2 watt	569-1004-682
R50	4.7k, ±10%, 1 4 watt	569-1002-472
R51	22Ω, ±10%, 1 2 watt	569-1004-220
R52	68Ω, ±10%, 1 2 watt	569-1004-680
R53	150Ω, ±10%, 1 2 watt	569-1004-151
R54	15Ω, ±10%, 1 2 watt	569-1004-150
R55	220Ω, ±10%, 1 2 watt	569-1004-221

SYMBOL NO.	DESCRIPTION	PART NO.
R56	15Ω, ±10%, 1/2 watt	569-1004-180
R57	0.33Ω, ±10%, 1/2 watt	569-2003-338
R58	Same as R57	
R59	2.2 kΩ, ±10%, 1/2 watt	569-1004-222
R60	1 kΩ, 1/2 watt	569-1004-102
R62	33 kΩ, ±10%, 1/2 watt	569-1004-333
R63	68 dΩ, ±10%, 1/2 watt	569-1004-683
R64	47 kΩ, ±10%, 1/2 watt	569-1004-473
R65	6.8 kΩ, ±10%, 1/2 watt	569-1004-682
R66	1 kΩ, 1/2 watt	569-1004-102
R67	220 kΩ, ±10%, 1/2 watt	569-1004-224
R68	2.2 kΩ, ±10%, 1/2 watt	569-1004-222
R69	10 kΩ, ±10%, 1/2 watt	569-1004-103
R70	1 kΩ, 1/2 watt	569-1004-102
R71	2.2 kΩ, ±10%, 1/2 watt	569-1004-222
R72	33 kΩ, ±10%, 1/2 watt	569-1004-333
R73	470 kΩ, ±10%, 1/2 watt	569-1004-474
R74	33 kΩ, ±10%, 1/2 watt	569-1004-333
R75	15 kΩ, ±10%, 1/2 watt	569-1004-153
R76	68 kΩ, ±10%, 1/2 watt	569-1004-683
R77	33 kΩ, ±10%, 1/2 watt	569-1004-333
R78	10 kΩ, ±10%, 1/2 watt	569-1004-103
R79	68 kΩ, ±10%, 1/2 watt	569-1004-683
R80	2.2 kΩ, ±10%, 1/2 watt	569-1004-222
R81	2.7 kΩ, ±10%, 1/4 watt	569-1002-272
R82	Same as R81	
R83	Same as R81	
R84	Same as R81	
R85	Same as R81	
R86	Same as R81	
R87	Same as R81	
R88	Same as R81	
R89	56Ω, ±10%, 1/2 watt	569-1004-560
R90	100Ω, ±10%, 1/2 watt	569-1004-101
R91	Same as R90	
R92	150Ω, ±10%, 1/2 watt	569-1004-151
	SWITCH	
S1	Switch, push button (S1-S9)	583-4008-004
	TRANSFORMERS	
T1	Transformer, P.C., 140-165 MHz	592-5009-016
T2	Same as T1	
T3	Same as T1	
T4	Same as T1	
T6	Transformer, P.C.	592-5030-021
T7	Transformer, P.C., 27 MHz	592-5015-004
T8	Transformer, P.C., 10.7 MHz	592-5009-011
T9	Same as T8	
T10	Same as T8	
T11	Transformer, P.C., 455 kHz	592-5006-006
T12	Transformer, P.C., 455 kHz	592-5006-008
T13	Transformer, audio	592-1015-001
T14	Transformer, power	592-3001-008
T15	Transformer, P.C., 455 kHz	592-5006-006
	INTEGRATED CIRCUITS (IC'S)	
U1	IC, CA3011	544-2002-004
U2	IC, SC5245P (MC1351P)	544-2003-002
U3	IC, MC723P	544-3001-002
U4	IC, MC790P	544-3001-003
U5	IC, MC724P	544-3001-001
	PC BOARD	
U6	PC board	035-0143-002
	COAX CABLE	
W30	Coaxial cable, 50 Ω, RG-188/U	071-0321-930
W31	Same as W30	
W32	Same as W30	
W41	Same as W30	

SYMBOL NO.	DESCRIPTION	PART NO
	CRYSTAL BLOCK	
X1	Crystal block, 8-pin	126-0110-108
X2	Crystal block, 8-pin	126-0110-109
	CRYSTAL (SECOND OSCILLATOR)	
Y1	Crystal, 11.155 MHz	519-0009-001
	FILTERS	
Z1	Filter, Murata SF 10.7 MHz	532-2001-011
Z2	Filter, 455 kHz	532-2003-001

Crystal Ordering Information

We recommend crystals manufactured by M-Tron Industries, Inc., Yankton, South Dakota 57078

Use the following information when ordering M-Tron crystals:

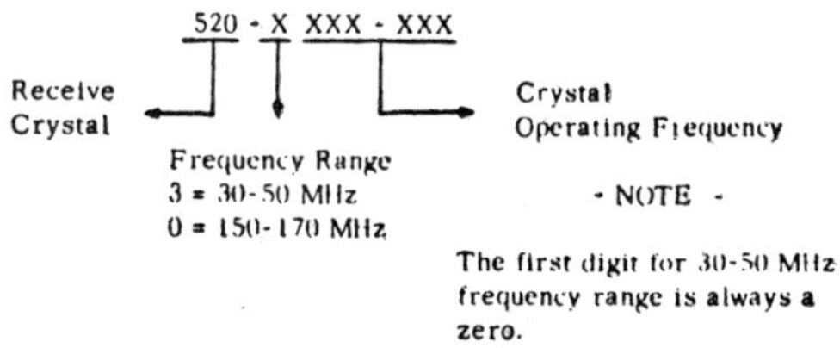

For Example: A crystal with an operating frequency of 35.080 MHz will have a part number of 520-3035-080.

A crystal with an operating frequency of 151.625 MHz will have a part number of 520-0151-625.

UHF—VHF DUO-SCAN

	ACCESSORY PACKAGE	023-3066-001
	Includes:	
E1	Antenna, collapsible	501-0010-001
EP1	6 ft. power cable	597-1002-011
EP3	Connector package	023-2209-001
ML4	Schematic	564-3001-392
ML8	Hardware envelope	023-2615-001
MP4	Rubber bumper	574-1006-003
W9	Battery cable	023-1652-001
	ASSEMBLIES	
A1	Cabinet Assembly	023-2920-002
A2	Rear-Panel Assembly	023-3171-001
	Includes:	
J1	CC tini-jax NTT312	515-2001-001
J3	Cord connector AC	597-1002-012
J4	Terminal bushing, red	515-4100-001
T14	AC power transformer	592-3001-008
U9	Rear panel	023-2965-003
A3	Front-Panel Assembly	023-3074-010
	Includes:	
HW3	Nut spanner	013-1313-001
MP1	Knob	013-1086-008
MP2	Front-panel	032-0307-001
MP3	Front-panel overlay	559-2058-113
R45	Potentiometer, 10K volume	562-0018-008
R61	Potentiometer 5K squelch	562-0018-007
LS7	Speaker Assembly	023-2927-002
	Includes:	
LS1	Speaker 4-ohm	589-1003-005
MP26	Foam ring	018-0960-001
	CAPACITORS	
C1	3 pF J 500V composition	510-9002-309

SYMBOL NO.	DESCRIPTION	PART NO.
C2	Same as C1	
C3	Same as C1	
C4	1000 pF M 1KV Y5S disc	510-3061-102
C5	Same as C4	
C6	6.8 pF 200V N750 ceramic	510-3020-689
C7	0.56 pF K 500V composition	510-9003-568
C8	8.2 pF J 200V N750 ceramic	510-3020-829
C9	1 pF J 500V composition	510-9002-109
C11	6.8 pF J 200V N750 ceramic	510-3020-689
C12	1000 pF M 1KV Y5S disc	510-3061-102
C13	1 pF J 500V composition	510-9002-109
C14	0.010 MF M 16V Y5S disc	510-3010-103
C16	Same as C14	
C40	1500 pF J 25V polystyrene	510-1101-152
C41	0.10 MF Z 10V Y5U disc	510-3008-104
C42	Same as C41	
C43	1 pF J 500V composition	510-9002-109
C44	220 pF J 100V 1DM15	510-0001-221
C45	22 pF J 200V N750 ceramic	510-3020-220
C46	0.010 MF M 16V Y5S disc	510-3010-103
C48	0.10 MF Z 10V Y5U disc	510-3008-104
C49	Same as C48	
C50	Same as C48	
C51	Same as C48	
C52	Same as C48	
C53	1500 pF J 25V polystyrene	510-1101-152
C54	0.10 MF Z 10V Y5U disc	510-3008-104
C55	Same as C54	
C56	0.022 MF M 16V Y5S disc	510-3010-223
C57	0.10 MF Z 10V Y5U disc	510-3008-104
C56	0.022 MF M 16V Y5S disc	510-3010-223
C57	0.10 MF Z 10V Y5U disc	510-3008-104
C58	Same as C57	
C59	Same as C57	
C60	1 MF M 35V dipped	510-2045-109
C61	56 pF J 200V NPO ceramic	510-3013-560
C62	270 pF J 100V 1DM15	510-0001-271
C63	0.56 MF M 35V tubular	510-2005-568
C64	4700 pF M 500V Y5U disc	510-3004-472
C65	0.56 MF M 35V tubular	510-2005-568
C66	330 MF 10V aluminum	510-4003-007
C67	0.010 MF M 16V Y5S disc	510-3010-103
C68	1000 pF M 1KV Y5S disc	510-3061-102
C69	Same as C68	
C71	2.2 MF M 35V dipped	510-2045-229
C73	0.10 MF Z 10V Y5U disc	510-3008-104
C74	2.2 MF M 35V dipped	510-2045-229
C75	Same as C74	
C76	Same as C74	
C77	470 pF J 500V 4DM15	510-0004-471
C78	220 pF J 100V 1DM15	510-0001-221
C79	1000 MF 16V aluminum	510-4006-005
C80	Same as C79	
C82	0.010 MF M 16V Y5S disc	510-3010-103
C83	0.10 MF M 16V Y5S disc	510-3010-104
C84	1000 MF 16V aluminum	510-4006-005
C101	10 MF 25V aluminum	510-4006-002
C102	56 pF J 200V N750 ceramic	510-3020-560
C103	110 pF J 200V N750 ceramic	510-3020-111
C104	1000 pF M 1KV Y5S disc	510-3061-102
C105	180 pF J 200V N750 ceramic	510-3020-181
C106	15 pF J 200V N750 ceramic	510-3020-150
C108	1000 pF M 1KV Y5S disc	510-3061-102
C109	68 pF J 100V 1DM10	510-0002-680
C111	1000 pF M 1KV Y5S disc	510-3061-102
C112		187-0106-095
C113	1000 pF M 1KV Y5S disc	510-3061-102
C114	150 pF J 500V 4DM15	510-0004-151
C115	47 pF J 200V N750 ceramic	510-3020-470
C116	1000 pF M 1KV Y5S disc	510-3061-102
C117	Same as C116	
C118		187-0106-095
C119	1000 pF M 1KV Y5S disc	510-3061-102
C120	160 pF J 100V 1DM10	510-0002-161
C121	10 pF J 200V N750 ceramic	510-3020-100
C122		187-0106-095
C123	1000 pF M 1KV Y5S disc	510-3061-102
C124	0.010 MF M 16V Y5S disc	510-3010-103
C125		187-0106-095
C126	1000 pF M 1KV Y5S disc	510-3061-102
C127	39 pF J 200V NPO ceramic	510-3013-390
C129	27 pF J 200V NPO ceramic	510-3013-270
C130		187-0106-095

SYMBOL NO.	DESCRIPTION	PART NO.
C131	3.3 pF J 500V composition	510-9002-339
	DIODES	
CR1	1N4149 75V 75MA SI SW	523-0006-002
CR2	Same as CR1	
CR3	Same as CR1	
CR4	Same as CR1	
CR5	Same as CR1	
CR6	Same as CR1	
CR7	Same as CR1	
CR8	Same as CR1	
CR9	1N67A 80V 30MA germanium	523-1000-067
CR10	Same as CR9	
CR11	1N4003 200V 1A rectifier	523-0001-002
CR12	Same as CR11	
CR13	Same as CR11	
CR14	Same as CR11	
CR15	10V J 1W zener	523-2003-100
CR16	1N67A 80V 30MA germanium	523-1000-067
CR17	Same as CR16	
CR101	MV2209 VVC	523-0009-049
CR102	1N881 silicon diode	523-1000-881
CR103	Same as CR102	
CR104	Same as CR102	
CR105	Same as CR102	
CR106	Same as CR102	
CR107	Same as CR102	
CR108	Same as CR102	
CR109	Same as CR102	
CR110	1N67A 80V 30MA germanium	523-1000-067
	INDUCTORS	
L101	0.14-0.21 UH variable inductor	542-1006-001
L102	RF inductor	542-1006-020
L103	20 UHY choke	542-3002-002
L104	Inductor	023-3165-004
L105	2 1/2 turn inductor	542-0007-001
L106	1.0 UH choke	542-3002-001
L107	Same as L106	
L108	0.22-0.37 UH variable inductor	542-1006-017
	TRANSISTORS	
Q1	SI NPN 175 MHz amp TO72	576-0003-029
Q2	Same as Q1	
Q7	Same as Q1	
Q8	SL NPN SWT XSTR	576-0003-028
Q9	SI NPN Gen. Purp. TO92	576-0003-011
Q10	SI NPN 60V 1.5A 1W TO5	576-0002-006
Q11	SI NPN 60V 4A 36W X75	576-0002-001
Q12	Same as Q11	
Q13	SI NPN Gen. Purp. TO92	576-0003-011
Q14	Same as Q13	
Q15	Same as Q13	
Q16	SL NPN SWT XSTR	576-0003-028
Q17	Same as Q16	
Q18	SI PNP 50 MHz amp TO92	576-0003-017
Q19	SL NPN SWT XSTR	576-0003-028
Q20	Same as Q19	
Q21	Same as Q19	
Q22	Same as Q19	
Q23	Same as Q19	
Q24	Same as Q19	
Q25	Same as Q19	
Q26	Same as Q19	
Q27	Same as Q19	
Q28	Same as Q19	
Q29	SI NPN 60V 4A 36W X75	576-0002-001
Q101	SI NPN Gen. Purp. TO92	576-0003-011
Q102	Same as Q101	
Q103	SI PNP 50 MHz amp TO92	576-0003-017
Q104	SI NPN 175 MHz amp TO72	576-0003-029
Q105	SI PNP 50 MHz amp TO92	576-0003-017
Q106	Same as Q105	
Q107	SI NPN 470 MHz mix TO72	576-0003-027
Q108	SI MOSFET 470 MHz	576-0006-222
Q109	SI NPN 175 MHz amp TO72	576-0003-029

SYMBOL NO.	DESCRIPTION	PART NO.
	RESISTORS	
R2	4.7K ohm K 1/2 W	569-1004-472
R3	27K ohm K 1/2 W	569-1004-273
R4	1.0K ohm K 1/2 W	569-1004-102
R5	560 ohm K 1/2 W	569-1004-561
R6	33K ohm K 1/2 W	569-1004-333
R7	8.2K ohm K 1/2 W	569-1004-822
R9	1.0K ohm K 1/2 W	569-1004-102
R28	2.2K ohm K 1/2 W	569-1004-222
R29	560 ohm K 1/2 W	569-1004-561
R31	Same as R29	
R32	1.0K ohm K 1/2 W	569-1004-102
R33	18K ohm K 1/4 W	569-1002-183
R34	27K ohm K 1/2 W	569-1004-273
R35	6.8K ohm K 1/4 W	569-1002-682
R36	2.2K ohm K 1/2 W	569-1004-222
R37	Same as R36	
R38	180 ohm K 1/2 W	569-1004-181
R39	100 ohm K 1/2 W	569-1004-101
R40	10K ohm K 1/2 W	569-1004-103
R41	Same as R40	
R42	100K ohm K 1/2 W	569-1004-104
R43	47K ohm K 1/2 W	569-1004-473
R44	1.0K ohm K 1/2 W	569-1004-102
R46	Same as R44	
R47	10K ohm K 1/2 W	569-1004-103
R48	33K ohm K 1/2 W	569-1004-333
R49	6.8K ohm K 1/2 W	569-1004-682
R50	4.7K ohm K 1/4 W	569-1002-472
R51	22 ohm K 1/2 W	569-1004-220
R52	68 ohm K 1/2 W	569-1004-680
R53	150 ohm K 1/2 W	569-1004-151
R54	15 ohm K 1/2 W	569-1004-150
R55	220 ohm K 1/2 W	569-1004-221
R56	15 ohm K 1/2 W	569-1004-150
R57	0.33 ohm K 1/2 W	569-2003-338
R58	Same as R57	
R59	2.2K ohm K 1/2 W	569-1004-222
R60	1.0K ohm K 1/2 W	569-1004-102
R63	68K ohm K 1/2 W	569-1004-683
R64	47K ohm K 1/2 W	569-1004-473
R65	6.8K ohm K 1/2 W	569-1004-682
R66	1.0K ohm K 1/2 W	569-1004-102
R67	220K ohm K 1/2 W	569-1004-224
R68	2.2K ohm K 1/2 W	569-1004-222
R69	10K ohm K 1/2 W	569-1004-103
R70	1.0K ohm K 1/2 W	569-1004-102
R71	2.2K ohm K 1/2 W	569-1004-222
R72	33K ohm K 1/2 W	569-1004-333
R73	470K ohm K 1/2 W	569-1004-474
R74	33K ohm K 1/2 W	569-1004-333
R75	15K ohm K 1/2 W	569-1004-153
R76	68K ohm K 1/2 W	569-1004-683
R77	33K ohm K 1/2 W	569-1004-333
R78	10K ohm K 1/2 W	569-1004-103
R79	68K ohm K 1/2 W	569-1004-683
R81	1.5K ohm K 1/4 W	569-1002-152
R82	Same as R81	
R83	Same as R81	
R84	Same as R81	
R85	Same as R81	
R86	Same as R81	
R87	Same as R81	
R88	Same as R81	
R89	56 ohm K 1/2 W	569-1004-560
R90	100 ohm K 1/2 W	569-1004-101
R91	Same as R90	
R92	150 ohm K 1/2 W	569-1004-151
R101	47K ohm K 1/4 W	569-1002-473
R102	1.0K ohm K 1/4 W	569-1002-102
R103	3.3K ohm K 1/4 W	569-1002-332
R104	3.9K ohm K 1/4 W	569-1002-392
R105	2K 1/8 W PC trim pot.	562-0004-202
R106	3.3K ohm K 1/4 W	569-1004-332
R107	1.0K ohm K 1/4 W	569-1002-102
R108	100 ohm K 1/4 W	569-1002-101
R110	3.3K ohm K 1/4 W	569-1002-332
R111	10K ohm K 1/4 W	569-1002-103
R112	82K ohm K 1/4 W	569-1002-823
R113	1.0K ohm K 1/4 W	569-1002-102
R114	10K ohm K 1/4 W	569-1002-103
R115	47K ohm K 1/4 W	569-1002-473
R116	39K ohm K 1/4 W	569-1002-393
R117	100 ohm K 1/4 W	569-1002-101
R118	10K ohm K 1/4 W	569-1002-103
R119	22K ohm K 1/4 W	569-1002-223
R109	1.0K ohm K 1/4 W	569-1002-102
R120	330 ohm K 1/2 W	569-1004-331
R121	22 ohm K 1/2 W	569-1004-220
R122	68K ohm K 1/2 W	569-1004-683
R123	47K ohm K 1/2 W	569-1004-473
R124	1.0K ohm K 1/4 W	569-1002-102
R125	Same as R124	
R126	100 ohm K 1/2 W	569-1004-101
R127	470K ohm K 1/4 W	569-1002-474
R128	1.5K ohm K 1/4 W	569-1002-152
R129	3.9K ohm K 1/4 W	569-1002-392
R131	15K ohm K 1/4 W	569-1002-153
R132	6.8K ohm K 1/2 W	569-1004-682
R133	1.0K ohm K 1/2 W	569-1004-102
R134	6.8K ohm K 1/4 W	569-1002-682
R135	Same as R134	
	SWITCH	
SW1	Push button switch (S1-S9)	583-4008-004
	SWITCH BUTTONS	
MP17	Switch button, white (9 required)	032-0306-010
	TRANSFORMERS	
T1	10MM 150 MHz	592-5009-016
T2	Same as T1	
T3	Same as T1	
T4	Same as T1	
T11	10MM 455 kHz IF	592-5006-006
T12	10MM 455 kHz IF	592-5006-008
T13	Input/driver	592-1015-001
T15	10MM 455 kHz IF	592-5006-006
T101	RF transformer	023-3165-001
T102	RF transformer	023-3165-002
T103	10.7 MHz	592-5009-011
T104	RF transformer	023-3165-003
	INTEGRATED CIRCUITS (IC'S)	
U1	IC	544-2002-004
U2	IC MC1351PQ FM	544-2003-002
U3	IC MC 723P	544-3001-002
U4	IC MC790P 2X J-K FF	544-3001-003
U5	IC MC724P 4X gate	544-3001-001
	PC BOARD	
U6	PC board	035-0180-002
	COAX CABLE	
W30	RG-188A/U coax cable	071-0321-930
W31	Same as W30	
W32	Same as W30	
W41	Same as W30	
	CRYSTAL SOCKET	
X1	Crystal socket	126-0110-108
X2	Crystal socket	126-0110-109
	CRYSTAL (SECOND OSCILLATOR)	
Y1	11.155 MHz, 32 pF HC-18/U	519-0009-001
Y2	11.168 MHz 32 pF HC-18/u	519-0009-005

SYMBOL NO.	DESCRIRTION	PART NO.
	FILTERS	
Z1	10.7 MHz filter	532-2001-011
Z2	455 kHz filter	532-2003-001

Crystal Ordering Information

We recommend crystals manufactured by M-Tron Industries, Inc., Yankton, South Dakota 57078

Use the following information when ordering M-Tron crystals:

520 - X XXX - XXX

Receive Crystal ← 520; X ↓ Frequency Range; XXX - XXX → Crystal Operating Frequency

Frequency Range
0 = 150-170 MHz
5 = 450-470 MHz

For Example: A crystal with an operating frequency of 151.625 MHz will have a part number of 520-0151-625.

A crystal with an operating frequency of 451.175 MHz will have a part number of 520-5451-175.

COMPONENT CODES

Capacitors	Resistors
1DM10, 15 - 1 coat dipped mica, size 10, 15	CC - carbon composition
4DM10, 15 - 4 coat dipped mica, size 10, 15	WW - wire wound
D - ±.5 pF, J - ±5%, M - ±20%	K - ±10%
Z - +80 -20%	MF - ±20% at 25°C, ±7.5% over temp. range

VHF MINI-SCAN

Parts List

The parts list is arranged in alphabetical and numerical order for ease in locating replacement part numbers.

Following is a list of codes used in the component descriptions.

D – ±0.5 pF
J – ±5%
K – ±10%
M – ±20%
Z – +80 –20%

1DM10 – 1 coat dipped mica, size 10

SYMBOL NO.	DESCRIPTION	PART NO.
	CAPACITORS	
C1	Capacitor, variable	187-0103-005
C2	10 pF K 250V X5R cera.	510-3171-100
C3	1000 pF K 40V Y5P plate	510-3160-102
C4	Same as C3	
C5	2.2 pF D 40V P100 cera.	510-3161-229
C6	0.56 pF J 500V comp.	510-9002-568
C7	6.8 pF J NPO 50V submin.	510-3113-689
C8	1000 pF K 40V Y5P plate	510-3160-102
C9	Same as C8	
C10	91 pF J 100V 1DM10	510-0002-910
C11	39 pF J 100V 1DM10	510-0002-390
C12	1000 pF K 40V Y5P plate	510-3160-102
C13	18 pF K 250V X5R cera.	510-3171-180
C14	Same as C13	
C15	1 pF J 500V composition	510-9002-109
C16	0.010 μF M 16V Y5S disc	510-3010-103
C17	0.1 μF K Y5R 50V submin.	510-3109-104
C18	180 pF J 100V 1DM10	510-0002-181
C19	6.8 pF J NPO 50V submin.	510-3113-689
C20	1 pF J 500V composition	510-9002-109
C21	200 pF J 100V 1DM10	510-0002-201
C22	22 pF J 100V 1DM10	510-0002-220
C23	0.010 μF M 16V Y5S disc	510-3010-103
C24	0.1 μF K Y5R 50V submin.	510-3109-104
C25	0.10 μF Z 10V Y5U disc	510-3008-104
C26	1000 pF K 40V Y5P plate	510-3160-102
C27	0.1 μF K Y5R 50V submin.	510-3109-104
C28	1000 pF K 40V Y5P plate	510-3160-102
C29	Same as C28	
C30	Same as C28	
C31	Same as C28	
C32	Capacitor, 0.0047 μF	510-3109-472
C33	1.0 μF K 35V submin. tub.	510-2075-109
C34	5.6 μF M 15V molded	510-2023-569
C35	0.022 μF M 16V Y5S disc	510-3010-223
C36	1.0 μF K 35V submin. tub.	510-2075-109
C37	100 μF K 10V tant.	510-2055-107
C38	1.0 μF K 35V submin. tub.	510-2075-109
C39	Capacitor, 0.0047 μF	510-3109-472
C40	100 μF K 10V tant.	510-2055-107
C41	Same as C40	
C42	0.010 μF M 16V Y5S disc	510-3010-103
C43	1000 pF K 40V Y5P plate	510-3160-102
C45	0.1 μF K Y5R 50V submin.	510-3109-104
C46	1000 pF M 500V Y5U disc	510-3004-102
C47	0.1 μF K Y5R 50V submin.	510-3109-104
C48	2.2 μF M 35V dipped	510-2045-229
C49	Same as C48	
C50	0.1 μF K Y5R 50V submin.	510-3109-104
C51	100 μF K 10V tant.	510-2055-107
C52	0.1 μF K Y5R 50V submin.	510-3109-104
C53	1000 pF K 40V Y5P plate	510-3160-102
	DIODES	
CR1	Channel indicator, light-emitting diodes	549-4001-002
CR2	Same as CR1	
CR3	Same as CR1	
CR4	Same as CR1	
CR5	1N67A 80V 30 mA germ.	523-1000-067
CR6	Same as CR5	
CR7	Same as CR5	
CR8	Same as CR5	
CR9	Same as CR5	
CR10	1N4148 SI diode	523-1500-883
CR11	Same as CR10	
CR12	Same as CR10	
CR13	Same as CR10	
	CHASSIS PART	
CH1	Scan lockout switch bracket	017-1814-001
	HARDWARE	
HW4	Nut, spanner	013-1313-001
	JACKS	
J2	External speaker jack	515-2002-001
J3	External charge jack	
	Includes:	
	Spring contact	016-2028-001
	Pin contact	013-1434-001
	Wire terminal	022-0069-010
	External antenna jack	
	Includes:	
	Brass bushing	013-1400-001
	Antenna ground lug	016-1988-001
	Ring nut 3/8-32 x 094	560-9066-016
	COILS & CHOKES	
L1	0.17-0.22 μH var. ind.	542-1012-011
L2	0.9-1.2 μH var. ind.	542-1012-002
L3	0.11-0.14 μH var. ind.	542-1006-013
L4	1.0 μH choke	542-3002-001
L5	13.5T coil 26AWG	542-0001-135
L6	10 mH RFchoke	542-8007-100
	SPEAKER	
LS1	Speaker, 8.2Ω	589-1012-002

SYMBOL NO.	DESCRIPTION	PART NO.
	MECHANICAL PARTS	
MP1	SCR cover mtg.	013-1399-003
MP2	Contact spring	580-1008-001
MP3	Knob squelch	547-0011-001
MP4	Knob, Volume	547-0011-001
MP5	Case	032-0364-001
MP6	Speaker cover	032-0365-001
MP7	Chassis rail	032-0363-002
MP8	Gasket speaker cover	018-1010-001
	TRANSISTORS	
Q1	SI NPN 175 MHz amp TO72	576-0003-029
Q2	SI NPN gen. purp. TO92	576-0003-011
Q3	SI NPN 175 MHz amp TO72	576-0003-029
Q4	SI NPN gen. purp. TO92	576-0003-011
Q5	Same as Q4	
Q6	SI NPN 50 MHz amp. TO92	576-0003-018
Q7	SI NPN gen. purp. TO92	576-0003-011
Q8	Same as Q7	
Q9	Same as Q7	
Q10	Same as Q7	
Q11	SI PNP 50 MHz amp. TO92	576-0003-012
Q12	SI NPN gen. parp. TO92	576-0003-011
Q13	Same as Q12	
Q14	Same as Q12	
Q15	Same as Q12	
Q16	Same as Q12	
Q17	Same as Q12	
Q18	Same as Q12	
Q19	Same as Q12	
Q20	Same as Q12	
Q21	SI PNP 50 MHz amp. TO92	576-0003-012
Q22	SI NPN gen. purp. TO92	576-0003-011
	RESISTORS	
R1	470K ohm K 1/8 W	569-1010-474
R2	120K ohm K 1/8 W	569-1010-124
R3	8.2K ohm K 1/8 W	569-1010-822
R4	47K ohm K 1/8 W	569-1010-473
R5	33K ohm K 1/8 W	569-1010-333
R6	100 ohm K 1/8 W	569-1010-101
R7	470 ohm K 1/8 W	569-1010-471
R8	47K ohm K 1/8 W	569-1010-473
R9	12K ohm K 1/8 W	569-1010-123
R10	470 ohm K 1/8 W	569-1010-471
R12	3.3K ohm K 1/8 W	569-1010-332
R13	220K ohm K 1/8 W	569-1010-224
R14	1.8K ohm K 1/8 W	569-1010-182
R15	47K ohm K 1/8 W	569-1010-473
R16	10K ohm K 1/8 W	569-1010-103
R17	5.6K ohm K 1/8 W	569-1010-562
R18	1.2K ohm K 1/8 W	569-1010-122
R19	27K ohm 1/8 W	569-1010-273
R20	18K ohm K 1/8 W	569-1010-183
R21	6.8K ohm K 1/8 W	569-1010-682
R22	2.2K ohm K 1/8 W	569-1010-222
R23	Same as R22	
R24	150K ohm K 1/8 W	569-1010-154
R25	10K ohm K 1/8 W	569-1010-103
R26	150K ohm K 1/8 W	569-1010-154
R27	10K ohm K 1/8 W	569-1010-103
R28	150K ohm K 1/8 W	569-1010-154
R29	10K ohm K 1/8 W	569-1010-103
R30	1.2K ohm K 1/8 W	569-1010-122
R31	180K ohm K 1/8 W	569-1010-184
R32	10K ohm K 1/8 W	569-1010-103
R33	150K ohm K 1/8 W	569-1010-154
R34	Same as R33	
R35	39K ohm K 1/8 W	569-1010-393
R36	15K ohm K 1/8 W	569-1010-153
R37	2.2K ohm K 1/8 W	569-1010-222
R38	2.2K ohm K 1/8 W	569-1010-222
R39	50K 1/4 W A 1/2 SPST volume control/on-off switch	562-0018-017

SYMBOL NO.	DESCRIPTION	PART NO.
R40	120K ohm K 1/8 W	569-1010-124
R41	56K ohm K 1/8 W	569-1010-563
R42	27 ohm K 1/8 W	569-1010-270
R43	1K ohm K 1/8 W	569-1010-102
R44	68 ohm K 1/8 W	569-1010-680
R45	1.2K ohm K 1/8 W	569-1010-122
R47	5K 1/4 W squelch control	562-0018-016
R48	18K ohm K 1/8 W	569-1010-183
R49	22K ohm K 1/8 W	569-1010-223
R50	10K ohm K 1/8 W	569-1010-103
R51	3.3K ohm K 1/8 W	569-1010-332
R52	1.2K ohm K 1/8 W	569-1010-122
R53	27K ohm 1/8 W	569-1010-273
R54	5.6K ohm K 1/8 W	569-1010-562
R55	15K ohm K 1/8 W	569-1010-153
R56	68K ohm K 1/8 W	569-1010-683
R57	100K ohm K 1/8 W	569-1010-104
R58	680K ohm K 1/8 W	569-1010-684
R59	1K ohm K 1/8 W	569-1010-102
R60	Same as R59	
R61	Same as R59	
R62	Same as R59	
R63	4.7K ohm K 1/8 W	569-1010-472
	SWITCHES	
S1	Scan/lockout switches	583-3001-201
S2	Same as S1	
S3	Same as S1	
S4	Same as S1	
	TRANSFORMERS	
T1	6-32 5-1/4 tap ST2	592-5022-008
T2	7MM 455 kHz IF	592-5020-003
	INTEGRATED CIRCUITS	
U1	IC CD4017AE counter	544-3001-117
U2	IC MFC40008 1/4 W audio	544-2003-006
	ANTENNA	
U3	Internal antenna	023-2694-002
	PRINTED-CIRCUIT BOARDS	
U4	Printed-circuit board	035-0213-001
	CRYSTAL SOCKET	
X1	Crystal socket	126-0110-110
	CRYSTALS	
Y1	11.155 MHz 32 pF HC-18/U	519-0009-001

Y2-Y5

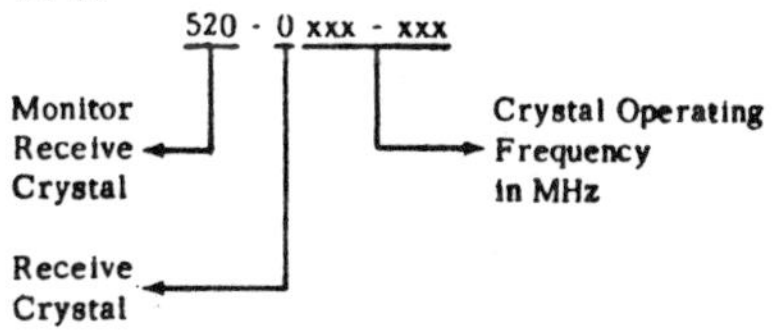

Example: 520-0151-625 is a receive crystal on a channel frequency of 151.625 MHz

Receive Frequency Scheme:

$$\text{Crystal Frequency} = \frac{\text{Channel Frequency} - 10.7\ \text{MHz}}{3}$$

SYMBOL NO.	DESCRIPTION	PART NO.
	CERAMIC FILTERS	
Z1	Ceramic filter 10.7-290	532-2001-011
Z2	Ceramic filter 455-15	532-2004-001
Z3	455 kHz ceramic resonator	532-2005-001

Index

T

U

V